1476343

Co
2/29/08

352-465-

2920

WELDING METALLURGY

SECOND EDITION

WELDING METALLURGY

SECOND EDITION

Sindo Kou

Professor and Chair
Department of Materials Science and Engineering
University of Wisconsin

A JOHN WILEY & SONS, INC., PUBLICATION

Library of Congress Cataloging-in-Publication Data

Kou, Sindo.
  Welding metallurgy / Sindo Kou.–2nd ed.
    p. cm.
"A Wiley-Interscience publication."
Includes bibliographical references and index.
  ISBN 0-471-43491-4
  1. Welding. 2. Metallurgy. 3. Alloys. I. Title.
  TS227 .K649 2002
  671.5′2–dc21

                          2002014327

Printed in the United States of America.

10 9 8 7 6 5

To Warren F. Savage
for his outstanding contributions to welding metallurgy

CONTENTS

PREFACE

Since the publication of the first edition of this book in 1987, there has been much new progress made in welding metallurgy. The purpose for the second edition is to update and improve the first edition. Examples of improvements include (1) much sharper photomicrographs and line drawings; (2) integration of the phase diagram, thermal cycles, and kinetics with the microstructure to explain microstructural development and defect formation in welds; and (3) additional exercise problems. Specific revisions are as follows.

In Chapter 1 the illustrations for all welding processes have been redrawn to show both the overall process and the welding area. In Chapter 2 the heat source efficiency has been updated and the melting efficiency added. Chapter 3 has been revised extensively, with the dissolution of atomic nitrogen, oxygen, and hydrogen in the molten metal considered and electrochemical reactions added. Chapter 4 has also been revised extensively, with the arc added, and with flow visualization, arc plasma dragging, and turbulence included in weld pool convection. Shot peening is added to Chapter 5.

Chapter 6 has been revised extensively, with solute redistribution and microsegregation expanded and the solidification path added. Chapter 7 now includes nonepitaxial growth at the fusion boundary and formation of non-dendritic equiaxed grains. In Chapter 8 solidification modes are explained with more illustrations. Chapter 9 has been expanded significantly to add ferrite formation mechanisms, new ferrite prediction methods, the effect of cooling rate, and factors affecting the austenite–ferrite transformation. Chapter 10 now includes the effect of both solid-state diffusion and dendrite tip under-cooling on microsegregation. Chapter 11 has been revised extensively to include the effect of eutectic reactions, liquid distribution, and ductility of the solidifying metal on solidification cracking and the calculation of fraction of liquid in multicomponent alloys.

Chapter 12 has been rewritten completely to include six different liquation mechanisms in the partially melted zone (PMZ), the direction and modes of grain boundary (GB) solidification, and the resultant GB segregation. Chapter 13 has been revised extensively to include the mechanism of PMZ cracking and the effect of the weld-metal composition on cracking.

Chapter 15 now includes the heat-affected zone (HAZ) in aluminum–lithium–copper welds and friction stir welds and Chapter 16 the HAZ of Inconel 718. Chapter 17 now includes the effect of multiple-pass welding on reheat cracking and Chapter 18 the grain boundary chromium depletion in a sensitized austenitic stainless steel.

The author thanks the National Science Foundation and NASA for supporting his welding research, from which this book draws frequently. He also thanks the American Welding Society and ASM International for permissions to use numerous copyrighted materials. Finally, he thanks C. Huang, G. Cao, C. Limmaneevichitr, H. D. Lu, K. W. Keehn, and T. Tantanawat for providing technical material, requesting permissions, and proofreading.

SINDO KOU

Madison, Wisconsin

WELDING METALLURGY

SECOND EDITION

PART I
Introduction

1 Fusion Welding Processes

Fusion welding processes will be described in this chapter, including gas welding, arc welding, and high-energy beam welding. The advantages and disadvantages of each process will be discussed.

1.1 OVERVIEW

1.1.1 Fusion Welding Processes

Fusion welding is a joining process that uses fusion of the base metal to make the weld. The three major types of fusion welding processes are as follows:

1. **Gas welding**:
   Oxyacetylene welding (OAW)
2. **Arc welding**:
   Shielded metal arc welding (SMAW)
   Gas–tungsten arc welding (GTAW)
   Plasma arc welding (PAW)
   Gas–metal arc welding (GMAW)
   Flux-cored arc welding (FCAW)
   Submerged arc welding (SAW)
   Electroslag welding (ESW)
3. **High-energy beam welding**:
   Electron beam welding (EBW)
   Laser beam welding (LBW)

Since there is no arc involved in the electroslag welding process, it is not exactly an arc welding process. For convenience of discussion, it is grouped with arc welding processes.

1.1.2 Power Density of Heat Source

Consider directing a 1.5-kW hair drier very closely to a 304 stainless steel sheet 1.6 mm ($\frac{1}{16}$ in.) thick. Obviously, the power spreads out over an area of roughly

50 mm (2 in.) diameter, and the sheet just heats up gradually but will not melt. With GTAW at 1.5 kW, however, the arc concentrates on a small area of about 6 mm ($\frac{1}{4}$ in.) diameter and can easily produce a weld pool. This example clearly demonstrates the importance of the power density of the heat source in welding.

The heat sources for the gas, arc, and high-energy beam welding processes are a gas flame, an electric arc, and a high-energy beam, respectively. The power density increases from a gas flame to an electric arc and a high-energy beam. As shown in Figure 1.1, as the power density of the heat source increases, the heat input to the workpiece that is required for welding decreases. The portion of the workpiece material exposed to a gas flame heats up so slowly that, before any melting occurs, a large amount of heat is already conducted away into the bulk of the workpiece. Excessive heating can cause damage to the workpiece, including weakening and distortion. On the contrary, the same material exposed to a sharply focused electron or laser beam can melt or even vaporize to form a deep *keyhole* instantaneously, and before much heat is conducted away into the bulk of the workpiece, welding is completed (1).

Therefore, the advantages of increasing the power density of the heat source are deeper weld penetration, higher welding speeds, and better weld quality with less damage to the workpiece, as indicated in Figure 1.1. Figure 1.2 shows that the weld strength (of aluminum alloys) increases as the heat input per unit length of the weld per unit thickness of the workpiece decreases (2). Figure 1.3*a* shows that angular distortion is much smaller in EBW than in

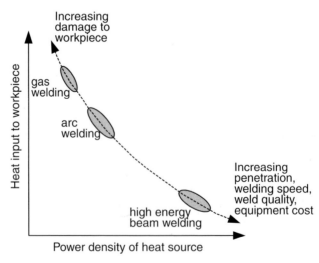

Figure 1.1   Variation of heat input to the workpiece with power density of the heat source.

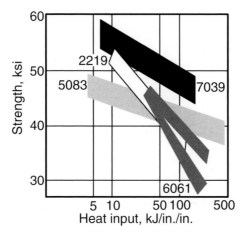

Figure 1.2 Variation of weld strength with heat input per unit length of weld per unit thickness of workpiece. Reprinted from Mendez and Eagar (2).

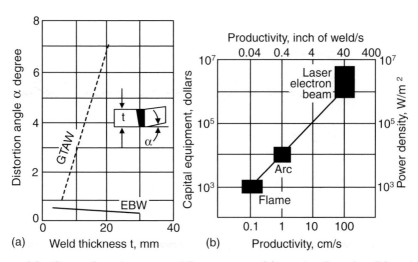

Figure 1.3 Comparisons between welding processes: (*a*) angular distortion; (*b*) capital equipment cost. Reprinted from Mendez and Eagar (2).

GTAW (2). Unfortunately, as shown in Figure 1.3*b*, the costs of laser and electron beam welding machines are very high (2).

1.1.3   Welding Processes and Materials

Table 1.1 summarizes the fusion welding processes recommended for carbon steels, low-alloy steels, stainless steels, cast irons, nickel-base alloys, and

TABLE 1.1 Overview of Welding Processes[a]

Material	Thickness[b]	SMAW	SAW	GMAW	FCAW	GTAW	PAW	ESW	OFW	EBW	LBW
Carbon steels	S	X	X	X		X			X	X	X
	I	X	X	X	X	X			X	X	X
	M	X	X	X	X				X	X	
	T	X	X	X	X			X	X	X	
Low-alloy steels	S	X	X	X		X			X	X	X
	I	X	X	X	X	X				X	X
	M	X	X	X	X					X	X
	T	X	X	X	X			X		X	
Stainless steels	S	X	X	X		X	X		X	X	X
	I	X	X	X	X	X	X			X	X
	M	X	X	X	X		X			X	X
	T	X	X	X	X			X		X	
Cast iron	I	X		X					X		
	M	X	X	X	X				X		
	T	X	X	X	X				X		
Nickel and alloys	S	X		X		X	X		X	X	X
	I	X		X		X	X			X	X
	M	X	X	X			X			X	X
	T	X	X	X				X			
Aluminum and alloys	S			X		X			X	X	X
	I			X		X				X	X
	M			X		X	X			X	
	T			X						X	

[a] Process code: SMAW, shielded metal arc welding; SAW, submerged arc welding; GMAW, gas-metal arc welding; FCAW, flux-cored arc welding; GTAW, gas-tungsten arc welding; PAW, plasma arc welding; ESW, electroslag welding; OFW, oxyfuel gas welding; EBW, electron beam welding; LBW, laser beam welding.

[b] Abbreviations: S, sheet, up to 3 mm ($1/8$ in.); I, intermediate, 3–6 mm ($1/8$–$1/4$ in.); M, medium, 6–19 mm ($1/4$–$3/4$ in.); T, thick, 19 mm ($3/4$ in.) and up; X, recommended.

Source: Welding Handbook (3).

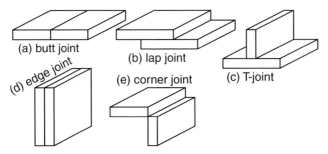

Figure 1.4   Five basic types of weld joint designs.

aluminum alloys (3). For one example, GMAW can be used for all the materials of almost all thickness ranges while GTAW is mostly for thinner workpieces. For another example, any arc welding process that requires the use of a flux, such as SMAW, SAW, FCAW, and ESW, is not applicable to aluminum alloys.

1.1.4   Types of Joints and Welding Positions

Figure 1.4 shows the basic weld joint designs in fusion welding: the butt, lap, T-, edge, and corner joints. Figure 1.5 shows the transverse cross section of some typical weld joint variations. The surface of the weld is called the face, the two junctions between the face and the workpiece surface are called the toes, and the portion of the weld beyond the workpiece surface is called the reinforcement. Figure 1.6 shows four welding positions.

1.2   OXYACETYLENE WELDING

1.2.1   The Process

Gas welding is a welding process that melts and joins metals by heating them with a flame caused by the reaction between a fuel gas and oxygen. Oxyacetylene welding (OAW), shown in Figure 1.7, is the most commonly used gas welding process because of its high flame temperature. A flux may be used to deoxidize and cleanse the weld metal. The flux melts, solidifies, and forms a slag skin on the resultant weld metal. Figure 1.8 shows three different types of flames in oxyacetylene welding: neutral, reducing, and oxidizing (4), which are described next.

1.2.2   Three Types of Flames

A. Neutral Flame   This refers to the case where oxygen (O_2) and acetylene (C_2H_2) are mixed in equal amounts and burned at the tip of the welding torch. A short inner cone and a longer outer envelope characterize a neutral flame

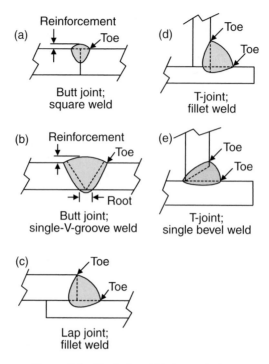

(a) Reinforcement
Toe
Butt joint;
square weld

(b) Reinforcement
Toe
Root
Butt joint;
single-V-groove weld

(c) Toe
Toe
Lap joint;
fillet weld

(d) Toe
Toe
T-joint;
fillet weld

(e) Toe
Toe
T-joint;
single bevel weld

Figure 1.5   Typical weld joint variations.

(Figure 1.8a). The inner cone is the area where the primary combustion takes place through the chemical reaction between O_2 and C_2H_2, as shown in Figure 1.9. The heat of this reaction accounts for about two-thirds of the total heat generated. The products of the primary combustion, CO and H_2, react with O_2 from the surrounding air and form CO_2 and H_2O. This is the secondary combustion, which accounts for about one-third of the total heat generated. The area where this secondary combustion takes place is called the outer envelope. It is also called the protection envelope since CO and H_2 here consume the O_2 entering from the surrounding air, thereby protecting the weld metal from oxidation. For most metals, a neutral flame is used.

B. Reducing Flame   When excess acetylene is used, the resulting flame is called a reducing flame. The combustion of acetylene is incomplete. As a result, a greenish acetylene feather between the inert cone and the outer envelope characterizes a reducing flame (Figure 1.8b). This flame is reducing in nature and is desirable for welding aluminum alloys because aluminum oxidizes easily. It is also good for welding high-carbon steels (also called carburizing flame in this case) because excess oxygen can oxidize carbon and form CO gas porosity in the weld metal.

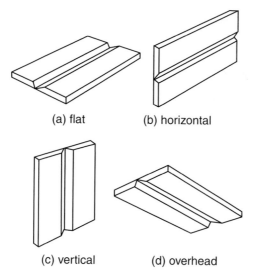

(a) flat   (b) horizontal

(c) vertical   (d) overhead

Figure 1.6 Four welding positions.

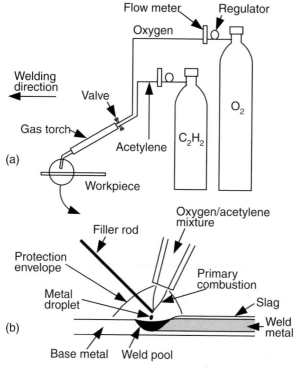

Figure 1.7 Oxyacetylene welding: (*a*) overall process; (*b*) welding area enlarged.

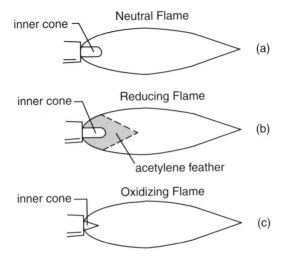

Figure 1.8   Three types of flames in oxyacetylene welding. Modified from *Welding Journal* (4). Courtesy of American Welding Society.

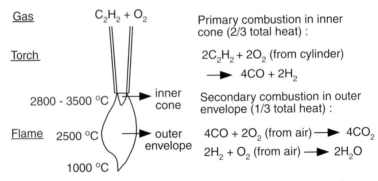

Figure 1.9   Chemical reactions and temperature distribution in a neutral oxyacetylene flame.

C. Oxidizing Flame   When excess oxygen is used, the flame becomes oxidizing because of the presence of unconsumed oxygen. A short white inner cone characterizes an oxidizing flame (Figure 1.8c). This flame is preferred when welding brass because copper oxide covers the weld pool and thus prevents zinc from evaporating from the weld pool.

1.2.3   Advantages and Disadvantages

The main advantage of the oxyacetylene welding process is that the equipment is simple, portable, and inexpensive. Therefore, it is convenient for maintenance and repair applications. However, due to its limited power density, the welding speed is very low and the total heat input per unit length of the weld is rather high, resulting in large heat-affected zones and severe distortion. The oxyacetylene welding process is not recommended for welding reactive metals such as titanium and zirconium because of its limited protection power.

1.3 SHIELDED METAL ARC WELDING

1.3.1 The Process

Shielded metal arc welding (SMAW) is a process that melts and joins metals by heating them with an arc established between a sticklike covered electrode and the metals, as shown in Figure 1.10. It is often called *stick welding*. The electrode holder is connected through a welding cable to one terminal of the power source and the workpiece is connected through a second cable to the other terminal of the power source (Figure 1.10a).

The core of the covered electrode, the core wire, conducts the electric current to the arc and provides filler metal for the joint. For electrical contact, the top 1.5 cm of the core wire is bare and held by the electrode holder. The electrode holder is essentially a metal clamp with an electrically insulated outside shell for the welder to hold safely.

The heat of the arc causes both the core wire and the flux covering at the electrode tip to melt off as droplets (Figure 1.10b). The molten metal collects in the weld pool and solidifies into the weld metal. The lighter molten flux, on the other hand, floats on the pool surface and solidifies into a slag layer at the top of the weld metal.

1.3.2 Functions of Electrode Covering

The covering of the electrode contains various chemicals and even metal powder in order to perform one or more of the functions described below.

A. Protection It provides a gaseous shield to protect the molten metal from air. For a *cellulose*-type electrode, the covering contains cellulose, $(C_6H_{10}O_5)_x$. A large volume of gas mixture of H_2, CO, H_2O, and CO_2 is produced when cellulose in the electrode covering is heated and decomposes. For a *limestone*-$(CaCO_3-)$ type electrode, on the other hand, CO_2 gas and CaO slag form when the limestone decomposes. The limestone-type electrode is a *low-hydrogen*-type electrode because it produces a gaseous shield low in hydrogen. It is often used for welding metals that are susceptible to hydrogen cracking, such as high-strength steels.

B. Deoxidation It provides deoxidizers and fluxing agents to deoxidize and cleanse the weld metal. The solid slag formed also protects the already solidified but still hot weld metal from oxidation.

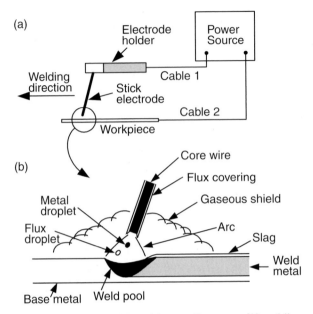

Figure 1.10   Shielded metal arc welding: (*a*) overall process; (*b*) welding area enlarged.

C. Arc Stabilization   It provides arc stabilizers to help maintain a stable arc. The arc is an ionic gas (a plasma) that conducts the electric current. Arc stabilizers are compounds that decompose readily into ions in the arc, such as potassium oxalate and lithium carbonate. They increase the electrical conductivity of the arc and help the arc conduct the electric current more smoothly.

D. Metal Addition   It provides alloying elements and/or metal powder to the weld pool. The former helps control the composition of the weld metal while the latter helps increase the deposition rate.

1.3.3 Advantages and Disadvantages

The welding equipment is relatively simple, portable, and inexpensive as compared to other arc welding processes. For this reason, SMAW is often used for maintenance, repair, and field construction. However, the gas shield in SMAW is not clean enough for reactive metals such as aluminum and titanium. The deposition rate is limited by the fact that the electrode covering tends to overheat and fall off when excessively high welding currents are used. The limited length of the electrode (about 35 cm) requires electrode changing, and this further reduces the overall production rate.

1.4   GAS–TUNGSTEN ARC WELDING

1.4.1   The Process

Gas–tungsten arc welding (GTAW) is a process that melts and joins metals by heating them with an arc established between a nonconsumable tungsten electrode and the metals, as shown in Figure 1.11. The torch holding the tungsten electrode is connected to a shielding gas cylinder as well as one terminal of the power source, as shown in Figure 1.11*a*. The tungsten electrode is usually in contact with a water-cooled copper tube, called the contact tube, as shown in Figure 1.11*b*, which is connected to the welding cable (cable 1) from the terminal. This allows both the welding current from the power source to enter the electrode and the electrode to be cooled to prevent overheating. The workpiece is connected to the other terminal of the power source through a different cable (cable 2). The shielding gas goes through the torch body and is directed by a nozzle toward the weld pool to protect it from the air. Protection from the air is much better in GTAW than in SMAW because an inert gas such as argon or helium is usually used as the shielding gas and because the shielding gas is directed toward the weld pool. For this reason, GTAW is

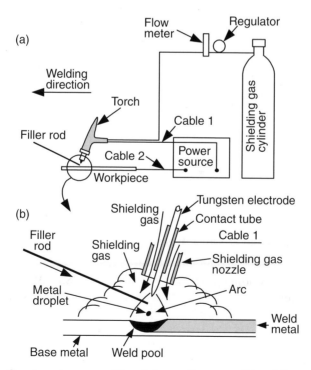

Figure 1.11   Gas–tungsten arc welding: (*a*) overall process; (*b*) welding area enlarged.

also called *tungsten–inert gas* (TIG) welding. However, in special occasions a noninert gas (Chapter 3) can be added in a small quantity to the shielding gas. Therefore, GTAW seems a more appropriate name for this welding process. When a filler rod is needed, for instance, for joining thicker materials, it can be fed either manually or automatically into the arc.

1.4.2 Polarity

Figure 1.12 shows three different polarities in GTAW (5), which are described next.

A. *Direct-Current Electrode Negative (DCEN)*   This, also called the *straight polarity*, is the most common polarity in GTAW. The electrode is connected to the negative terminal of the power supply. As shown in Figure 1.12a, electrons are emitted from the tungsten electrode and accelerated while traveling through the arc. A significant amount of energy, called the work function, is required for an electron to be emitted from the electrode. When the electron enters the workpiece, an amount of energy equivalent to the work function is released. This is why in GTAW with DCEN more power (about two-thirds) is located at the work end of the arc and less (about one-third) at the electrode end. Consequently, a relatively narrow and deep weld is produced.

B. *Direct-Current Electrode Positive (DCEP)*   This is also called the *reverse polarity*. The electrode is connected to the positive terminal of the power source. As shown in Figure 1.12b, the heating effect of electrons is now at the tungsten electrode rather than at the workpiece. Consequently, a shallow weld is produced. Furthermore, a large-diameter, water-cooled electrodes must be used in order to prevent the electrode tip from melting. The positive ions of the shielding gas bombard the workpiece, as shown in Figure 1.13, knocking off oxide films and producing a clean weld surface. Therefore, DCEP can be

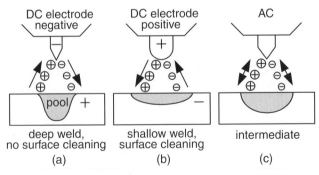

Figure 1.12   Three different polarities in GTAW.

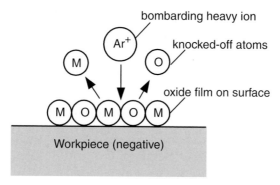

Figure 1.13   Surface cleaning action in GTAW with DC electrode positive.

used for welding thin sheets of strong oxide-forming materials such as aluminum and magnesium, where deep penetration is not required.

C. Alternating Current (AC)   Reasonably good penetration and oxide cleaning action can both be obtained, as illustrated in Figure 1.12c. This is often used for welding aluminum alloys.

1.4.3   Electrodes

Tungsten electrodes with 2% cerium or thorium have better electron emissivity, current-carrying capacity, and resistance to contamination than pure tungsten electrodes (3). As a result, arc starting is easier and the arc is more stable. The electron emissivity refers to the ability of the electrode tip to emit electrons. A lower electron emissivity implies a higher electrode tip temperature required to emit electrons and hence a greater risk of melting the tip.

1.4.4   Shielding Gases

Both argon and helium can be used. Table 1.2 lists the properties of some shielding gases (6). As shown, the *ionization potentials* for argon and helium are 15.7 and 24.5 eV (electron volts), respectively. Since it is easier to ionize argon than helium, arc initiation is easier and the voltage drop across the arc is lower with argon. Also, since argon is heavier than helium, it offers more effective shielding and greater resistance to cross draft than helium. With DCEP or AC, argon also has a greater oxide cleaning action than helium. These advantages plus the lower cost of argon make it more attractive for GTAW than helium.

TABLE 1.2   Properties of Shielding Gases Used for Welding

Gas	Chemical Symbol	Molecular Weight (g/mol)	Specific Gravity with Respect to Air at 1 atm and 0°C	Density (g/L)	Ionization Potential (eV)
Argon	Ar	39.95	1.38	1.784	15.7
Carbon dioxide	CO_2	44.01	1.53	1.978	14.4
Helium	He	4.00	0.1368	0.178	24.5
Hydrogen	H_2	2.016	0.0695	0.090	13.5
Nitrogen	N_2	28.01	0.967	1.25	14.5
Oxygen	O_2	32.00	1.105	1.43	13.2

Source: Reprinted from Lyttle (6).

Because of the greater voltage drop across a helium arc than an argon arc, however, higher power inputs and greater sensitivity to variations in the arc length can be obtained with helium. The former allows the welding of thicker sections and the use of higher welding speeds. The latter, on the other hand, allows a better control of the arc length during automatic GTAW.

1.4.5   Advantages and Disadvantages

Gas–tungsten arc welding is suitable for joining thin sections because of its limited heat inputs. The feeding rate of the filler metal is somewhat independent of the welding current, thus allowing a variation in the relative amount of the fusion of the base metal and the fusion of the filler metal. Therefore, the control of dilution and energy input to the weld can be achieved without changing the size of the weld. It can also be used to weld butt joints of thin sheets by fusion alone, that is, without the addition of filler metals or *autogenous* welding. Since the GTAW process is a very clean welding process, it can be used to weld reactive metals, such as titanium and zirconium, aluminum, and magnesium.

However, the deposition rate in GTAW is low. Excessive welding currents can cause melting of the tungsten electrode and results in brittle tungsten inclusions in the weld metal. However, by using preheated filler metals, the deposition rate can be improved. In the hot-wire GTAW process, the wire is fed into and in contact with the weld pool so that resistance heating can be obtained by passing an electric current through the wire.

1.5   PLASMA ARC WELDING

1.5.1   The Process

Plasma arc welding (PAW) is an arc welding process that melts and joins metals by heating them with a constricted arc established between a tungsten electrode and the metals, as shown in Figure 1.14. It is similar to GTAW, but an orifice gas as well as a shielding gas is used. As shown in Figure 1.15, the arc in PAW is constricted or collimated because of the converging action of the orifice gas nozzle, and the arc expands only slightly with increasing arc length (5). Direct-current electrode negative is normally used, but a special variable-polarity PAW machine has been developed for welding aluminum, where the presence of aluminum oxide films prevents a keyhole from being established.

1.5.2   Arc Initiation

The tungsten electrode sticks out of the shielding gas nozzle in GTAW (Figure 1.11*b*) while it is recessed in the orifice gas nozzle in PAW (Figure 1.14*b*). Consequently, arc initiation cannot be achieved by striking the electrode tip against the workpiece as in GTAW. The control console (Figure 1.14*a*) allows a pilot arc to be initiated, with the help of a high-frequency generator, between the electrode tip and the water-cooled orifice gas nozzle. The arc is then gradually

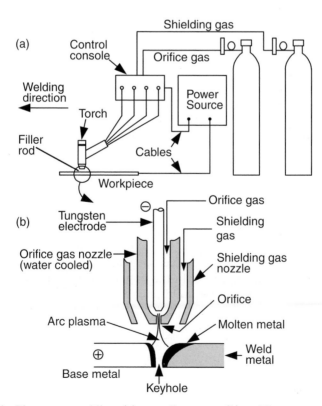

Figure 1.14   Plasma arc welding: (*a*) overall process; (*b*) welding area enlarged and shown with keyholing.

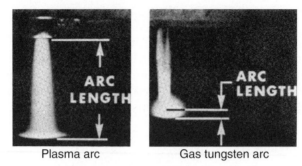

Plasma arc   Gas tungsten arc

Figure 1.15   Comparison between a gas–tungsten arc and a plasma arc. From *Welding Handbook* (5). Courtesy of American Welding Society.

transferred from between the electrode tip and the orifice gas nozzle to between the electrode tip and the workpiece.

1.5.3 Keyholing

In addition to the *melt-in mode* adopted in conventional arc welding processes (such as GTAW), the *keyholing mode* can also be used in PAW in certain ranges of metal thickness (e.g., 2.5–6.4 mm). With proper combinations of the orifice gas flow, the travel speed, and the welding current, keyholing is possible. Keyholing is a positive indication of full penetration and it allows the use of significantly higher welding speeds than GTAW. For example, it has been reported (7) that PAW took one-fifth to one-tenth as long to complete a 2.5-m-long weld in 6.4-mm-thick 410 stainless steel as GTAW. Gas–tungsten arc welding requires multiple passes and is limited in welding speed. As shown in Figure 1.16, 304 stainless steel up to 13 mm ($\frac{1}{2}$ in.) thick can be welded in a single pass (8). The wine-cup-shaped weld is common in keyholing PAW.

1.5.4 Advantages and Disadvantages

Plasma arc welding has several advantages over GTAW. With a collimated arc, PAW is less sensitive to unintentional arc length variations during manual welding and thus requires less operator skill than GTAW. The short arc length in GTAW can cause a welder to unintentionally touch the weld pool with the electrode tip and contaminate the weld metal with tungsten. However, PAW does not have this problem since the electrode is recessed in the nozzle. As already mentioned, the keyhole is a positive indication of full penetration, and it allows higher welding speeds to be used in PAW.

However, the PAW torch is more complicated. It requires proper electrode tip configuration and positioning, selection of correct orifice size for the application, and setting of both orifice and shielding gas flow rates. Because of the

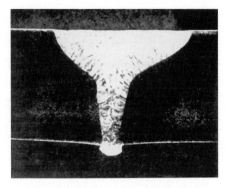

Figure 1.16 A plasma arc weld made in 13-mm-thick 304 stainless steel with keyholing. From Lesnewich (8).

need for a control console, the equipment cost is higher in PAW than in GTAW. The equipment for variable-polarity PAW is much more expensive than that for GTAW.

1.6 GAS–METAL ARC WELDING

1.6.1 The Process

Gas–metal arc welding (GMAW) is a process that melts and joins metals by heating them with an arc established between a continuously fed filler wire electrode and the metals, as shown in Figure 1.17. Shielding of the arc and the molten weld pool is often obtained by using inert gases such as argon and helium, and this is why GMAW is also called the *metal–inert gas* (MIG) welding process. Since noninert gases, particularly CO_2, are also used, GMAW seems a more appropriate name. This is the most widely used arc welding process for aluminum alloys. Figure 1.18 shows gas–metal arc welds of 5083 aluminum, one made with Ar shielding and the other with 75% He–25% Ar shielding (9). Unlike in GTAW, DCEP is used in GMAW. A stable arc, smooth metal transfer with low spatter loss and good weld penetration can be obtained. With DCEN or AC, however, metal transfer is erratic.

1.6.2 Shielding Gases

Argon, helium, and their mixtures are used for nonferrous metals as well as stainless and alloy steels. The arc energy is less uniformly dispersed in an Ar arc than in a He arc because of the lower thermal conductivity of Ar. Consequently, the Ar arc plasma has a very high energy core and an outer mantle of lesser thermal energy. This helps produce a stable, axial transfer of metal

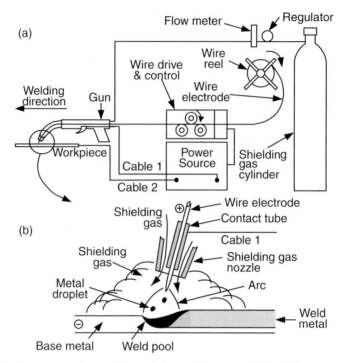

Figure 1.17   Gas–metal arc welding: (*a*) overall process; (*b*) welding area enlarged.

Figure 1.18   Gas–metal arc welds in 6.4-mm-thick 5083 aluminum made with argon (left) and 75% He–25% Ar (right). Reprinted from Gibbs (9). Courtesy of American Welding Society.

droplets through an Ar arc plasma. The resultant weld transverse cross section is often characterized by a papillary- (nipple-) type penetration pattern (10) such as that shown in Figure 1.18 (left). With pure He shielding, on the other hand, a broad, parabolic-type penetration is often observed.

With ferrous metals, however, He shielding may produce spatter and Ar shielding may cause undercutting at the fusion lines. Adding O_2 (about 3%) or CO_2 (about 9%) to Ar reduces the problems. Carbon and low-alloy steels are often welded with CO_2 as the shielding gas, the advantages being higher welding speed, greater penetration, and lower cost. Since CO_2 shielding produces a high level of spatter, a relatively low voltage is used to maintain a short buried arc to minimize spatter; that is, the electrode tip is actually below the workpiece surface (10).

1.6.3 Modes of Metal Transfer

The molten metal at the electrode tip can be transferred to the weld pool by three basic transfer modes: globular, spray, and short-circuiting.

A. Globular Transfer Discrete metal drops close to or larger than the electrode diameter travel across the arc gap under the influence of gravity. Figure 1.19*a* shows globular transfer during GMAW of steel at 180 A and with Ar–2% O_2 shielding (11). Globular transfer often is not smooth and produces spatter. At relatively low welding current globular transfer occurs regardless of the type of the shielding gas. With CO_2 and He, however, it occurs at all usable welding currents. As already mentioned, a short buried arc is used in CO_2-shielded GMAW of carbon and low-alloy steels to minimize spatter.

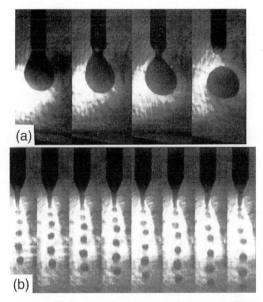

Figure 1.19 Metal transfer during GMAW of steel with Ar–2% O_2 shielding: (*a*) globular transfer at 180 A and 29 V shown at every 3×10^{-3} s; (*b*) spray transfer at 320 A and 29 V shown at every 2.5×10^{-4} s. Reprinted from Jones et al. (11). Courtesy of American Welding Society.

B. Spray Transfer   Above a critical current level, small discrete metal drops travel across the arc gap under the influence of the electromagnetic force at much higher frequency and speed than in the globular mode. Figure 1.19*b* shows spray transfer during GMAW of steel at 320 A and with Ar–2% O_2 shielding (11). Metal transfer is much more stable and spatter free. The critical current level depends on the material and size of the electrode and the composition of the shielding gas. In the case of Figure 1.19, the critical current was found to be between 280 and 320 A (11).

C. Short-Circuiting Transfer   The molten metal at the electrode tip is transferred from the electrode to the weld pool when it touches the pool surface, that is, when short circuiting occurs. Short-circuiting transfer encompasses the lowest range of welding currents and electrode diameters. It produces a small and fast-freezing weld pool that is desirable for welding thin sections, out-of-position welding (such as overhead-position welding), and bridging large root openings.

1.6.4   Advantages and Disadvantages

Like GTAW, GMAW can be very clean when using an inert shielding gas. The main advantage of GMAW over GTAW is the much higher deposition rate, which allows thicker workpieces to be welded at higher welding speeds. The dual-torch and twin-wire processes further increase the deposition rate of GMAW (12). The skill to maintain a very short and yet stable arc in GTAW is not required. However, GMAW guns can be bulky and difficult-to-reach small areas or corners.

1.7   FLUX-CORE ARC WELDING

1.7.1   The Process

Flux-core arc welding (FCAW) is similar to GMAW, as shown in Figure 1.20*a*. However, as shown in Figure 1.20*b*, the wire electrode is flux cored rather than solid; that is, the electrode is a metal tube with flux wrapped inside. The functions of the flux are similar to those of the electrode covering in SMAW, including protecting the molten metal from air. The use of additional shielding gas is optional.

1.8   SUBMERGED ARC WELDING

1.8.1   The Process

Submerged arc welding (SAW) is a process that melts and joins metals by heating them with an arc established between a consumable wire electrode

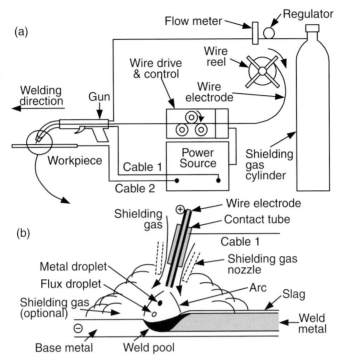

Figure 1.20 Flux-core arc welding: (*a*) overall process; (*b*) welding area enlarged.

and the metals, with the arc being shielded by a molten slag and granular flux, as shown in Figure 1.21. This process differs from the arc welding processes discussed so far in that the arc is submerged and thus invisible. The flux is supplied from a hopper (Figure 1.21*a*), which travels with the torch. No shielding gas is needed because the molten metal is separated from the air by the molten slag and granular flux (Figure 1.21*b*). Direct-current electrode positive is most often used. However, at very high welding currents (e.g., above 900 A) AC is preferred in order to minimize arc blow. Arc blow is caused by the electromagnetic (Lorentz) force as a result of the interaction between the electric current itself and the magnetic field it induces.

1.8.2 Advantages and Disadvantages

The protecting and refining action of the slag helps produce clean welds in SAW. Since the arc is submerged, spatter and heat losses to the surrounding air are eliminated even at high welding currents. Both alloying elements and metal powders can be added to the granular flux to control the weld metal composition and increase the deposition rate, respectively. Using two or more electrodes in tandem further increases the deposition rate. Because of its high

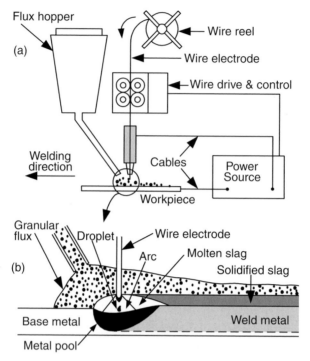

Figure 1.21 Submerged arc welding: (*a*) overall process; (*b*) welding area enlarged.

deposition rate, workpieces much thicker than that in GTAW and GMAW can be welded by SAW. However, the relatively large volumes of molten slag and metal pool often limit SAW to flat-position welding and circumferential welding (of pipes). The relatively high heat input can reduce the weld quality and increase distortions.

1.9 ELECTROSLAG WELDING

1.9.1 The Process

Electroslag welding (ESW) is a process that melts and joins metals by heating them with a pool of molten slag held between the metals and continuously feeding a filler wire electrode into it, as shown in Figure 1.22. The weld pool is covered with molten slag and moves upward as welding progresses. A pair of water-cooled copper shoes, one in the front of the workpiece and one behind it, keeps the weld pool and the molten slag from breaking out. Similar to SAW, the molten slag in ESW protects the weld metal from air and refines it. Strictly speaking, however, ESW is not an arc welding process, because the arc exists only during the initiation period of the process, that is, when the arc

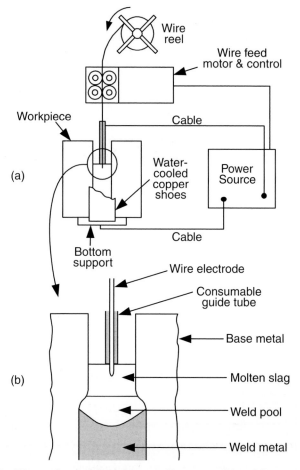

Figure 1.22   Electroslag welding: (*a*) overall process; (*b*) welding area enlarged.

heats up the flux and melts it. The arc is then extinguished, and the resistance heating generated by the electric current passing through the slag keeps it molten. In order to make heating more uniform, the electrode is often oscillated, especially when welding thicker sections. Figure 1.23 is the transverse cross section of an electroslag weld in a steel 7 cm thick (13). Typical examples of the application of ESW include the welding of ship hulls, storage tanks, and bridges.

1.9.2   Advantages and Disadvantages

Electroslag welding can have extremely high deposition rates, but only one single pass is required no matter how thick the workpiece is. Unlike SAW or other arc welding processes, there is no angular distortion in ESW because the

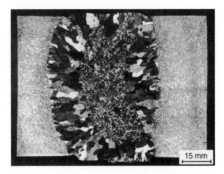

Figure 1.23 Transverse cross section of electroslag weld in 70-mm-thick steel. Reprinted from Eichhorn et al. (13). Courtesy of American Welding Society.

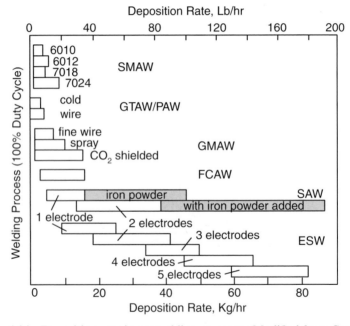

Figure 1.24 Deposition rate in arc welding processes. Modified from Cary (14).

weld is symmetrical with respect to its axis. However, the heat input is very high and the weld quality can be rather poor, including low toughness caused by the coarse grains in the fusion zone and the heat-affected zone. Electroslag welding is restricted to vertical position welding because of the very large pools of the molten metal and slag.

Figure 1.24 summarizes the deposition rates of the arc welding processes discussed so far (14). As shown, the deposition rate increases in the order of

GTAW, SMAW, GMAW and FCAW, SAW, and ESW. The deposition rate can be much increased by adding iron powder in SAW or using more than one wire in SAW, ESW, and GMAW (not shown).

1.10 ELECTRON BEAM WELDING

1.10.1 The Process

Electron beam welding (EBW) is a process that melts and joins metals by heating them with an electron beam. As shown in Figure 1.25a, the cathode of the electron beam gun is a negatively charged filament (15). When heated up to its thermionic emission temperature, this filament emits electrons. These electrons are accelerated by the electric field between a negatively charged bias electrode (located slightly below the cathode) and the anode. They pass through the hole in the anode and are focused by an electromagnetic coil to a point at the workpiece surface. The beam currents and the accelerating voltages employed for typical EBW vary over the ranges of 50–1000 mA and 30–175 kV, respectively. An electron beam of very high intensity can vaporize the metal and form a vapor hole during welding, that is, a keyhole, as depicted in Figure 1.25b.

Figure 1.26 shows that the beam diameter decreases with decreasing ambient pressure (1). Electrons are scattered when they hit air molecules, and the lower the ambient pressure, the less they are scattered. This is the main reason for EBW in a vacuum chamber.

The electron beam can be focused to diameters in the range of 0.3–0.8 mm and the resulting power density can be as high as 10^{10} W/m² (1). The very high

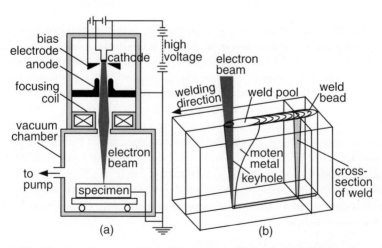

Figure 1.25 Electron beam welding: (a) process; (b) keyhole. Modified from Arata (15).

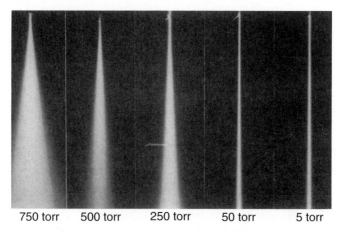

750 torr   500 torr   250 torr   50 torr   5 torr

Figure 1.26   Dispersion of electron beam at various ambient pressures (1). Reprinted from *Welding Handbook* (1). Courtesy of American Welding Society.

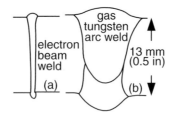

Figure 1.27   Welds in 13-mm-thick 2219 aluminum: (*a*) electron beam weld; (*b*) gas–tungsten arc weld. From Farrell (16).

power density makes it possible to vaporize the material and produce a deep-penetrating keyhole and hence weld. Figure 1.27 shows a single-pass electron beam weld and a dual-pass gas–tungsten arc weld in a 13-mm-thick (0.5-in.) 2219 aluminum, the former being much narrower (16). The energy required per unit length of the weld is much lower in the electron beam weld (1.5 kJ/cm, or 3.8 kJ/in.) than in the gas–tungsten arc weld (22.7 kJ/cm, or 57.6 kJ/in.).

Electron beam welding is not intended for incompletely degassed materials such as rimmed steels. Under high welding speeds gas bubbles that do not have enough time to leave deep weld pools result in weld porosity. Materials containing high-vapor-pressure constituents, such as Mg alloys and Pb-containing alloys, are not recommended for EBW because evaporation of these elements tends to foul the pumps or contaminate the vacuum system.

1.10.2   Advantages and Disadvantages

With a very high power density in EBW, full-penetration keyholing is possible even in thick workpieces. Joints that require multiple-pass arc welding can

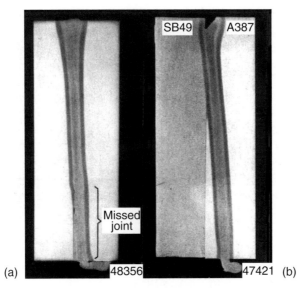

Figure 1.28 Missed joints in electron beam welds in 150-mm-thick steels: (*a*) 2.25Cr–1Mo steel with a transverse flux density of 3.5 G parallel to joint plane; (*b*) SB (C–Mn) steel and A387 (2.25Cr–1Mo) steel. Reprinted from Blakeley and Sanderson (17). Courtesy of American Welding Society.

be welded in a single pass at a high welding speed. Consequently, the total heat input per unit length of the weld is much lower than that in arc welding, resulting in a very narrow heat-affected zone and little distortion. Reactive and refractory metals can be welded in vacuum where there is no air to cause contamination. Some dissimilar metals can also be welded because the very rapid cooling in EBW can prevent the formation of coarse, brittle intermetallic compounds. When welding parts varying greatly in mass and size, the ability of the electron beam to precisely locate the weld and form a favorably shaped fusion zone helps prevent excessive melting of the smaller part.

However, the equipment cost for EBW is very high. The requirement of high vacuum (10^{-3}–10^{-6} torr) and x-ray shielding is inconvenient and time consuming. For this reason, medium-vacuum (10^{-3}–25 torr) EBW and nonvacuum (1 atm) EBW have also been developed. The fine beam size requires precise fit-up of the joint and alignment of the joint with the gun. As shown in Figure 1.28, residual and dissimilar metal magnetism can cause beam deflection and result in missed joints (17).

1.11 LASER BEAM WELDING

1.11.1 The Process

Laser beam welding (LBW) is a process that melts and joins metals by heating them with a laser beam. The laser beam can be produced either by a solid-state laser or a gas laser. In either case, the laser beam can be focused and directed by optical means to achieve high power densities. In a solid-state laser, a single crystal is doped with small concentrations of transition elements or rare earth elements. For instance, in a *YAG laser* the crystal of yttrium–aluminum–garnet (YAG) is doped with neodymium. The electrons of the dopant element can be selectively excited to higher energy levels upon exposure to high-intensity flash lamps, as shown in Figure 1.29a. Lasing occurs when these excited electrons return to their normal energy state, as shown in Figure 1.29b. The power level of solid-state lasers has improved significantly, and continuous YAG lasers of 3 or even 5 kW have been developed.

In a *CO_2 laser*, a gas mixture of CO_2, N_2, and He is continuously excited by electrodes connected to the power supply and lases continuously. Higher

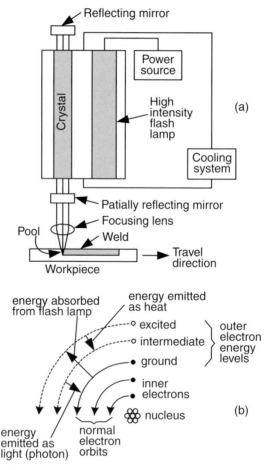

Figure 1.29   Laser beam welding with solid-state laser: (*a*) process; (*b*) energy absorption and emission during laser action. Modified from *Welding Handbook* (1).

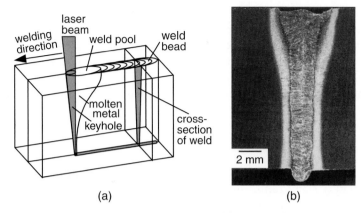

(a)   (b)

Figure 1.30   Laser beam welding with CO_2 laser: (*a*) process; (*b*) weld in 13-mm-thick A633 steel. (*b*) Courtesy of E.A. Metzbower.

power can be achieved by a CO_2 laser than a solid-state laser, for instance, 15 kW. Figure 1.30*a* shows LBW in the keyholing mode. Figure 1.30*b* shows a weld in a 13-mm-thick A633 steel made with a 15-kW CO_2 laser at 20 mm/s (18).

Besides solid-state and gas lasers, semiconductor-based *diode lasers* have also been developed. Diode lasers of 2.5 kW power and 1 mm focus diameter have been demonstrated (19). While keyholing is not yet possible, conduction-mode (surface melting) welding has produced full-penetration welds with a depth–width ratio of 3:1 or better in 3-mm-thick sheets.

1.11.2   Reflectivity

The very high reflectivity of a laser beam by the metal surface is a well-known problem in LBW. As much as about 95% of the CO_2 beam power can be reflected by a polished metal surface. Reflectivity is slightly lower with a YAG laser beam. Surface modifications such as roughening, oxidizing, and coating can reduce reflectivity significantly (20). Once keyholing is established, absorption is high because the beam is trapped inside the hole by internal reflection.

1.11.3   Shielding Gas

A plasma (an ionic gas) is produced during LBW, especially at high power levels, due to ionization by the laser beam. The plasma can absorb and scatter the laser beam and reduce the depth of penetration significantly. It is therefore necessary to remove or suppress the plasma (21). The shielding gas for protecting the molten metal can be directed sideways to blow and deflect the plasma away from the beam path. Helium is often preferred to argon as the shielding gas for high-power LBW because of greater penetration depth (22).

Since the ionization energy of helium (24.5 eV) is higher than that of argon (15.7 eV), helium is less likely to be ionized and become part of the plasma than argon. However, helium is lighter than air and is thus less effective in displacing air from the beam path. Helium–10% Ar shielding has been found to improve penetration over pure He at high-speed welding where a light shielding gas may not have enough time to displace air from the beam path (23).

1.11.4   Lasers in Arc Welding

As shown in Figure 1.31, laser-assisted gas metal arc welding (LAGMAW) has demonstrated significantly greater penetration than conventional GMAW (24). In addition to direct heating, the laser beam acts to focus the arc by heating its path through the arc. This increases ionization and hence the conductivity of the arc along the beam path and helps focus the arc energy along the path. It has been suggested that combining the arc power with a 5-kW CO_2 laser, LAGMAW has the potential to achieve weld penetration in mild steel equivalent to that of a 20–25-kW laser (24). Albright et al. (25) have shown that a lower power CO (not CO_2) laser of 7 W and 1 mm diameter can initiate, guide, and focus an Ar–1% CO gas–tungsten arc.

1.11.5   Advantages and Disadvantages

Like EBW, LBW can produce deep and narrow welds at high welding speeds, with a narrow heat-affected zone and little distortion of the workpiece. It can

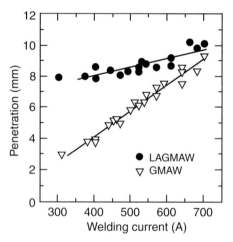

Figure 1.31 Weld penetration in GMAW and laser-assisted GMAW using CO_2 laser at 5.7 kW. Reprinted from Hyatt et al. (24). Courtesy of American Welding Society.

be used for welding dissimilar metals or parts varying greatly in mass and size. Unlike EBW, however, vacuum and x-ray shielding are not required in LBW. However, the very high reflectivity of a laser beam by the metal surface is a major drawback, as already mentioned. Like EBW, the equipment cost is very high, and precise joint fit-up and alignment are required.

REFERENCES

1. *Welding Handbook*, Vol. 3, 7th ed., American Welding Society, Miami, FL, 1980, pp. 170–238.
2. Mendez, P. F., and Eagar, T. W., *Advanced Materials and Processes*, **159:** 39, 2001.
3. *Welding Handbook,* Vol. 1, 7th ed., American Welding Society, Miami, FL, 1976, pp. 2–32.
4. Welding Workbook, Data Sheet 212a, *Weld. J.*, **77:** 65, 1998.
5. *Welding Handbook*, Vol. 2, 7th ed., American Welding Society, Miami, FL, 1978, pp. 78–112, 296–330.
6. Lyttle, K. A., in *ASM Handbook*, Vol. 6, ASM International, Materials Park, OH, 1993, p. 64.
7. Schwartz, M. M., *Metals Joining Manual*, McGraw-Hill, New York, 1979, pp. 2–1 to 3–40.
8. Lesnewich, A., in *Weldability of Steels*, 3rd ed., Eds. R. D. Stout and W. D. Doty, Welding Research Council, New York, 1978, p. 5.
9. Gibbs, F. E., *Weld. J.*, **59:** 23, 1980.
10. Fact Sheet—Choosing Shielding for GMA Welding, *Weld. J.*, **79:** 18, 2000.
11. Jones, L. A., Eagar, T. W., and Lang, J. H., *Weld. J.*, **77:** 135s, 1998.
12. Blackman, S. A., and Dorling, D. V., *Weld. J.*, **79:** 39, 2000.
13. Eichhorn, F., Remmel, J., and Wubbels, B., *Weld. J.*, **63:** 37, 1984.
14. Cary, H. B., *Modern Welding Technology*, Prentice-Hall, Englewood Cliffs, NJ, 1979.
15. Arata, Y., *Development of Ultra High Energy Density Heat Source and Its Application to Heat Processing*, Okada Memorial Japan Society, 1985.
16. Farrell, W. J., *The Use of Electron Beam to Fabricate Structural Members*, Creative Manufacturing Seminars, ASTME Paper SP 63-208, 1962–1963.
17. Blakeley, P. J., and Sanderson, A., *Weld. J.*, **63:** 42, 1984.
18. Metzbower, E. A., private communication, Naval Research Laboratory, Washington, DC.
19. Bliedtner, J., Heyse, Th., Jahn, D., Michel, G., Muller, H., and Wolff, D., *Weld. J.*, **80:** 47, 2001.
20. Xie, J., and Kar, A., *Weld. J.*, **78:** 343s, 1999.
21. Mazumder, J., in *ASM Handbook*, Vol. 6, ASM International, Materials Park, OH, 1993, p. 874.
22. Rockstroh, T., and Mazumder, J., *J. Appl. Phys.*, **61:** 917, 1987.

23. Seaman, F., *Role of Shielding Gas in Laser Welding*, Technical Paper MR77-982, Society of Manufacturing Engineers, Dearborn, MI, 1977.

24. Hyatt, C. V., Magee, K. H., Porter, J. F., Merchant, V. E., and Matthews, J. R., *Weld. J.*, **80:** 163s, 2001.

25. Albright, C. E., Eastman, J., and Lempert, W., *Weld. J.*, **80:** 55, 2001.

26. Ushio, M., Matsuda, F., and Sadek, A. A., in " *International Trends in Welding Science and Technology*, Eds. S. A. David and J. M. Vitek, ASM International, Materials Park, OH, March 1993, p. 408.

FURTHER READING

1. Arata, Y., *Development of Ultra High Energy Density Heat Source and Its Application to Heat Processing*, Okada Memorial Society for the Promotion of Welding, Japan, 1985.

2. Schwartz, M. M., *Metals Joining Manual*, McGraw-Hill, New York, 1979.

3. *Welding Handbook*, Vols. 1–3, 7th ed., American Welding Society, Miami, FL, 1980.

4. Duley, W. W., *Laser Welding*, Wiley, New York, 1999.

5. *ASM Handbook*, Vol. 6, ASM International, Materials Park, OH, 1993.

PROBLEMS

1.1   It has been suggested that compared to SMAW, the cooling rate is higher in GMAW and it is, therefore, more likely for heat-affected zone cracking to occur in hardenable steels. What is the main reason for the cooling rate to be higher in GMAW than SMAW?

1.2   The diameter of the electrodes to be used in SMAW depends on factors such as the workpiece thickness, the welding position, and the joint design. Large electrodes, with their corresponding high currents, tend to produce large weld pools. When welding in the overhead or vertical position, do you prefer using larger or smaller electrodes?

1.3   In arc welding, the magnetic field induced by the welding current passing through the electrode and the workpiece can interact with the arc and cause "arc blow." Severe arc blow can cause excessive weld spatter and incomplete fusion. When arc blow is a problem in SMAW, do you expect to minimize it by using DC or AC for welding?

1.4   In the hot-wire GTAW process, shown in Figure P1.4, the tip of the filler metal wire is dipped in the weld pool and the wire itself is resistance heated by means of a second power source between the contact tube of the wire and the workpiece. In the case of steels, the deposition rate can

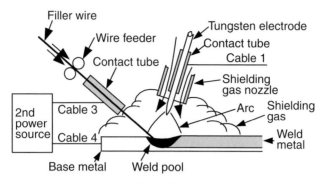

Figure P1.4

be more than doubled this way. Do you prefer using an AC or a DC power source for heating the wire? Do you expect to apply this process to aluminum and copper alloys?

1.5   In GTAW the welding cable is connected to the tungsten electrode through a water-cooled copper contact tube, as shown in Figure 1.11. Why is the tube positioned near the lower end of the electrode instead of the top?

1.6   Measurements of the axial temperature distribution along the GTAW electrode have shown that the temperature drops sharply from the electrode tip toward the contact tube. Why? For instance, with a 2.4-mm-diameter W–ThO_2 electrode at 150 A, the temperature drops from about 3600 K at the tip to about 2000 K at 5 mm above the tip. Under the same condition but with a W–CeO_2 electrode, the temperature drops from about 2700 K at the tip to about 1800 K at 5 mm above the tip (26). Which electrode can carry more current before melting and why?

1.7   Experimental results show that in EBW the penetration depth of the weld decreases as the welding speed increases. Explain why. Under the same power and welding speed, do you expect a much greater penetration depth in aluminum or steel and why?

1.8   How does the working distance in EBW affect the depth–width ratio of the resultant weld?

1.9   Consider EBW in the presence of a gas environment. Under the same power and welding speed, rank and explain the weld penetration for Ar, He, and air. The specific gravities of Ar, He, and air with respect to air are 1.38, 0.137, and 1, respectively, at 1 atm, 0°C.

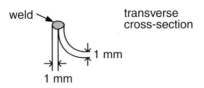

Figure P1.10

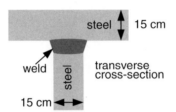

Figure P1.11

1.10 Which arc welding process could have been used for joining the edge weld of thin-gauge steel shown in Figure P1.10 and why?

1.11 Two 15-cm-thick steel plates were joined together in a single pass, as shown in Figure P1.11. Which welding process could have been used and why?

2 Heat Flow in Welding

Heat flow during welding, as will be shown throughout Parts II–IV of this book, can strongly affect phase transformations during welding and thus the resultant microstructure and properties of the weld. It is also responsible for weld residual stresses and distortion, as will be discussed in Chapter 5.

2.1 HEAT SOURCE

2.1.1 Heat Source Efficiency

A. Definition The *heat source efficiency* η is defined as

$$\eta = \frac{Qt_{weld}}{Q_{nominal}t_{weld}} = \frac{Q}{Q_{nominal}} \tag{2.1}$$

where Q is the rate of heat transfer from the heat source to the workpiece, $Q_{nominal}$ the nominal power of the heat source, and t_{weld} the welding time. A portion of the power provided by the heat source is transferred to the workpiece and the remaining portion is lost to the surroundings. Consequently, $\eta < 1$. If the heat source efficiency η is known, the heat transfer rate to the workpiece, Q, can be easily determined from Equation (2.1).

In arc welding with a constant voltage E and a constant current I, the arc efficiency can be expressed as

$$\eta = \frac{Qt_{weld}}{EIt_{weld}} = \frac{Q}{EI} \tag{2.2}$$

Equation (2.2) can also be applied to electron beam welding, where η is the heat source efficiency. In laser beam welding, $Q_{nominal}$ in Equation (2.1) is the power of the laser beam, for instance, 2500 W.

It should be noted that in the welding community the term *heat input* often refers to $Q_{nominal}$, or EI in the case of arc welding, and the term *heat input per unit length of weld* often refers to the ratio $Q_{nominal}/V$, or EI/V, where V is the welding speed.

B. Measurements The heat source efficiency can be measured with a calorimeter. The heat transferred from the heat source to the workpiece is in

37

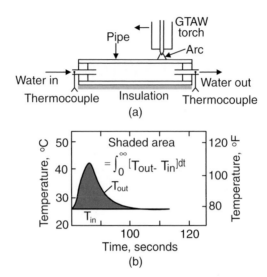

Figure 2.1   Measurement of arc efficiency in GTAW: (*a*) calorimeter; (*b*) rise in cooling water temperature as a function of time. Modified from Kou et al. (1, 2).

turn transferred from the workpiece to the calorimeter, which can be determined as described below.

Kou et al. (1, 2) used simple tubular calorimeters to determine the arc efficiency in GTAW of aluminum, as shown in Figure 2.1*a*. The calorimeter can be a round cross section if the workpiece is a pipe or a rectangular cross section if the workpiece is a sheet. The temperature rise in the cooling water ($T_{out} - T_{in}$) can be measured using thermocouples or thermistors. Heat transfer from the workpiece to the calorimeter is as follows (1–3):

$$Qt_{weld} = \int_0^\infty WC(T_{out} - T_{in})\, dt \approx WC\int_0^\infty (T_{out} - T_{in})dt \qquad (2.3)$$

where W is the mass flow rate of water, C the specific heat of water, T_{out} the outlet water temperature, T_{in} the inlet water temperature, and t time. The integral corresponds to the shaded area in Figure 2.1*b*. The arc efficiency η can be determined from Equations (2.2) and (2.3).

Giedt et al. (4) used the Seebeck envelop calorimeter shown in Figure 2.2*a* to measure the arc efficiency in GTAW. The name Seebeck came from the Seebeck thermoelectric effect of a thermocouple, namely, a voltage is produced between two thermocouple junctions of different temperatures. The torch can be quickly withdrawn after welding, and the calorimeter lid can be closed to minimize heat losses to the surrounding air. As shown in Figure 2.2*b*, heat transfer from the workpiece to the calorimeter can be determined by measuring the temperature difference ΔT and hence gradient

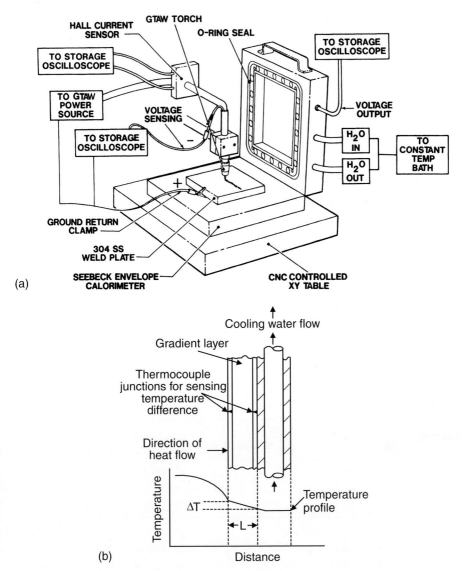

Figure 2.2 Measurement of arc efficiency in GTAW: (*a*) calorimeter; (*b*) layer of temperature gradient. Reprinted from Giedt et al. (4). Courtesy of American Welding Society.

across a "gradient layer" of material of known thermal conductivity k and thickness L:

$$Qt_{\text{weld}} = A\int_0^\infty k\frac{\Delta T}{L}\, dt \qquad (2.4)$$

where A is the area for heat flow and $\Delta T/L$ the temperature gradient. The arc efficiency η can be determined from Equations (2.4) and (2.2). This type of calorimeter was later used to determine the arc efficiencies in PAW, GMAW, and SAW (5–8).

Figure 2.3 shows the results of arc efficiency measurements in GTAW and PAW (2, 5, 7, 9), and Figure 2.4 shows similar results in GMAW and SAW (7, 10). These results were obtained using the two types of calorimeters described above except for the results of Lu and Kou for GMAW (10, 11), which are described in what follows.

In GMAW the arc, metal droplets, and cathode heating all contribute to the efficiency of the heat source. It has been observed in GMAW of aluminum and steel with Ar shielding that current flow or electron emission occurs not

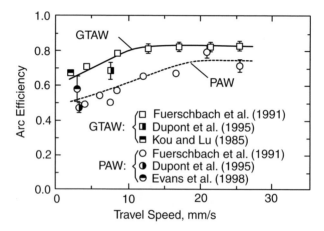

Figure 2.3   Arc efficiencies in GTAW and PAW.

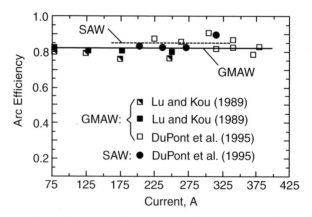

Figure 2.4   Arc efficiencies in GMAW and SAW.

uniformly over the workpiece surface but over localized areas on the workpiece surface called cathode spots (12, 13). The localized heating, called cathode heating, causes the surface oxide to dissociate and leaves a clean metal surface (12). Cathode heating is attributed to field-type emission of electrons. Unlike thermionic emission at the tungsten electrode in DC electrode-negative GTAW, field emission electrons do not cool the cathode (6).

Lu and Kou (10, 11) used a combination of three calorimeters to estimate the amounts of heat transfer from the arc, filler metal droplets, and cathode heating to the workpiece in GMAW of aluminum. Figure 2.5a shows the measurement of heat transfer from droplets (11). The arc is established between a GMAW torch and a GTA torch and the droplets collect in the calorimeter below. From the water temperature rise and the masses and specific heats of the water and the copper basin, the heat transfer from droplets can be deter-

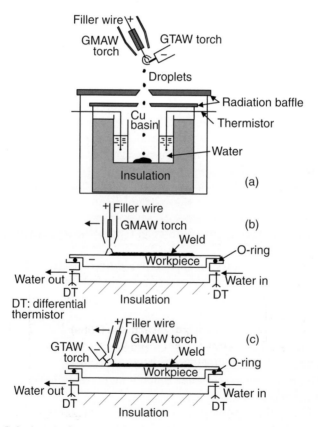

Figure 2.5   Calorimeter for measuring heat inputs in GMAW: (*a*) metal droplets; (*b*) total heat input; (*c*) combined heat inputs from arc and metal droplets. Reprinted from Lu and Kou (10, 11). Courtesy of American Welding Society.

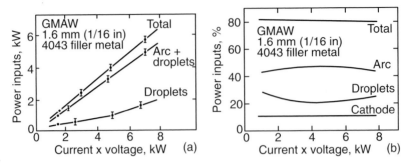

Figure 2.6   Power inputs during GMAW of aluminum: (*a*) measured results; (*b*) breakdown of total power input. Reprinted from Lu and Kou (10). Courtesy of American Welding Society.

mined. Figure 2.5*b* shows the measurement of the total heat transfer to the workpiece. Figure 2.5*c* shows the measurement of the combined heat transfer from the arc and the droplets, with cathode heating shifted to the tungsten electrode of a nearby GTAW torch (10). The results are shown in Figure 2.6*a*. By subtracting the combined heat transfer from the arc and droplets from the total heat transfer to the workpiece, heat transfer from cathode heating was determined. Figure 2.6*b* shows the breakdown of the total heat transfer to the workpiece into those from the arc, droplets, and cathode heating. Within the range of the power studied, the overall efficiency was about 80%, with about 45% from the arc, 25% from droplets, and 10% from cathode heating.

The heat source efficiency can be very low in LBW because of the high reflectivity of metal surfaces to a laser beam, for instance, 98% for CO_2 laser on polished aluminum. The reflectivity can be found by determining the ratio of the reflected beam power to the incident beam power. Xie and Kar (14) show that roughening the surface with sandpapers and oxidizing the surface by brief exposure to high temperatures can reduce the reflectivity significantly.

C. Heat Source Efficiencies in Various Welding Processes   Figure 2.7 summarizes the heat source efficiencies measured in several welding processes. A few comments are made as follows:

LBW: The heat source efficiency is very low because of the high reflectivity of metal surfaces but can be significantly improved by surface modifications, such as roughening, oxidizing, or coating.

PAW: The heat source efficiency is much higher than LBW because reflectivity is not a problem.

GTAW: Unlike in PAW there are no heat losses from the arc plasma to the water-cooled constricting torch nozzle. With DCEN, the imparting electrons are a major source of heat transfer to the workpiece. They release the work function as heat and their kinetic energy is also converted into

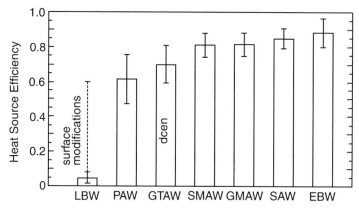

Figure 2.7   Heat source efficiencies in several welding processes.

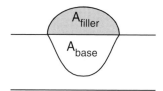

Figure 2.8   Transverse cross section of weld showing areas representing contributions from base metal and filler metal.

heat at the workpiece. In AC GTAW, however, electrons bombard the workpiece only during the electrode-negative half cycle, and the arc efficiency is thus lower. In GTAW with DCEP, the arc efficiency is even lower because the electrons bombard the electrode rather than the workpiece.

GMAW, SMAW: Unlike in GTAW, heat transfer to the electrode can be transferred back to the workpiece through metal droplets, thus improving the arc efficiency.

SAW: Unlike in GMAW or SMAW, the arc is covered with a thermally insulating blanket of molten slag and granular flux, thus reducing heat losses to the surroundings and improving the arc efficiency.

EBW: The keyhole in EBW acts like a "black body" trapping the energy from the electron beam. As a result, the efficiency of the electron beam is very high.

2.1.2   Melting Efficiency

The ability of the heat source to melt the base metal (as well as the filler metal) is of practical interest to the welder. Figure 2.8 shows the cross-sectional area representing the portion of the weld metal contributed by the base metal, A_{base}, and that contributed by the filler metal, A_{filler}. One way to define the *melting efficiency* of the welding arc, η_m, is as follows:

$$\eta_m = \frac{(A_{base}Vt_{weld})H_{base} + (A_{filler}Vt_{weld})H_{filler}}{\eta EIt_{weld}} \quad (2.5)$$

where V is the welding speed, H_{base} the energy required to raise a unit volume of base metal to the melting point and melt it, and H_{filler} the energy required to raise a unit volume of filler metal to the melting point and melt it. The quantity inside the parentheses represents the volume of material melted while the denominator represents the heat transfer from the heat source to the workpiece.

Figures 2.9a and b show the transverse cross section of two steel welds differing in the melting efficiency (7). Here, $EI = 3825$ W and $V = 10$ mm/s for the shallower weld of lower melting efficiency (Figure 2.9a) and $EI = 10170$ W and $V = 26$ mm/s for the deeper weld of higher melting efficiency (Figure 2.9b). Note that the ratio EI/V is equivalent in each case.

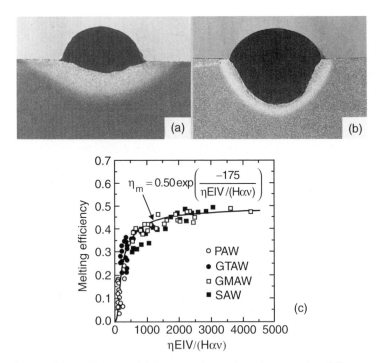

Figure 2.9 Melting efficiency: (*a*) lower at lower heat input and welding speed; (*b*) higher at higher heat input and welding speed; (*c*) variation with dimensionless parameter $\eta EIV/H\alpha v$. Reprinted from DuPont and Marder (7). Courtesy of American Welding Society.

Fuerschbach and Knorovsky (5) proposed the following empirical equation for the melting efficiency:

$$\eta_m = A\exp\left(\frac{-B}{\eta EIV/H\alpha v}\right) \qquad (2.6)$$

where A and B are constants, $H = H_{base} + H_{filler}$, α is the thermal diffusivity, and v is the kinematic viscosity of the weld pool. The results of DuPont and Marder (7) shown in Figure 2.9c confirms the validity of Equation (2.6). As the dimensionless parameter $\eta EIV/H\alpha v$ increases, the melting efficiency increases rapidly first and then levels off. If the arc efficiency η is known, $\eta EIV/H\alpha v$ is also known and the melting efficiency can be predicted from Figure 2.9. With the help of the following equation for determining A_{filler}, A_{base} can then be calculated from Equation (2.5):

$$A_{filler}Vt_{weld} = \pi R^2_{filler}V_{filler}t_{weld} \qquad (2.7)$$

or

$$A_{filler} = \frac{\pi R^2_{filler}V_{filler}}{V} \qquad (2.8)$$

In the above equations R_{filler} and V_{filler} are the radius and feeding speed of the filler metal, respectively. The left-hand side of Equation (2.7) is the volume of the weld metal contributed by the filler metal while the right-hand side is the volume of filler metal used during welding.

It should be noted that the melting efficiency cannot be increased indefinitely by increasing the welding speed without increasing the power input. To do so, the power input must be increased along with the welding speed. It should also be noted that in the presence of a surface-active agent such as sulfur in steel, the weld pool can become much deeper even though the welding parameters and physical properties in Equation (2.6) remain unchanged (Chapter 4).

2.1.3   Power Density Distribution of Heat Source

A. Effect of Electrode Tip Angle   In GTAW with DCEN, the shape of the electrode tip affects both the shape and power density distribution of the arc. As the electrode tip becomes blunter, the diameter of the arc decreases and the power density distribution increases, as illustrated in Figure 2.10. Glickstein (15) showed, in Figure 2.11, that the arc becomes more constricted as the conical tip angle of the tungsten electrode increases.

Savage et al. (16) observed that, under the same welding current, speed, and arc gap, the weld depth–width ratio increases with increasing vertex angle of

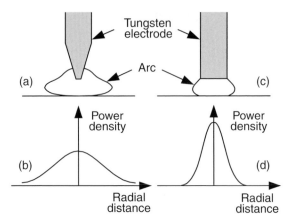

Figure 2.10   Effect of electrode tip angle on shape and power density distribution of gas–tungsten arc.

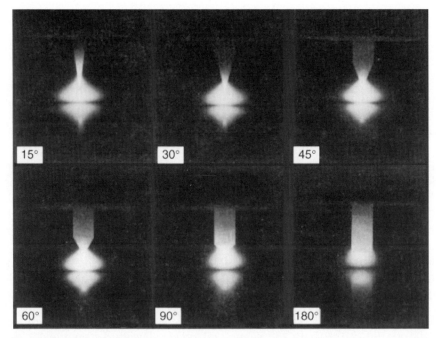

Figure 2.11   Effect of electrode tip angle on shape of gas–tungsten arc. Reprinted from Glickstein (15).

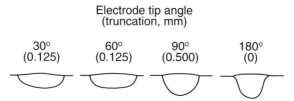

Figure 2.12   Effect of electrode tip geometry on shape of gas–tungsten arc welds in stainless steel (pure Ar, 150 A, 2.0 s, spot-on-plate). Reprinted from Key (17). Courtesy of American Welding Society.

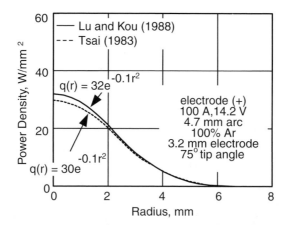

Figure 2.13   Measured power density distributions. Reprinted from Lu and Kou (3). Courtesy of American Welding Society.

the conical tip of the tungsten electrode. Key (17) reported a similar effect of the tip angle, at least with Ar shielding, as shown in Figure 2.12.

B. Measurements   Several investigators have measured the power density distribution (and current density distribution) in the gas tungsten arc by using the split-anode method (2, 18–20). Figure 2.13 shows the results of Tsai (20) and Lu and Kou (3). For simplicity, the Gaussian-type distribution is often used as an approximation (21–23).

2.2   ANALYSIS OF HEAT FLOW IN WELDING

Figure 2.14 is a schematic showing the welding of a stationary workpiece (24). The origin of the coordinate system moves with the heat source at a constant speed V in the negative-x direction. Except for the initial and final transients

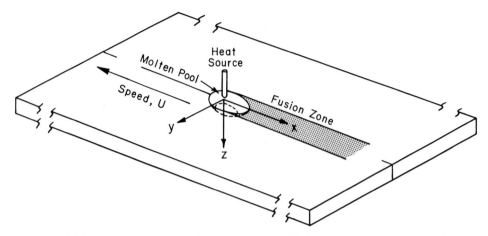

Figure 2.14 Coordinate system (x, y, z) moving with heat source. From Kou and Le (24).

of welding, heat flow in a workpiece of sufficient length is steady, or quasi-stationary, with respect to the moving heat source. In other words, for an observer moving with the heat source, the temperature distribution and the pool geometry do not change with time. This steady-state assumption was first used by Rosenthal (25) to simplify the mathematical treatment of heat flow during welding.

2.2.1 Rosenthal's Equations

Rosenthal (25) used the following simplifying assumptions to derive analytical equations for heat flow during welding:

1. steady-state heat flow,
2. point heat source,
3. negligible heat of fusion,
4. constant thermal properties,
5. no heat losses from the workpiece surface, and
6. no convection in the weld pool.

A. Rosenthal's Two-Dimensional Equation Figure 2.15 is a schematic showing the welding of thin sheets. Because of the small thickness of the workpiece, temperature variations in the thickness direction are assumed negligible and heat flow is assumed two dimensional. Rosenthal (25) derived the following equation for two-dimensional heat flow during the welding of thin sheets of infinite width:

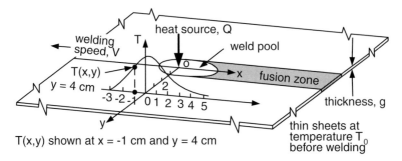

Figure 2.15 Two-dimensional heat flow during welding of thin workpiece.

$$\frac{2\pi(T-T_0)kg}{Q} = \exp\left(\frac{Vx}{2\alpha}\right)K_0\left(\frac{Vr}{2\alpha}\right)$$

(2.9)

where T = temperature

T_0 = workpiece temperature before welding

k = workpiece thermal conductivity

g = workpiece thickness

Q = heat transferred from heat source to workpiece

V = travel speed

α = workpiece thermal diffusivity, namely, $k/\rho C$, where ρ and C are density and specific heat of the workpiece, respectively

K_0 = modified Bessel function of second kind and zero order, as shown in Figure 2.16 (26)

r = radial distance from origin, namely, $(x^2 + y^2)^{1/2}$

Equation (2.9) can be used to calculate the temperature $T(x, y)$ at any location in the workpiece (x, y) with respect to the moving heat source, for instance, at $x = -1$ cm and $y = 4$ cm shown in Figure 2.15. The temperatures at other locations along $y = 4$ cm can also be calculated, and the temperature distribution along $y = 4$ cm can thus be determined. Table 2.1 lists the thermal properties for several materials (27).

B. Rosenthal's Three-Dimensional Equation The analytical solution derived by Rosenthal for three-dimensional heat flow in a semi-infinite workpiece during welding, Figure 2.17, is as follows (25):

$$\frac{2\pi(T-T_0)kR}{Q} = \exp\left[\frac{-V(R-x)}{2\alpha}\right]$$

(2.10)

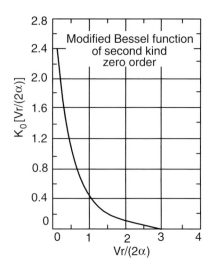

Figure 2.16 Modified Bessel function of second kind and zero order (26).

TABLE 2.1 Thermal Properties for Several Materials

Material	Thermal Diffusivity, α (m²/s)	Volume Thermal Capacity, ρC_s (J/m³ K)	Thermal Conductivity, k (J/m s K)	Melting Point (K)
Aluminum	8.5×10^{-5}	2.7×10^{6}	229.0	933
Carbon steel	9.1×10^{-6}	4.5×10^{6}	41.0	1800
9% Ni steel	1.1×10^{-5}	3.2×10^{6}	35.2	1673
Austenitic stainless steel	5.3×10^{-6}	4.7×10^{6}	24.9	1773
Inconel 600	4.7×10^{-6}	3.9×10^{6}	18.3	1673
Ti alloy	9.0×10^{-6}	3.0×10^{6}	27.0	1923
Copper	9.6×10^{-5}	4.0×10^{6}	384.0	1336
Monel 400	8.0×10^{-6}	4.4×10^{6}	35.2	1573

Source: Gray et al. (27).

where R is the radial distance from the origin, namely, $(x^2 + y^2 + z^2)^{1/2}$. For a given material and a given welding condition, an isotherm T on a plane at a given x has a radius of R. In other words, Equation (2.10) implies that on the transverse cross section of the weld all isotherms, including the fusion boundary and the outer boundary of the heat-affected zone, are semicircular in shape. Equation (2.10) can be used to calculate the steady-state temperature $T(x, y, z)$, with respect to the moving heat source, at any location in the workpiece (x, y, z), for instance, at $x = 1$ cm, $y = 4$ cm, and $z = 0$ cm, as shown in Figure 2.17. The temperatures at other locations along $y = 4$ cm can also be

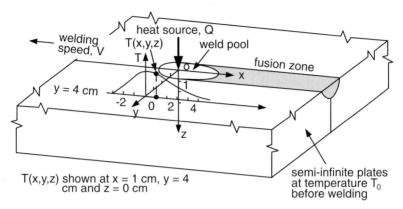

Figure 2.17 Three-dimensional heat flow during welding of semi-infinite workpiece.

calculated, and the temperature distribution along $y = 4$ cm can thus be determined.

C. Thermal Cycles and Temperature Distributions Equations (2.9) and (2.10) can be used to calculate the temperature distribution in the workpiece during welding. The temperature distribution in the welding direction, for instance, the T–x curves in Figures 2.15 and 2.17, are of particular interest. They can be readily converted into temperature–time plots, namely, the *thermal cycles*, by converting distance x into time t through $t = (x - 0)/V$. Figures 2.18 and 2.19 show calculated thermal cycles and temperature distributions at the top surface ($z = 0$) of a thick 1018 steel at two different sets of heat input and welding speed. The infinite peak temperature at the origin of the coordinate system is the result of the singularity problem in Rosenthal's solutions caused by the point heat source assumption.

It should be mentioned, however, that Rosenthal's analytical solutions, though based on many simplifying assumptions, are easy to use and have been greatly appreciated by the welding industry.

2.2.2 Adams' Equations

Adams (28) derived the following equations for calculating the peak temperature T_p at the workpiece surface ($z = 0$) at a distance Y away from the fusion line (measured along the normal direction):

$$\frac{1}{T_p - T_0} = \frac{4.13VYg\rho C}{Q} + \frac{1}{T_m - T_0} \qquad (2.11)$$

for two-dimensional heat flow and

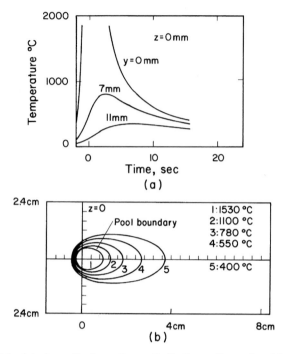

Figure 2.18   Calculated results from Rosenthal's three-dimensional heat flow equation: (*a*) thermal cycles; (*b*) isotherms. Welding speed: 2.4 mm/s; heat input: 3200 W; material: 1018 steel.

$$\frac{1}{T_p - T_0} = \frac{5.44\pi k \alpha}{QV}\left[2 + \left(\frac{VY}{2\alpha}\right)^2\right] + \frac{1}{T_m - T_0} \qquad (2.12)$$

for three-dimensional heat flow.

Several other analytical solutions have also been derived for two-dimensional (29–34) and three-dimensional (31, 35–37) welding heat flow. Because of the many simplifying assumptions used, analytical solutions have had limited success.

2.2.3   Computer Simulation

Many computer models have been developed to study two-dimensional heat flow (e.g., refs. 38–43) and three-dimensional heat flow (e.g., refs. 44–55) during welding. Most assumptions of Rosenthal's analytical solutions are no longer required. Figure 2.20 shows the calculated results of Kou and Le (24) for the GTAW of 3.2-mm-thick sheets of 6061 aluminum alloy. The agreement with observed fusion boundaries and thermal cycles appears good.

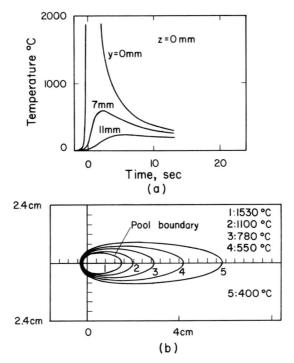

Figure 2.19   Calculated results similar to those in Figure 2.18 but with welding speed of 6.2 mm/s and heat input of 5000 W.

2.3   EFFECT OF WELDING PARAMETERS

2.3.1   Pool Shape

As the heat input Q and welding speed V both increase, the weld pool becomes more elongated, shifting from elliptical to teardrop shaped. Figure 2.21 shows the weld pools traced from photos taken during autogenous GTAW of 304 stainless steel sheets 1.6 mm thick (56). Since the pools were photographed from the side at an inclined angle (rather than vertically), the scale bar applies only to lengths in the welding direction. In each pool the cross indicates the position of the electrode tip relative to the pool. The higher the welding speed, the greater the length–width ratio becomes and the more the geometric center of the pool lags behind the electrode tip.

Kou and Le (57) quenched the weld pool during autogenous GTAW of 1.6-mm 309 stainless steel sheets and observed the sharp pool end shown in Figure 2.22. The welding current was 85 A, voltage 10 V, and speed 4.2 mm/s [10 in./min (ipm)]. The sharp end characteristic of a teardrop-shaped weld pool is evident.

The effect of the welding parameters on the pool shape is more significant in stainless steel sheets than in aluminum sheets. The much lower thermal con-

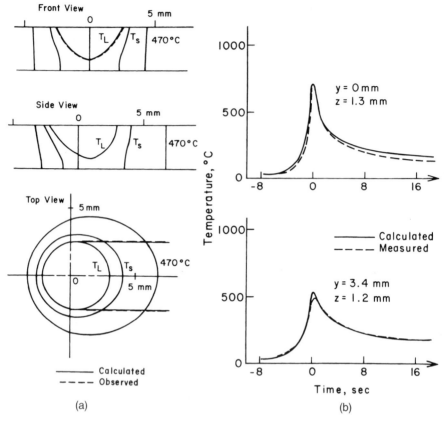

Figure 2.20   Computer simulation of GTAW of 3.2-mm-thick 6061 aluminum, 110 A, 10 V, and 4.23 mm/s: (*a*) fusion boundaries and isotherms; (*b*) thermal cycles. From Kou and Le (24).

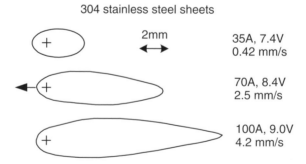

Figure 2.21   Weld pool shapes in GTAW of 304 stainless steel sheets. Courtesy of Smolik (56).

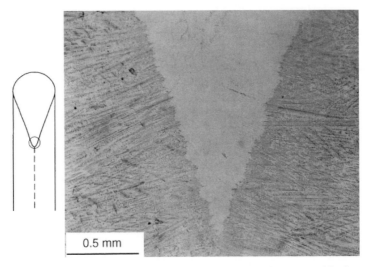

Figure 2.22   Sharp pool end in GTAW of 309 stainless steel preserved by ice quenching during welding. From Kou and Le (57).

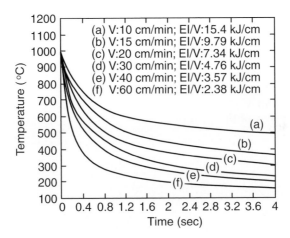

Figure 2.23   Variation in cooling rates with heat input per unit length of weld (EI/V). Reprinted from Lee et al. (58).

ductivity of stainless steels makes it more difficult for the weld pool to dissipate heat and solidify.

2.3.2   Cooling Rate and Temperature Gradient

The ratio EI/V represents the amount of heat input per unit length of weld. Lee et al. (58) measured the cooling rate in GTAW of 2024 aluminum by sticking a thermocouple into the weld pool. Figure 2.23 shows that increasing EI/V

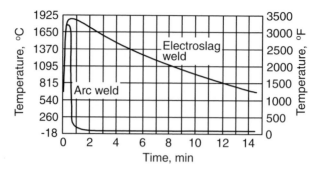

Figure 2.24 Thermal cycles of electroslag and arc welds. Reprinted from Liu et al. (60).

decreases the cooling rate (the slope). Kihara et al. (59) showed that the cooling rate decreases with increasing EI/V and preheating. Figure 2.24 shows that the cooling rate in ESW, which is known to have a very high Q/V, is much smaller than that in arc welding (60). The effects of the heat input, welding speed, and preheat on the cooling rate and temperature gradient can be illustrated by considering the following example.

Example: Bead-on-plate welding of a thick steel plate is carried out using GTAW at 200 A, 10 V, and 2 mm/s. Based on Rosenthal's three-dimensional equation, calculate the 500°C cooling rates along the x axis of the workpiece for zero and 250°C preheating. The arc efficiency is 70% and the thermal conductivity is 35 W/m°C.

Along the x axis of the workpiece,

$$y = z = 0 \quad \text{and} \quad R = x \tag{2.13}$$

Therefore, Equation (2.10) becomes

$$T - T_0 = \frac{Q}{2\pi k x} \tag{2.14}$$

Therefore, the temperature gradient is

$$\boxed{\left(\frac{\partial T}{\partial x}\right)_t = \frac{Q}{2\pi x}\frac{-1}{x^2} = -2\pi k \frac{(T-T_0)^2}{Q}} \tag{2.15}$$

From the above equation and

$$\left(\frac{\partial x}{\partial t}\right)_T = V \tag{2.16}$$

the cooling rate is

$$\left(\frac{\partial T}{\partial t}\right)_x = \left(\frac{\partial T}{\partial x}\right)_t\left(\frac{\partial x}{\partial t}\right)_T = -2\pi k V\frac{(T-T_0)^2}{Q} \tag{2.17}$$

Without preheating the workpiece before welding,

$$\left(\frac{\partial T}{\partial t}\right)_x = (-2\pi \times 35\,\text{W/m}\,°\text{C} \times 2\times10^{-3}\,\text{m/s})\frac{(500°\text{C}-25°\text{C})^2}{0.7\times200\,\text{A} \times 10\,\text{V}} = 71°\text{C/s} \tag{2.18}$$

With 250°C preheating,

$$\left(\frac{\partial T}{\partial t}\right)_x = (-2\pi \times 35\,\text{W/m}\,°\text{C} \times 2\times10^{-3}\,\text{m/s})\frac{(500°\text{C}-250°\text{C})^2}{0.7\times200\,\text{A} \times 10\,\text{V}} = 20°\text{C/s} \tag{2.19}$$

It is clear that the cooling rate is reduced significantly by preheating. Preheating is a common practice in welding high-strength steels because it reduces the risk of heat-affected zone cracking. In multiple-pass welding the interpass temperature is equivalent to the preheat temperature T_0 in single-pass welding.

Equation (2.17) shows that the cooling rate decreases with increasing Q/V, and Equation (2.15) shows that the temperature gradient decreases with increasing Q.

2.3.3 Power Density Distribution

Figure 2.25 shows the effect of the power density distribution of the heat source on the weld shape (24). Under the same heat input and welding speed, weld penetration decreases with decreasing power density of the heat source. As an approximation, the power density distribution at the workpiece surface is often considered Gaussian, as shown by the following equation:

$$q = \frac{3Q}{\pi a^2}\exp\left[\frac{r^2}{-a^2/3}\right] \tag{2.20}$$

where q is the power density, Q the rate of heat transfer from the heat source to the workpiece, and a the effective radius of the heat source.

2.3.4 Heat Sink Effect of Workpiece

Kihara et al. (59) showed that the cooling rate increases with the thickness of the workpiece. This is because a thicker workpiece acts as a better heat sink to cool the weld down. Inagaki and Sekiguchi (61) showed that, under the

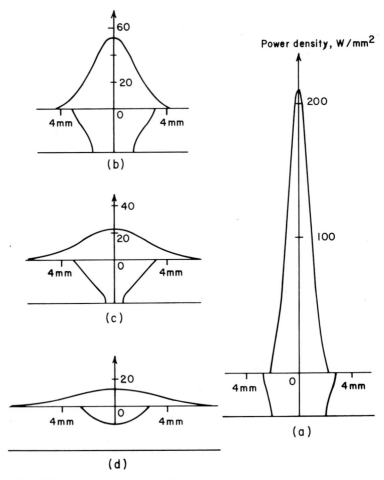

Figure 2.25   Effect of power density distribution on weld shape in GTAW of 3.2-mm 6061 aluminum with 880 W and 4.23 mm/s. From Kou and Le (24).

same heat input and plate thickness, the cooling time is shorter for fillet welding (a T-joint between two plates) than for bead-on-plate welding because of the greater heat sink effect in the former.

2.4   WELD THERMAL SIMULATOR

2.4.1   The Equipment

The thermal cycles experienced by the workpiece during welding can be duplicated in small specimens convenient for mechanical testing by using a weld thermal simulator called Gleeble, a registered trademark of Dynamic Systems.

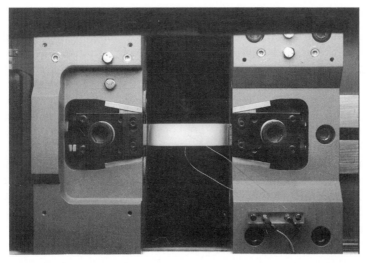

Figure 2.26   A weld simulator specimen held between two water-cooled jaws and resistance heated by electric current passing through it. Courtesy of Dynamic Systems Inc. (63).

These simulators evolved from an original device developed by Nippes and Savage in 1949 (62). Figure 2.26 shows a specimen being resistance heated by the electric current passing through the specimen and the water-cooled jaws holding it (63). A thermocouple spot welded to the middle of the specimen is connected to a feedback control system that controls the amount of electric current passing through the specimen such that a specific thermal cycle can be duplicated.

2.4.2   Applications

There are many applications for weld thermal simulators. For instance, a weld thermal simulator can be used in conjunction with a high-speed dilatometer to help construct continuous-cooling transformation diagrams useful for studying phase transformations in welding and heat treating of steels.

By performing high-speed tensile testing during weld thermal simulation, the elevated-temperature ductility and strength of metals can be evaluated. This is often called the *hot-ductility test*. Nippes and Savage (64, 65), for instance, used this test to investigate the heat-affected zone fissuring in austenitic stainless steels.

Charpy impact test specimens can also be prepared from specimens (1 × 1 cm in cross section) subjected to various thermal cycles. This synthetic-specimen or simulated-microstructure technique has been employed by numerous investigators to study the heat-affected-zone toughness.

2.4.3 Limitations

Weld thermal simulators, though very useful, have some limitations and draw-backs. First, extremely high cooling rates during electron and laser beam welding cannot be reproduced, due to the limited cooling capacity of the sim-ulators. Second, because of the surface heat losses, the temperature at the surface can be lower than that at the centerline of the specimen, especially if the peak temperature is high and the thermal conductivity of the specimen is low (66). Third, the temperature gradient is much lower in the specimen than in the weld heat-affected zone, for instance, 10°C/mm, as opposed to 300°C/mm near the fusion line of a stainless steel weld. This large difference in the temperature gradient tends to make the specimen microstructure differ from the heat-affected-zone microstructure. For example, the grain size tends to be significantly larger in the specimen than in the heat-affected zone, espe-cially at high peak temperatures such as 1100°C and above.

REFERENCES

1. Kou, S., and Le, Y., *Metall. Trans. A*, **15A:** 1165, 1984.
2. Kou, S., and Lu, M. J., in *Welding Metallurgy*, 1st ed., S. Kou, 1987, Wiley, New York, p. 32; Lu, M. J., Ph.D. Thesis, Department of Materials Science and Engineering, University of Wisconsin, Madison, WI, 1988.
3. Lu, M., and Kou, S., *Weld. J.*, **67:** 29s, 1988.
4. Giedt, W. H., Tallerico, L. N., and Fuerschbach, P. W., *Weld. J.*, **68:** 28s, 1989.
5. Fuerschbach, P. W., and Knorovsky, G. A., *Weld. J.*, **70:** 287s, 1991.
6. Fuerschbach, P. W., *Weld. J.*, **77:** 76s, 1998.
7. Dupont, J. N., and Marder, A. R., *Weld. J.*, **74:** 406s, 1995.
8. Dupont, J. N., and Marder, A. R., *Metall. Mater. Trans.*, **27B:** 481s, 1996.
9. Evans, D. M., Huang, D., McClure, J. C., and Nunes, A. C., *Weld. J.*, **77:** 53s, 1998.
10. Lu, M., and Kou, S., *Weld. J.*, **68:** 452s, 1989.
11. Lu, M., and Kou, S., *Weld. J.*, **68:** 382s, 1989.
12. Essers, W. G., and Van Gompel, M. R. M., *Weld. J.*, **63:** 26s, 1984.
13. Hasegawa, M., and Goto, H., IIW Document 212-212-71, International Welding Institute, London, 1971.
14. Xie, J., and Kar, A., *Weld. J.*, **78:** 343s, 1999.
15. Glickstein, S. S., *Weld. J.*, **55:** 222s, 1976.
16. Savage, W. F., Strunk, S. S., and Ishikawa, Y., *Weld. J.*, **44:** 489s, 1965.
17. Key, J. F., *Weld. J.*, **59:** 365s, 1980.
18. Nestor, O. H., *J. Appl. Phys.*, **33:** 1638, 1962.
19. Schoeck, P. A., in *Modern Developments in Heat Transfer*, Warren Ibele, Academic, New York, 1963, p. 353.
20. Tsai, N., Ph.D. Thesis, Department of Materials Science and Engineering, MIT, Cambridge, MA, 1983.

21. Kou, S., and Sun, D. K., *Metall. Trans. A*, **16A:** 203, 1985.

22. Oreper, G., Eagar, T. W., and Szekely, J., *Weld. J.*, **62:** 307s, 1983.

23. Pavelic, V., Tan Bakuchi, L. R., Uyechara, O. A., and Myers, P. S., *Weld. J.*, **48:** 295s, 1969.

24. Kou, S., and Le, Y., *Metall. Trans. A*, **14A:** 2245, 1983.

25. Rosenthal, D., *Weld. J.*, **20:** 220s, 1941.

26. Abramowitz, M., and Stegun, I.A., Eds., *Handbook of Mathematical Functions*, National Bureau of Standards, Washington, DC, 1964.

27. Gray, T. F. G., Spence, J., and North, T. H., *Rational Welding Design*, Newnes-Butterworth, London, 1975.

28. Adams, C. M., Jr., *Weld. J.*, **37:** 210s, 1958.

29. Grosh, R. J., Trabant, E. A., and Hawskins, G. A., *Q. Appl. Math.*, **13:** 161, 1955.

30. Swift-Hook, D. T., and Gick, A. E. F., *Weld. J.*, **52:** 492s, 1973.

31. Jhaveri, P., Moffatt, W. G., and Adams C. M., Jr., *Weld. J.*, **41:** 12s, 1962.

32. Myers, P. S., Uyehara, O. A., and Borman, G. L., *Weld. Res. Council Bull.*, **123:** 1967.

33. Ghent, H. W., Hermance, C. E., Kerr, H. W., and Strong, A. B., in *Proceedings of the Conference on Arc Physics and Weld Pool Behavior*, Welding Institute, Arbington Hall, Cambridge, 1979, p. 389.

34. Trivedi, R., and Shrinivason, S. R., *J. Heat Transfer*, **96:** 427, 1974.

35. Grosh, R. J., and Trabant, E. A., *Weld. J.*, **35:** 396s, 1956.

36. Malmuth, N. D., Hall, W. F., Davis, B. I., and Rosen, C. D., *Weld. J.*, **53:** 388s, 1974.

37. Malmuth, N. D., *Int. J. Heat Mass Trans.*, **19:** 349, 1976.

38. Kou, S., Kanevsky, T., and Fyfitch, S., *Weld. J.*, **62:** 175s, 1982.

39. Pavelic, V., and Tsao, K. C., in *Proceedings of the Conference on Arc Physics and Weld Pool Behavior*, Vol. 1, Welding Institute, Arbington Hall, Cambridge, 1980, p. 251.

40. Kou, S., and Kanevsky, T., in *Proceedings of the Conference on New Trends of Welding Research in the United States*, Ed. S. David, ASM International, Materials Park, OH, 1982, p. 77.

41. Friedman, E., *Trans. ASME; J. Pressure Vessel Techn.*, Series J, No. 3, **97:** 206, 1965.

42. Grill, A., *Metall. Trans. B*, **12B:** 187, 1981.

43. Grill, A., *Metall. Trans. B*, **12B:** 667, 1981.

44. Ushio, M., Ishmimura, T., Matsuda, F., and Arata, Y., *Trans. Japan Weld. Res. Inst.*, **6:** 1, 1977.

45. Kou, S., in *Proceedings of the Conference on Modeling of Casting and Welding Processes*, Metall. Society of AIME, Warrendale, PA, 1980.

46. Friedman, E., and Glickstein, S. S., *Weld. J.*, **55:** 408s, 1976.

47. Friedman, E., in *Numerical Modeling of Manufacturing Processes*, Ed. R. F. Jones, Jr., American Society of Mechanical Engineers, New York, 1977, p. 35.

48. Lewis, R. W., Morgan, K., and Gallagher, R. H., in *Numerical Modeling of Manufacturing Processes*, Ed. R. F. Jones, Jr., American Society of Mechanical Engineers, New York, 1977, p. 67.

49. Hsu, M. B., in *Numerical Modeling of Manufacturing Processes*, Ed. R. F. Jones, Jr., American Society of Mechanical Engineers, New York, 1977, p. 97.

50. Glickstein, S. S., and Friedman, E., *Weld. J.*, **60:** 110s, 1981.

51. Krutz, G. W., and Segerlind, L. J., *Weld. J.*, **57:** 211s, 1978.

52. Paley, Z., and Hibbert, P. D., *Weld. J.*, **54:** 385s, 1975.

53. Hibbitt, H. D., and Marcal, P. V., *Comput. Struct.*, **3:** 1145, 1973.

54. Mazumder, J., and Steen, W. M., *J. Appl. Phys.*, **51:** 941, 1980.

55. Tsai, N. S., and Eagar, T. W., *Weld. J.*, **62:** 82s, 1983.

56. Smolik, G., private communication, Idaho National Engineering Laboratories, Idaho Falls, Idaho, 1984.

57. Kou, S., and Le, Y., *Metall. Trans. A*, **13A:** 1141, 1982.

58. Lee, J. Y., Park, J. M., Lee, C. H., and Yoon, E. P., in *Synthesis/Processing of Light-weight Metallic Materials II*, Eds. C. M. Ward-Close, F. H. Froes, S. S. Cho, and D. J. Chellman, The Minerals, Metals and Materials Society, Warrendale, PA, 1996, p. 49.

59. Kihara, H., Suzuki, H., and Tamura, H., *Researches on Weldable High-Strength Steels*, 60th Anniversary Series, Vol. 1, Society of Naval Architects of Japan, Tokyo, 1957.

60. Liu, S., Brandi, S. D., and Thomas, R. D., in *ASM Handbook*, Vol. 6, ASM International, Materials Park, OH, 1993, p. 270.

61. Inagaki, M., and Sekiguchi, H., *Trans. Nat. Res. Inst. Metals, Tokyo, Japan*, **2**(2)**:** 102 (1960).

62. Nippes, E. F., and Savage, W. F., *Weld. J.*, **28:** 534s, 1949.

63. HAZ 1000, Duffers Scientific, Troy, NY, now Dynamic System, Inc.

64. Nippes, E. F., Savage, W. F., Bastian, B. J., Mason, H. F., and Curran, R. M., *Weld. J.*, **34:** 183s, 1955.

65. Nippes, E. F., Savage, W. F., and Grotke, G. E., *Weld. Res. Council Bull.*, **33:** February 1957.

66. Widgery, D. J., in *Weld Thermal Simulators for Research and Problem Solving*, Welding Institute, Cambridge, 1972, p. 14.

FURTHER READING

1. Rykalin, N. N., *Calculation of Heat Flow in Welding*, Trans. Z. Paley and C. M. Adams, Jr., Document 212-350-74, 1974, International Institute of Welding, London.

2. Rosenthal, D., *Weld. J.*, **20:** 220s (1941).

3. Meyers, P. S., Uyehara, O. A., and Borman, G. L., *Weld. Res. Council Bull.* **123:** 1967.

4. Kou, S., *Transport Phenomena and Materials Processing*, Wiley, New York, 1996.

PROBLEMS

2.1   In one welding experiment, 50-mm-thick steel plates were joined using electroslag welding. The current and voltage were 480 A and 34 V, respectively. The heat losses to the water-cooled copper shoes and by radiation from the surface of the slag pool were 1275 and 375 cal/s, respectively. Calculate the heat source efficiency.

2.2   It has been reported that the heat source efficiency in electroslag welding increases with increasing thickness of the workpiece. Explain why.

2.3   **(a)** Consider the welding of 25.4-mm-thick steel plates. Do you prefer to apply Rosenthal's two- or three-dimensional heat flow equation for full-penetration electron beam welds? What about bead-on-plate gas–tungsten arc welds?

(b) Suppose you are interested in studying the solidification structure of the weld metal and you wish to calculate the temperature distribution in the weld pool. Do you expect Rosenthal's equations to provide reliable thermal information in the pool? Why or why not?

(c) In multipass welding do you expect a higher or lower cooling rate in the first pass than in the subsequent passes? Why?

2.4   Large aluminum sheets 1.6 mm thick are butt welded using GTAW with alternating current. The current, voltage, and welding speed are 100 A, 10 V, and 2 mm/s, respectively. Calculate the peak temperatures at distance of 1.0 and 2.0 mm from the fusion boundary. Assume 50% arc efficiency.

2.5   Bead-on-plate welding of a thick-section carbon steel is carried out using 200 A, 20 V, and 2 mm/s. The preheat temperature and arc efficiency are 100°C and 60%, respectively. Calculate the cross-sectional area of the weld bead.

2.6   (a) Do you expect to have difficulty in achieving steady-state heat flow during girth (or circumferential) welding of tubes by keeping constant heat input and welding speed? Explain why. What is the consequence of the difficulty? (b) Suggest two methods that help achieve steady-state heat flow during girth welding.

2.7   A cold-rolled AISI 1010 low-carbon steel sheet 0.6 mm thick was tested for surface reflectivity in CO_2 laser beam welding under the following different surface conditions: (a) as received; (b) oxidized in air furnace at 1000°C for 20 s; (c) oxidized in air furnace at 1000°C for 40 s; (d) covered with steel powder. In which order does the reflectivity rank in these surface conditions and why?

2.8   It was observed in YAG laser beam welding of AISI 409 stainless steel that under the same power the beam size affected the depth–width ratio of the resultant welds significantly. Describe and explain the effect.

2.9   Calculate the thermal cycle at the top surface of a very thick carbon steel plate at 5 mm away from the centerline of the weld surface. The power of the arc is 2 kW, the arc efficiency 0.7, the travel speed 2 mm/s, and the preheat temperature 100°C.

2.10 Is the transverse cross section of the weld pool at a fixed value of x perfectly round according to Rosenthal's three-dimensional heat flow equation? Explain why or why not based on the equation. What does your answer tell you about the shape of the transverse cross section of a weld based on Rosenthal's three-dimensional equation?

3 Chemical Reactions in Welding

Basic chemical reaction during fusion welding will be described in this chapter, including gas–metal reactions and slag–metal reactions. The effect of these chemical reactions on the weld metal composition and mechanical properties will be discussed.

3.1 OVERVIEW

3.1.1 Effect of Nitrogen, Oxygen, and Hydrogen

Nitrogen, oxygen, and hydrogen gases can dissolve in the weld metal during welding. These elements usually come from air, the consumables such as the shielding gas and flux, or the workpiece such as the moist or dirt on its surface. Nitrogen, oxygen, and hydrogen can affect the soundness of the resultant weld significantly. Some examples of the effect of these gases are summarized in Table 3.1.

3.1.2 Techniques for Protection from Air

As described in Chapter 1, various techniques can be used to protect the weld pool during fusion welding. These techniques are summarized in Table 3.2. Figure 3.1 shows the weld oxygen and nitrogen levels expected from several different arc welding processes (1, 2). As will be explained below, techniques provide different degrees of weld metal protection.

A. GTAW and GMAW Gas–tungsten arc welding is the cleanest arc welding process because of the use of inert shielding gases (Ar or He) and a short, stable arc. Gas shielding through the torch is sufficient for welding most materials. However, as shown in Figure 3.2, additional gas shielding to protect the solidified but still hot weld is often provided both behind the torch and under the weld of highly reactive metals such as titanium (3, 4). Welding can also be conducted inside a special gas-filled box. Although also very clean, GMAW is not as clean as GTAW due to the less stable arc associated with the use of consumable electrodes. Furthermore, the greater arc length in GMAW reduces the protective effects of the shielding gas. Carbon dioxide is sometimes employed as shielding gas in GMAW. Under the high temperature of the arc,

TABLE 3.1   Effect of Nitrogen, Oxygen, and Hydrogen on Weld Soundness

	Nitrogen	Oxygen	Hydrogen
Steels	Increases strength but reduces toughness	Reduces toughness but improves it if acicular ferrite is promoted	Induces hydrogen cracking
Austenitic or duplex stainless steels	Reduces ferrite and promotes solidification cracking		
Aluminum		Forms oxide films that can be trapped as inclusions	Forms gas porosity and reduces both strength and ductility
Titanium	Increases strength but reduces ductility	Increases strength but reduces ductility	

TABLE 3.2   Protection Techniques in Common Welding Processes

Protection Technique	Fusion Welding Process
Gas	Gas tungsten arc, gas metal arc, plasma arc
Slag	Submerged arc, electroslag
Gas and slag	Shielded metal arc, flux-cored arc
Vacuum	Electron beam
Self-protection	Self-shielded arc

decomposition into CO and O is favored, potentially increasing the weld oxygen level.

B. SMAW   The flow of gas in SMAW is not as well directed toward the weld pool as the flow of inert gas in GTAW or GMAW. Consequently, the protection afforded the weld metal is less effective, resulting in higher weld oxygen and nitrogen levels. Carbon dioxide, produced by the decomposition of carbonate or cellulose in the electrode covering, can potentially increase the weld oxygen level.

C. Self-Shielded Arc Welding   Self-shielded arc welding uses strong nitride formers such as Al, Ti, and Zr in the electrode wire alone to protect against nitrogen. Since these nitride formers are also strong deoxidizers, the weld

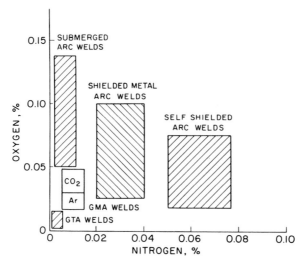

Figure 3.1 Oxygen and nitrogen levels expected from several arc welding processes. From Rein (1); reproduced from Eagar (2).

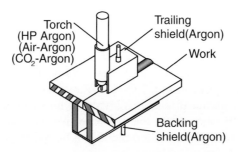

Figure 3.2 Gas–tungsten arc welding of titanium with additional gas shielding. Reprinted from Harwig et al. (3). Courtesy of American Welding Society.

oxygen levels are slightly lower than in SMAW. Unfortunately, as can be seen in Figure 3.1, the weld nitrogen content is still rather high.

D. SAW The weld oxygen level in submerged arc welds can vary significantly, depending on the composition of the flux; the very high oxygen levels associated with acidic fluxes containing large percentages of SiO_2, according to Eagar (2), are the result of SiO_2 decomposition. This is consistent with the large increase in the weld metal silicon content when acidic fluxes are used (2). If atmospheric contamination were the reason for the weld metal oxygen content, the nitrogen content would also have been high.

3.2 GAS–METAL REACTIONS

The gas–metal reactions here refer to chemical reactions that take place at the interface between the gas phase and the liquid metal. They include the dissolution of nitrogen, oxygen, and hydrogen in liquid metal and the evolution of carbon monoxide.

3.2.1 Thermodynamics of Reactions

In steelmaking, exposure of molten steel to molecular nitrogen, N_2, can result in dissolution of nitrogen atoms in the molten steel, that is,

$$\frac{1}{2} N_2(g) = \underline{N} \tag{3.1}$$

where the underlining bar denotes dissolution in molten metal. From thermodynamics (5), the equilibrium concentration of dissolved nitrogen, $[\underline{N}]$, at any given temperature T can be determined from the following relationship:

$$\ln K_N^d = \ln\left(\frac{[\underline{N}]}{\sqrt{p_{N_2}}}\right) = \frac{-\Delta G^\circ}{RT} \tag{3.2}$$

where K_N^d is the equilibrium constant for reaction (3.1) based on dissolution from a diatomic gas N_2, p_{N_2} the partial pressure (in atmospheres) of N_2 above the molten metal, ΔG° the standard free energy of formation (in calories per mole), and R the gas constant 1.987 cal/(K mol). Table 3.3 shows the values of ΔG° for several chemical reactions involving nitrogen, oxygen, and hydrogen (6–9). From Equation (3.2), the well-known *Sievert law* (10) for the dissolution of a diatomic gas in molten metal can be written as

TABLE 3.3 Free Energy of Reactions Involving Nitrogen, Oxygen, and Hydrogen

Gas	Reaction	Free Energy of Reaction, ΔG° (cal/mol)	Reference
Nitrogen	$\frac{1}{2}N_2(g) = N(g)$	$86596.0 - 15.659T$ (K)	6
	$N(g) = \underline{N}(\text{wt \% in steel})$	$-85736.0 + 21.405T$	
	$\frac{1}{2}N_2(g) = \underline{N}(\text{wt \% in steel})$	$860.0 + 5.71T$	7
Oxygen	$\frac{1}{2}O_2(g) = O(g)$	$60064 - 15.735T$	6
	$O(g) = \underline{O}(\text{wt \% in steel})$	$-88064 + 15.045\ T$	
	$\frac{1}{2}O_2(g) = \underline{O}(\text{wt \% in steel})$	$-28000 - 0.69T$	8
Hydrogen	$\frac{1}{2}H_2(g) = H(g)$	$53500.0 - 14.40T$	9
	$H(g) = \underline{H}(\text{ppm in steel})$	$-44780.0 + 3.38T$	9
	$\frac{1}{2}H_2(g) = \underline{H}(\text{ppm in steel})$	$8720.0 - 11.02T$	8

$$[\underline{N}] = K_N^d \sqrt{p_{N_2}} \qquad (3.3)$$

In arc welding, however, a portion of the N_2 molecules can dissociate (or even ionize) under the high temperature of the arc plasma. The atomic N so produced can dissolve in the molten metal as follows:

$$N = \underline{N} \qquad (3.4)$$

$$\ln K_N^m = \ln\left(\frac{[\underline{N}]}{p_N}\right) = \frac{-\Delta G^\circ}{RT} \qquad (3.5)$$

where K_N^m is the equilibrium constant for reaction (3.4) based on dissolution from a monatomic gas N and p_N the partial pressure (in atmospheres) of N above the molten metal.

It is interesting to compare dissolution of nitrogen in molten steel from molecular nitrogen to that from atomic nitrogen. Consider an arbitrary temperature of 1600°C for molten steel for the purpose of discussion. Based on the free energy of reaction ΔG° shown in Table 3.3, for molecular nitrogen a pressure of $p_{N_2} = 1$ atm is required in order to have $[\underline{N}] = 0.045$ wt %. For atomic nitrogen, however, only a pressure of $p_N = 2 \times 10^{-7}$ atm is required to dissolve the same amount of nitrogen in molten steel.

Similarly, for the dissolution of oxygen and hydrogen from $O_2(g)$ and $H_2(g)$,

$$\frac{1}{2}O_2(g) = \underline{O} \qquad (3.6)$$

$$\frac{1}{2}H_2(g) = \underline{H} \qquad (3.7)$$

However, as in the case of nitrogen, a portion of the O_2 and H_2 molecules can dissociate (or even ionize) under the high temperature of the arc plasma. The atomic O and H so produced can dissolve in the molten metal as follows:

$$O = \underline{O} \qquad (3.8)$$

$$H = \underline{H} \qquad (3.9)$$

DebRoy and David (11) showed, in Figure 3.3, the dissolution of monoatomic, rather than diatomic, nitrogen and hydrogen dominates in molten iron. As they pointed out, several investigators have concluded that the species concentration in the weld metal can be significantly *higher* than those calculated from dissolution of diatomic molecules. Dissociation of such molecules to neutral atoms and ions in the arc leads to enhanced dissolution in the molten metal.

In the case of hydrogen the calculated results are consistent with the earlier ones of Gedeon and Eagar (9) shown in Figure 3.4. The calculation is based

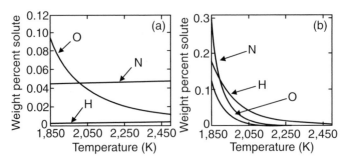

Figure 3.3 Equilibrium concentration of nitrogen, oxygen, and hydrogen in liquid iron as a function of temperature: (a) $N_2(g)$ with $p_{N_2} = 1$ atm, $O_2(g)$ with $p_{O_2} = 10^{-9}$ atm, $H_2(g)$ with $p_{H_2} = 1$ atm; (b) $N(g)$ with $p_N = 10^{-6}$ atm, $O(g)$ with $p_O = 10^{-8}$ atm, $H(g)$ with $p_H = 5 \times 10^{-2}$ atm. From DebRoy and David (11).

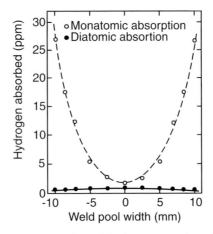

Figure 3.4 Equilibrium concentration of hydrogen as a function of weld pool location. Reprinted from Gedeon and Eagar (9). Courtesy of American Welding Society.

on a dissociation temperature of 2500°C, 0.01 atm hydrogen added to the argon shielding gas, and the pool surface temperature distribution measured by Krause (12). As shown, the majority of hydrogen absorption appears to take place around the outer edge of the weld pool, and monatomic hydrogen absorption dominates the contribution to the hydrogen content. This contradicts predictions based on Sievert's law that the maximum absorption occurs near the center of the pool surface where the temperature is highest. However, as they pointed out, the dissolution process alone does not determine the hydrogen content in the resultant weld metal. Rejection of the dissolved hydrogen atoms by the solidification front and diffusion of the hydrogen atoms from the weld pool must also be considered. It is interesting to note that

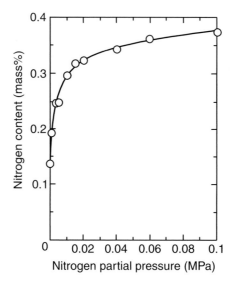

Figure 3.5   Effect of nitrogen partial pressure in Ar–N$_2$ shielding gas on nitrogen content in welds of duplex stainless steel. Reprinted from Sato et al. (14).

Hooijmans and Den Ouden (13) suggested that a considerable amount of hydrogen absorbed by the liquid metal during welding leaves the weld metal immediately after the extinction of the arc.

3.2.2   Nitrogen

For metals that neither dissolve nor react with nitrogen, such as copper and nickel, nitrogen can be used as the shielding gas during welding. On the other hand, for metals that either dissolve nitrogen or form nitrides (or both), such as Fe, Ti, Mn, and Cr, the protection of the weld metal from nitrogen should be considered.

A. Sources of Nitrogen   The presence of nitrogen in the welding zone is usually a result of improper protection against air. However, nitrogen is some-times added purposely to the inert shielding gas. Figure 3.5 shows the weld nitrogen content of a duplex stainless steel as a function of the nitrogen partial pressure in the Ar–N$_2$ shielding gas (14). Nitrogen is an *austenite stabilizer* for austenitic and duplex stainless steels. Increasing the weld metal nitrogen content can decrease the ferrite content (Chapter 9) and increase the risk of solidification cracking (Chapter 11).

B. Effect of Nitrogen   The presence of nitrogen in the weld metal can sig-nificantly affect its mechanical properties. Figure 3.6 shows the needlelike structure of iron nitride (Fe$_4$N) in a ferrite matrix (15). The sharp ends of such

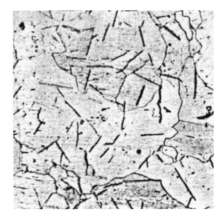

Figure 3.6   Iron nitride in a ferrite matrix (×500). From Seferian (15).

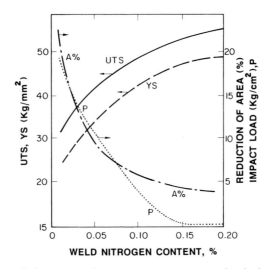

Figure 3.7   Effect of nitrogen on the room temperature mechanical properties of mild steel welds. From Seferian (15).

a brittle nitride act as ideal sites for crack initiation. As shown in Figure 3.7, the ductility and the impact toughness of the weld metal decrease with increasing weld metal nitrogen (15). Figure 3.8 shows that nitrogen can decrease the ductility of Ti welds (4).

C. Protection against Nitrogen   In the self-shielded arc welding process, strong nitride formers (such as Ti, Al, Si, and Zr) are often added to the filler wire (16). The nitrides formed enter the slag and nitrogen in the weld metal is thus reduced. As already shown in Figure 3.1, however, the nitrogen con-

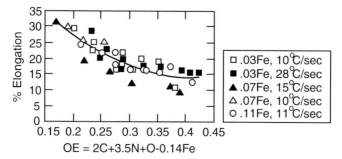

Figure 3.8 Effect of oxygen equivalence (OE) on ductility of titanium welds. Reprinted from Harwig et al. (4). Courtesy of American Welding Society.

tents of self-shielded arc welds can still be rather high, and other arc welding processes such as GTAW, GMAW, or SAW should be used if weld nitrogen contamination is to be minimized.

3.2.3   Oxygen

A. Sources of Oxygen   Oxygen in the weld metal can come from the air, the use of excess oxygen in oxyfuel welding, and the use of oxygen- or CO_2-containing shielding gases. It can also come from the decomposition of oxides (especially SiO_2 and MnO and FeO) in the flux and from the slag–metal reactions in the weld pool, which will be discussed subsequently.

In GMAW of steels the addition of oxygen or carbon dioxide to argon (e.g., Ar–2% O_2) helps stabilize the arc, reduce spatter, and prevent the filler metal from drawing away from (or not flowing out to) the fusion line (17). Carbon dioxide is widely used as a shielding gas in FCAW, the advantages being low cost, high welding speed, and good weld penetration. Baune et al. (18) pointed out that CO_2 can decompose under the high temperature of the welding arc as follows:

$$CO_2(g) = CO(g) + \frac{1}{2}O_2(g) \tag{3.10}$$

$$CO(g) = C(s) + \frac{1}{2}O_2(g) \tag{3.11}$$

B. Effect of Oxygen   Oxygen can oxidize the carbon and other alloying elements in the liquid metal, modifying their prevailing role, depressing hardenability, and producing inclusions. The oxidation of carbon is as follows:

$$\underline{C} + \underline{O} = CO(g) \tag{3.12}$$

TABLE 3.4   Effect of Oxygen–Acetylene Ratio on Weld Metal Composition and Properties of Mild Steel

	Before	$a = O_2/C_2H_2 > 1$				$a \leq 1$	
		$a = 1.14$	$a = 1.33$	$a = 2$	$a = 2.37$	$a = 1$	$a = 0.82$
C	0.155	0.054	0.054	0.058	0.048	0.15	1.56
Mn	0.56	0.38	0.265	0.29	0.18	0.29	0.375
Si	0.03						
S	0.030						
P	0.018						
O	—	0.04	0.07	0.09	—	0.02	0.01
N	—	0.015	0.023	0.030	—	0.012	0.023
Impact value, kg/cm^2	—	5.5	1.40	1.50	1.30	6.9	2.3
Hardness, HB	—	130	132	115	100	140	320
Grain size	—	6	5	4	4	4	5

Source: Seferian (15).

The oxidation of other alloying elements, which will be discussed subsequently in slag–metal reactions, forms oxides that either go into the slag or remain in the liquid metal and become inclusion particles in the resultant weld metal.

Table 3.4 shows the effect of gas composition in oxyacetylene welding of mild steel on the weld metal composition and properties (15). When too much oxygen is used, the weld metal has a high oxygen level but low carbon level. On the other hand, when too much acetylene is used, the weld metal has a low oxygen level but high carbon level (the flame becomes carburizing). In either case, the weld mechanical properties are poor. When the oxygen–acetylene ratio is close to 1, both the impact toughness and strength (proportional to hardness) are reasonably good.

If oxidation results in excessive inclusion formation in the weld metal or significant loss of alloying elements to the slag, the mechanical properties of the weld metal can deteriorate. Figure 3.9 shows that the strength, toughness, and ductility of mild steel welds can all decrease with increasing oxygen contamination (15). In some cases, however, fine inclusion particles can act as nucleation sites for acicular ferrite to form and improve weld metal toughness (Chapter 9). For aluminum and magnesium alloys, the formation of insoluble oxide films on the weld pool surface during welding can cause incomplete fusion. Heavy oxide films prevent a keyhole from being established properly in conventional PAW of aluminum, and more advanced DC variable-polarity PAW has to be used. In the latter, oxide films are cleaned during the electrode-positive part of the current cycle.

Bracarense and Liu (19) discovered in SMAW that the metal transfer *droplet size* can increase gradually during welding, resulting in increasing Mn

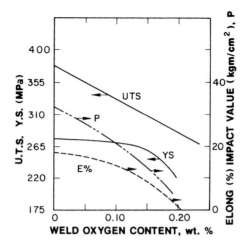

Figure 3.9 Effect of the oxygen content on the mechanical properties of mild steel welds. From Seferian (15).

and Si transfer to the weld pool and hence increasing weld metal hardness along the weld length, as shown in Figure 3.10. As the electrode is heated up more and more during welding, the droplet size increases gradually (becomes more globular). This reduces the surface area (per unit volume) for oxygen to react with Mn and Si and hence improves the efficiency of Mn and Si transfer to the weld pool.

3.2.4 Hydrogen

A. Steels The presence of hydrogen during the welding of high-strength steels can cause hydrogen cracking (Chapter 17).

A.1. Sources of Hydrogen Hydrogen in the welding zone can come from several different sources: the combustion products in oxyfuel welding; decomposition products of cellulose-type electrode coverings in SMAW; moisture or grease on the surface of the workpiece or electrode; and moisture in the flux, electrode coverings, or shielding gas.

As mentioned previously in Chapter 1, in SMAW high-cellulose electrodes contain much cellulose, $(C_6H_{10}O_5)_x$, in the electrode covering. The covering decomposes upon heating during welding and produces a gaseous shield rich in H_2, for instance, 41% H_2, 40% CO, 16% H_2O, and 3% CO_2 in the case of E6010 electrodes (20). On the other hand, low-hydrogen electrodes contain much $CaCO_3$ in the electrode covering. The covering decomposes during welding and produces a gaseous shield low in H_2, for example, 77% CO, 19% CO_2, 2% H_2, and 2% H_2O in the case of E6015 electrodes. As such, to reduce weld metal hydrogen, low-hydrogen electrodes should be used.

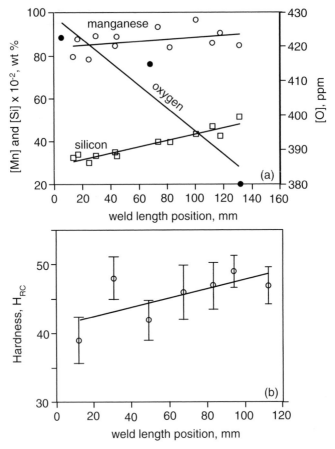

Figure 3.10   Variations of weld metal along the length of a weld made with an E7018 electrode: (*a*) composition; (*b*) hardness. Reprinted from Bracarense and Liu (19). Courtesy of American Welding Society.

A.2. Measuring Hydrogen Content   Various methods have been developed for measuring the hydrogen content in the weld metal of steels. The mercury method and the gas chromatography method are often used. Figure 3.11 shows the *mercury method* (21). A small test specimen ($13 \times 25 \times 127$ mm, or $\frac{1}{2} \times 1 \times 5$ in.) is welded in a copper fixture. The welded test specimen is then immersed in mercury contained in a eudiometer tube. As hydrogen diffuses out of the welded test specimen, the mercury level in the eudiometer tube continues to drop. From the final mercury level, *H* (in millimeters), the amount of hydrogen that diffuses out of the specimen, that is, the so-called diffusible hydrogen, can be measured. This method, however, can take days because of the slow diffusion of hydrogen at room temperature. In the *gas chromatography method* (22), the specimen is transferred to a leak-tight chamber after

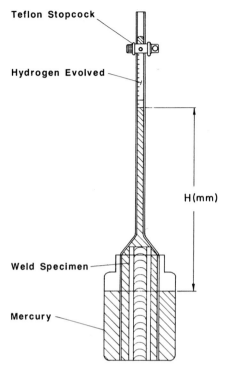

Teflon Stopcock

Hydrogen Evolved

H(mm)

Weld Specimen

Mercury

Figure 3.11   Mercury method for measuring diffusible hydrogen in welds. Reprinted from Shutt and Fink (21). Courtesy of American Welding Society.

welding, which can be heated to accelerate the hydrogen evolution from the specimen. After that, the chamber can be connected to a gas chromatograph analyzer to measure the total amount of hydrogen present. The advantages are that it can separate other gases present and measure only hydrogen and it takes hours instead of days. One disadvantage is the relatively high cost of the equipment.

Newer methods have also been developed. Albert et al. (23) developed a new sensor for detecting hydrogen. The sensor is a conducting polymer film coated with Pd on one side to be exposed to a hydrogen-containing gas and the other side to air. The current going through the sensor is directionally proportional to the hydrogen content in the gas. Figure 3.12 shows the hydrogen contents measured by the new sensor as well as gas chromatography (GC). These are gas–tungsten arc welds of a 0.5Cr–0.5Mo steel made with Ar–H_2 as the shielding gas. Smith et al. (24) developed a new hydrogen sensor that generates results in less than 1 h and allows analysis to be done on the actual welded structure. The sensor is a thin porous film of tungsten oxide, which changes color upon reacting with hydrogen.

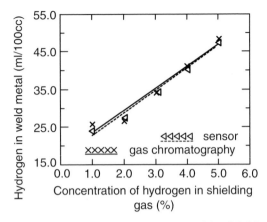

Figure 3.12 Hydrogen content in gas–tungsten arc welds of 0.5Cr–0.5Mo steel as a function of volume percent of hydrogen in Ar–H_2 shielding gas. Reprinted from Albert et al. (23). Courtesy of American Welding Society.

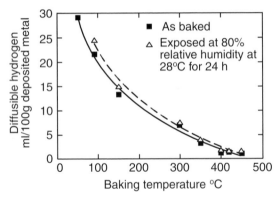

Figure 3.13 Effect of electrode baking temperature on weld metal diffusible hydrogen levels. Reprinted from Fazackerley and Gee (25).

A.3. Hydrogen Reduction Methods The weld hydrogen content can be reduced in several ways. First, avoid hydrogen-containing shielding gases, including the use of hydrocarbon fuel gases, cellulose-type electrode coverings, and hydrogen-containing inert gases. Second, dry the electrode covering and flux to remove moisture and clean the filler wire and workpiece to remove grease. Figure 3.13 shows the effect of the electrode baking temperature on the weld metal hydrogen content (25). Third, adjust the composition of the consumables if feasible. Figure 3.14 shows that CO_2 in the shielding gas helps reduce hydrogen in the weld metal (22), possibly because of reaction between the two gases. Increasing the CaF_2 content in the electrode covering or the flux

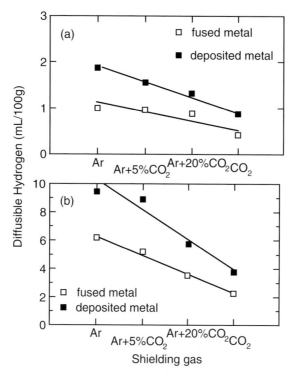

Figure 3.14 Effect of shielding gases on weld metal hydrogen content: (*a*) GMAW; (*b*) FCAW. Reprinted from Mirza and Gee (22).

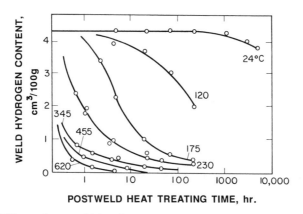

Figure 3.15 Effect of postweld heating on the weld metal hydrogen content of mild steel. Reprinted from Flanigan (27).

has been reported to reduce the weld hydrogen content. This reduction in hydrogen has been ascribed to the reaction between hydrogen and CaF_2 (26). Fourth, as shown in Figure 3.15, use postweld heating to help hydrogen diffuse out of the weld (27).

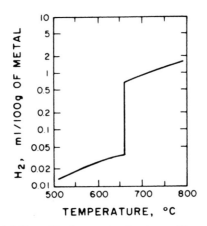

Figure 3.16   Solubility of hydrogen in aluminum. From Eastwood (28).

B. Aluminum   Hydrogen can cause porosity in aluminum welds (Chapter 10). The oxide films on the surface of the workpiece or electrode can absorb moisture from the air and introduce hydrogen into molten aluminum during welding. Grease on the surface of the workpiece or electrode and moisture in shielding gas can also be the sources of hydrogen. Figure 3.16 shows the solubility of hydrogen in aluminum (28). Since the solubility of hydrogen is much *higher* in liquid aluminum than in solid aluminum, hydrogen is rejected into the weld pool by the advancing solid–liquid interface. Consequently, hydrogen porosity is often observed in aluminum welds. Devletian and wood (29) have reviewed the factors affecting porosity in aluminum welds.

B.1. Effect of Hydrogen Porosity   As shown in Figure 3.17, excessive hydrogen porosity can severely reduce both the strength and ductility of aluminum welds (30). It has also been reported to reduce the fatigue resistance of aluminum welds (31).

B.2. Reducing Hydrogen Porosity   To reduce hydrogen porosity, the surface of Al–Li alloys has been scrapped, milled, or even thermovacuum degassed to remove hydrogen present in the form of hydrides or hydrated oxides (32, 33). Similarly, Freon (CCl_2F_2) has been added to the shielding gas to reduce hydrogen in aluminum welds. The weld pool has been magnetically stirred to help hydrogen bubbles escape and thus reduce hydrogen porosity (34). Keyhole plasma arc welding, with variable-polarity direct current, has been used to reduce hydrogen porosity in aluminum welds (35). The cleaning action of the DCEP cycle helps remove hydrated oxides and hydrides. The keyhole, on the other hand, helps eliminate entrapment of oxides and foreign materials in the weld, by allowing contaminants to enter the arc stream instead of being trapped in the weld. Consequently, the welds produced are practically porosity free.

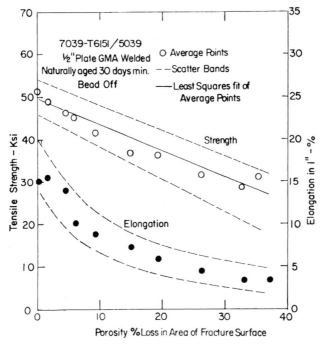

Figure 3.17 Effect of porosity on tensile properties of aluminum welds. From Shore (30).

Consider the reduction of severe gas porosity in the welding of a high-strength, lightweight Al–5Mg–2Li alloy (33, 36). The presence of Li in the alloy promotes the formation of lithium hydride during heat treatment as well as the hydration of the surface oxide at room temperature. Surface cleaning and thermovacuum treatment before welding, which help reduce the hydrogen content of the workpiece surface, have been reported to reduce the porosity level of the weld metal. Reduction in porosity has also been achieved by using an alternating magnetic field to stir the weld pool and by using variable-polarity keyhole PAW (33).

Consider also the reduction of severe gas porosity in the welding of PM (powder metallurgy) parts of Al–8.0Fe–1.7Ni alloy (33, 36). Oxidation and subsequent hydration of the aluminum powder during and after powder production by air atomization result in a high surface moisture content. When the powder is consolidated into PM parts, the moisture is trapped inside the parts. Because of the difficulty in removing the moisture from deep inside the workpiece, thermovacuum treatments at temperatures as high as 595°C have been found necessary. Unfortunately, the use of such a high temperature causes unacceptable degradation in the base-metal strength. Therefore, atomization and consolidation techniques that minimize powder oxidation and hydration are required for producing porosity-free welds.

C. Copper   Hydrogen can also cause problems in copper welding. It can react with oxygen to form steam, thus causing porosity in the weld metal. It can also diffuse to the HAZ and react with oxygen to form steam along the grain boundaries. This can cause microfissuring in the HAZ. These problems can be minimized if deoxidized copper is used for welding.

3.3   SLAG–METAL REACTIONS

3.3.1   Thermochemical Reactions

The thermochemical slag–metal reactions here refer to thermochemical reactions that take place at the interface between the molten slag and the liquid metal. Examples of such reactions are decomposition of metal oxides in the flux, oxidation of alloying elements in the liquid metal by the oxygen dissolved in the liquid metal, and desulfurization of the weld metal.

A. Decomposition of Flux   In studying SAW, Chai and Eagar (37) suggested that in the high-temperature environment near the welding plasma, all oxides are susceptible to decomposition and produce oxygen. It was found that the stability of metal oxides during welding decreases in the following order: (i) CaO, (ii) K_2O, (iii) Na_2O and TiO_2, (iv) Al_2O_3, (v) MgO, and (vi) SiO_2 and MnO (FeO was not included but can be expected to be rather unstable, too). For instance, SiO_2 and MnO can decompose as follows (2):

$$(SiO_2) = SiO(g) + \frac{1}{2}O_2(g) \tag{3.13}$$

$$(MnO) = Mn(g) + \frac{1}{2}O_2(g) \tag{3.14}$$

It was concluded that in fluxes of low FeO content (<10% FeO), SiO_2 and MnO are the primary sources of oxygen contamination and the stability of metal oxides in welding is not directly related to their thermodynamic stability. It was also concluded that CaF_2 reduces the oxidizing potential of welding fluxes due to dilution of the reactive oxides by CaF_2 rather than to reactivity of the CaF_2 itself and significant losses of Mn may occur by evaporation from the weld pool due to the high vapor pressure of Mn.

B. Oxidation by Oxygen in Metal

$$\underline{Mn} + \underline{O} = (MnO) \tag{3.15}$$

$$\underline{Si} + 2\underline{O} = (SiO_2) \tag{3.16}$$

$$\underline{Ti} + 2\underline{O} = (TiO_2) \tag{3.17}$$

$$2\underline{Al} + 3\underline{O} = (Al_2O_3) \tag{3.18}$$

C. *Desulfurization of Liquid Metal*

$$\underline{S} + (CaO) = (CaS) + \underline{O} \tag{3.19}$$

3.3.2   Effect of Flux on Weld Metal Composition

Burck et al. (38) welded 4340 steel by SAW with manganese silicate fluxes, keeping SiO_2 constant at 40 wt % and adding CaF_2, CaO, and FeO separately at the expense of MnO. Figure 3.18 shows the effect of such additions on the extent of oxygen transfer from the flux to the weld metal, expressed in Δ(weld metal oxygen). A positive Δ quantity means transfer of an element (oxygen in this case) from the flux to the weld metal, while a negative Δ quantity means loss of the element from the weld metal to the flux. The FeO additions, at the expense of MnO, increase the extent of oxygen transfer to the weld metal. This is because FeO is less stable than MnO and thus decomposes and produces oxygen in the arc more easily than MnO. The CaO additions at the expense of MnO decrease the extent of oxygen transfer to the weld metal because CaO is more stable than MnO. The CaF_2 additions at the expense of MnO also decrease the extent of oxygen transfer to the weld metal but more significantly. It is worth noting that Chai and Eagar (37) reported previously that

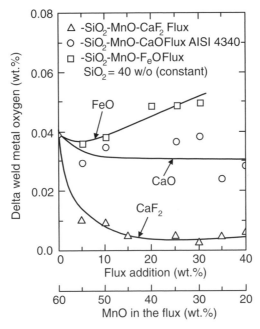

Figure 3.18   Effect of flux additions to manganese silicate flux on extent of oxygen transfer to the weld metal in submerged arc welding of 4340 steel. Reprinted from Burck et al. (38). Courtesy of American Welding Society.

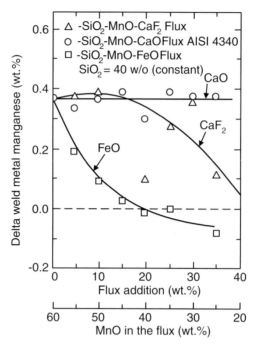

Figure 3.19   Effect of flux additions to manganese silicate flux on extent of manganese transfer to the weld metal in submerged arc welding of 4340 steel. Reprinted from Burck et al. (38). Courtesy of American Welding Society.

CaF_2 reduces oxygen transfer by acting as a diluent rather than an active species.

Figure 3.19 shows the effect of flux additions on the manganese change of the weld metal (38). It is surprising that the CaO additions at the expense of MnO do not decrease the extent of manganese transfer from the flux to the weld metal. From the steelmaking data shown in Figure 3.20, it appears that the CaO additions do not reduce the activity of MnO (39), as indicated by the dots along the constant MnO activity of about 0.30. The additions of FeO and CaF_2 at the expense of MnO decrease the extent of manganese transfer from the flux to the weld metal, as expected. Beyond 20% FeO, Δ(weld metal manganese) becomes negative, namely, Mn is lost from the weld metal to the slag. This is likely to be caused by the oxidation of Mn by the oxygen introduced into the liquid metal from FeO, namely, $\underline{Mn} + \underline{O} = (MnO)$. The flux additions also affect the extents of loss of alloying elements such as Cr, Mo, and Ni.

Therefore, the flux composition can affect the weld metal composition and hence mechanical properties rather significantly. The loss of alloying elements can be made up by the addition of ferroalloy powder (e.g., Fe–50% Si and Fe–80% Mn) to SAW fluxes or SMAW electrode coverings. In doing so, the

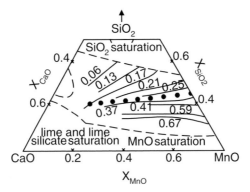

Figure 3.20 Activity of MnO in CaO–MnO–SiO$_2$ melts at 1500°C. Reprinted from Burck et al. (38). Courtesy of American Welding Society.

alloying element recovery, that is, the percentage of the element transferred across the arc and into the weld metal, should be considered. The recovery varies significantly from element to element. In SMAW, for example, it can be about 100% for Ni and Cr, 75% for Mn, 70% for Nb, 45% for Si, and 5% for Ti (40).

3.3.3 Types of Fluxes, Basicity Index, and Weld Metal Properties

The use of proper welding fluxes during fusion welding helps control the composition of the weld metal as well as protect it from air. Welding fluxes can be categorized into the following three groups according to the types of main constituents (26):

   (a) Halide-type fluxes: for example, CaF$_2$–NaF, CaF$_2$–BaCl$_2$–NaF, KCl–NaCl–Na$_3$AlF$_6$, and BaF$_2$–MgF$_2$–CaF$_2$–LiF.
   (b) Halide–oxide-type fluxes: for example, CaF$_2$–CaO–Al$_2$O$_3$, CaF$_2$–CaO–SiO$_2$, CaF$_2$–CaO–Al$_2$O$_3$–SiO$_2$, and CaF$_2$–CaO–MgO–Al$_2$O$_3$.
   (c) Oxide-type fluxes: for example, MnO–SiO$_2$, FeO–MnO–SiO$_2$, and CaO–TiO$_2$–SiO$_2$.

The halide-type fluxes are oxygen free and are used for welding titanium and aluminum alloys (26, 41). The halide–oxide-type fluxes, which are slightly oxidizing, are often used for welding high-alloy steels. The oxide-type fluxes, which are mostly oxidizing, are often used for welding low-carbon or low-alloy steels. When oxide-type fluxes are used for welding a reactive metal such as titanium, the weld metal can be contaminated with oxygen.

The oxides in a welding flux can be roughly categorized into the following three groups (26):

(a) Acidic oxides, in the order of decreasing acidity: SiO_2, TiO_2, P_2O_5, V_2O_5.
(b) Basic oxides, in the order of decreasing basicity: K_2O, Na_2O, CaO, MgO, BaO, MnO, FeO, PbO, Cu_2O, NiO.
(c) Amphoteric oxides: Al_2O_3, Fe_2O_3, Cr_2O_3, V_2O_3, ZnO.

Oxides that are donors of free oxide ions, O^{2-}, are considered as basic oxides, CaO being the most well known example. Oxides that are acceptors of O^{2-} are considered as acidic oxides, SiO_2 being the most well known example. Oxides that are neutral are considered as amphoteric oxides.

3.3.4   Basicity Index

The concept of the basicity index (BI) was adopted in steelmaking to explain the ability of the slag to remove sulfur from the molten steel. It was later broadened to indicate the flux oxidation capability. The BI of a flux (especially an oxide-type one) can be defined in the following general form (42):

$$BI = \frac{\sum(\% \ basic \ oxides)}{\sum(\% \ nonbasic \ oxides)} \qquad (3.20)$$

The concept of the BI was applied to welding. Tuliani et al. (43) used the following well-known formula for the fluxes in SAW:

$$BI = \frac{\begin{array}{c} CaF_2 + CaO + MgO + BaO + SrO \\ + Na_2O + K_2O + Li_2O + 0.5(MnO + FeO) \end{array}}{SiO_2 + 0.5(Al_2O_3 + TiO_2 + ZrO_2)} \qquad (3.21)$$

where components are in weight fractions. Using the above expression, the flux is regarded as acidic when BI < 1, as neutral when 1.0 < BI < 1.2, and as basic when BI > 1.2. The formula correlates well with the oxygen content in submerged arc welds.

Eagar and Chai (44, 45), however, modified Equation (3.21) by considering CaF_2 as neutral rather than basic and omitting the CaF_2 term. As shown in Figure 3.21, the formula correlates well with the oxygen content in submerged arc welds (44, 45). The oxygen content decreases as the basicity index increases up to about 1.25 and reaches a constant value around 250 ppm at larger basicity values.

Baune et al. (18, 46) modified Equation (3.21) for FCAW electrodes by using the composition of the solidified slag after welding rather than the composition of the flux before welding. Furthermore, mole fraction was used rather than weight fraction, and FeO was replaced by Fe_2O_3. The composition of the solidified slag was thought to provide more information about the extent of

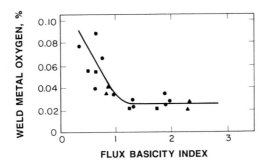

Figure 3.21   Weld metal oxygen content in steel as a function of flux basicity in submerged arc welding. From Chai and Eagar (45).

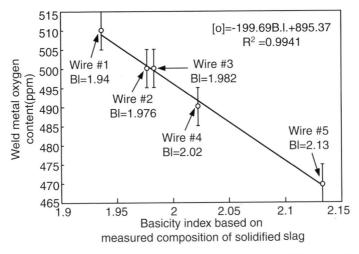

Figure 3.22   Weld metal oxygen content in steel as a function of flux basicity in flux-core arc welding. Reprinted from Baune et al. (46). Courtesy of American Welding Society.

the slag–metal reactions during welding than the composition of the flux before welding. The Fe_2O_3 was thought to be the iron oxide that forms in a welding slag. As shown in Figure 3.22, the weld oxygen content correlates well with the new basicity index (46).

Oxide inclusions in steel welds can affect the formation of acicular ferrite, which improves the weld metal toughness (47–50). It has been reported that acicular ferrite forms in the range of about 200–500 ppm oxygen (51–53). This will be discussed later in Chapter 9.

For SAW of high-strength, low-alloy steels with the CaF_2–CaO–SiO_2 flux system, Dallam et al. (47) used the following simple formula for the basicity index:

$$BI = \frac{CaO}{SiO_2} \tag{3.22}$$

Figure 3.23 shows the effect of the basicity index on sulfur transfer (47). As the basicity index increases from 0 to 5, Δsulfur becomes increasingly negative, namely, more undesirable sulfur is transferred from the weld metal to the slag (desulfurization). This is because CaO is a strong desulfurizer, as shown previously by Equation (3.19).

Excessive weld metal oxygen and hence oxide inclusions can deteriorate weld metal mechanical properties. As shown in Figure 3.24, some inclusion

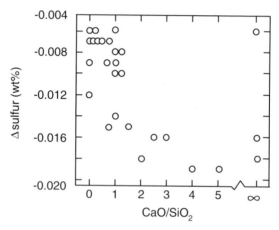

Figure 3.23 Desulfurization of high-strength, low-alloy steel welds as a function of basicity index CaO/SiO$_2$ of CaF$_2$–CaO–SiO$_2$ type flux. Reprinted from Dallam et al. (47). Courtesy of American Welding Society.

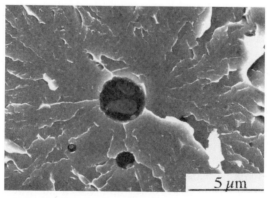

Figure 3.24 Fracture initiation at an inclusion in flux-cored arc weld of high-strength, low-alloy steel. Reprinted from Bose et al. (54).

particles can act as a fracture initiation site (54). Besides oxide inclusions, oxygen in the weld pool can also react with carbon to form CO gas during solidification. As shown in Figure 3.25, this can result in gas porosity in steel welds (55). The addition of deoxidizers such as Al, Ti, Si, and Mn in the filler metal helps reduce the amount of porosity. Figure 3.26 shows that the toughness of the weld metal decreases with increasing oxygen content (56). However, if the content of the acicular ferrite in the weld metal increases with the weld oxygen content, the weld metal toughness may in fact increase (Chapter 9).

Basic fluxes, however, can have some drawbacks. They are often found to have a greater tendency to absorb moisture, which can result in hydrogen embrittlement unless they are dried before welding. The slag detachability may

Figure 3.25   Wormhole porosity in weld metal. From Jackson (55).

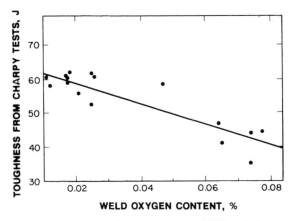

Figure 3.26   Relationship between the toughness at 20°C and the oxygen content of steel welds. Reprinted from North et al. (56). Courtesy of American Welding Society.

not be very good in a fully basic flux. This makes slag removal more difficult, especially in multiple-pass or narrow-groove welding. In the case of FCAW, basic fluxes have also been observed to generate unstable arcs (57).

3.3.5 Electrochemical Reactions

Kim et al. (58, 59) studied the effect of electrochemical reactions on the weld metal composition in SAW, and significant composition differences were observed when the electrode polarity was varied in DC welding. The following *anodic oxidation reactions* were proposed:

$$M(\text{metal}) + nO^{2-}(\text{slag}) = MO_n(\text{slag}) + 2ne^- \qquad (3.23)$$

$$O^{2-}(\text{slag}) = O(\text{metal}) + 2e^- \qquad (3.24)$$

These reactions occur at the electrode tip–slag interface in the electrode-positive polarity or the weld pool–slag interface in the electrode-negative polarity. Therefore, oxidation losses of alloying elements and pickup of oxygen are expected at the anode.

The following *cathodic reduction reactions* were also proposed:

$$M^{2+}(\text{slag}) + 2e^- = M(\text{metal}) \qquad (3.25)$$

$$Si^{4+}(\text{slag}) + 4e^- = Si(\text{metal}) \qquad (3.26)$$

$$O(\text{metal}) + 2e^- = O^{2-}(\text{slag}) \qquad (3.27)$$

The first two reactions are the reduction of metallic cations from the slag, and the third reaction is the removal (refining) of oxygen from the metal. These reactions occur at the electrode tip–slag interface in the electrode-negative polarity or the weld pool–slag interface in the electrode-positive polarity. The current density is much higher at the electrode tip–slag interface than at the weld pool–slag interface. Therefore, reactions at the electrode tip may exert a greater influence on the weld metal composition than those at the weld pool.

A carbon steel containing 0.18% C, 1.25% Mn, and 0.05% Si was submerged arc welded with a low-carbon steel wire of 0.06% C, 1.38% Mn, and 0.05% Si and a flux of 11.2% SiO_2, 18.14% Al_2O_3, 33.2% MgO, 25.3% CaF_2, 6.9% CaO, and 1.2% MnO. Figure 3.27 shows the oxygen contents of the melted electrode tips and the detached droplets for both polarities, the 20 ppm oxygen content of the wire being included as a reference (58). A significant oxygen pickup in the electrode tips is evident for both polarities, suggesting that the excess oxygen came from decomposition of oxide components in the flux and the surrounding atmosphere. The anodic electrode tip has about twice as much oxygen as the cathodic electrode tip, suggesting the significant effect of electrochemical reactions. The difference in the oxygen content is due to

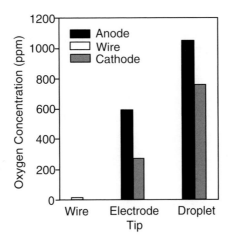

Figure 3.27 Oxygen contents of the welding wire, melted electrode tips, and detached droplets for both electrode-positive and electrode-negative polarities. Reprinted from Kim et al. (58). Courtesy of American Welding Society.

oxygen pickup at the anode and oxygen removal (refining) at the cathode. In either polarity the detached droplets contain more oxygen than the melted electrode tip. This suggests that the electrochemical reactions cease after the droplets separate from the electrode tips, but the droplets pick up more oxygen from decomposed oxides while falling through the arc plasma. The higher oxygen content in the droplets separated from the anodic electrode tip is due to the higher oxygen content of the melted anodic electrode tip.

Figure 3.28 shows the silicon change in the weld pool for both polarities (58). Since there is plenty of oxygen from the decomposition of oxides, loss of silicon from the weld pool due to the thermochemical reaction $\underline{Si} + 2\underline{O} = (SiO_2)$ is likely. But the silicon loss is recovered by the cathodic reduction $(Si^{4+}) + 4e^- = \underline{Si}$ and worsened by the anodic oxidation $\underline{Si} + 2(O^{2-}) = (SiO_2) + 4e^-$. Consequently, the silicon loss is more significant at the anode than at the cathode.

Figure 3.29 shows the manganese change in the weld pool for both polarities (58). The manganese loss from the weld pool to the slag is significant in both cases. This is because manganese is the richest alloying element in the weld pool (judging from the compositions of the workpiece and the filler metal) and MnO is among the poorest oxides in the flux. As such, the thermochemical reaction $\underline{Mn} + \underline{O} = (MnO)$ can shift to the right easily and cause much manganese loss. Since manganese is known to have a high vapor pressure (Chapter 4), evaporation from the liquid can be another reason for much manganese loss. However, the manganese loss is partially recovered by the cathodic reduction $(Mn^{2+}) + 2e^- = \underline{Mn}$ and worsened by the anodic oxidation $\underline{Mn} + (O^{2-}) = (MnO) + 2e^-$. Consequently, the manganese loss is more significant at the anode than at the cathode.

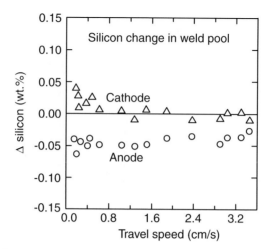

Figure 3.28   Gain or loss of weld metal silicon due to reactions in weld pool for electrode-positive and electrode-negative polarities as a function of welding speed. Reprinted from Kim et al. (58). Courtesy of American Welding Society.

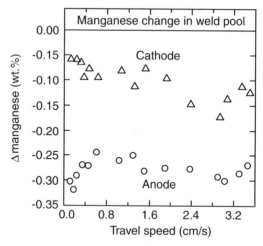

Figure 3.29   Loss of weld metal manganese due to reactions in weld pool for electrode-positive and electrode-negative polarities as a function of welding speed. Reprinted from Kim et al. (58). Courtesy of American Welding Society.

REFERENCES

1. Rein, R. H., in *Proceedings of a Workshop on Welding Research Opportunities*, Ed. B. A. McDonald, Office of Naval Research, Washington, DC, 1974, p. 92.

2. Eagar, T. W., in *Weldments: Physical Metallurgy and Failure Phenomena, Proceedings of the Fifth Bolton Landing Conference*, General Electric Co., Schenectady, NY, 1979, p. 31.

3. Harwig, D. D., Fountain, C., Ittiwattana, W., and Castner, H., *Weld. J.*, **79:** 305s, 2000.

4. Harwig, D. D., Ittiwattana, W., and Castner, H., *Weld. J.*, **80:** 126s, 2001.

5. Darken, L. S., and Gurry, R. W., *Physical Chemistry of Metals*, McGraw-Hill, New York, 1953.

6. Elliott, J. F., and Gleiser, M., *Thermochemistry for Steelmaking*, Vol. I, Addison Wesley, Reading, MA, 1960.

7. Pehlke, R. D., and Elliott, J. F., *Trans. Metall. Soc. AIME.*, **218:** 1091, 1960.

8. Elliott, J. F., Gleiser, M., and Ramakrishna, V., *Thermochemistry for Steelmaking*, Vol. II, Addison Wesley, Reading, MA, 1963.

9. Gedeon, S. A., and Eagar, T. W., *Weld. J.*, **69:** 264s, 1990.

10. Pehlke, R. D., *Unit Processes in Extractive Metallurgy*, Elsevier, New York, 1979.

11. DebRoy, T., and David, S. A., *Rev. Modern Phys.*, **67:** 85, 1995.

12. Krause, H. G., *Weld. J.*, **66:** 353s, 1987.

13. Hooijmans, J. W., and Den Ouden, G., *Weld. J.*, **76:** 264s, 1997.

14. Sato, Y. S., Kokawa, H., and Kuwana, T., in *Trends in Welding Research*, Eds. J. M. Vitek, S. A. David, J. A. Johnson, H. B. Smartt, and T. DebRoy, ASM International, Materials Park, OH, June 1998, p. 131.

15. Seferian, D., *The Metallurgy of Welding*, Wiley, New York, 1962.

16. Atarashi, N., *Welding Metallurgy*, Maruzen, Tokyo, 1972 (in Japanese).

17. *Welding Handbook*, Vol. 2, 7th ed., American Welding Society, Miami, FL, 1978, p. 134.

18. Baune, E., Bonnet, C., and Liu, S., *Weld. J.*, **79:** 57s, 2000.

19. Bracarense, A. Q., and Liu, S., *Weld. J.*, **72:** 529s, 1993.

20. Christensen, N., *Welding Metallurgy*, Lecture Notes, Colorado School of Mines, 1979.

21. Shutt, R. C., and Fink, D. A., *Weld. J.*, **64:** 19, 1985.

22. Mirza, R. M., and Gee, R., *Sci. Technol. Weld. Join.*, **4:** 104, 1999.

23. Albert, S. K., Remash, C., Murugesan, N., Gill, T. P. S., Periaswami, G., and Kulkarni, S. D., *Weld. J.*, **76:** 251s, 1997.

24. Smith, R. D., Benson, D. K., Maroef, I., Olson, D. L., and Wildeman, T. R., *Weld. J.*, **80:** 122s, 2001.

25. Fazackerley, W., and Gee, R., in *Trends in Welding Research*, Eds. H. B. Smartt, J. A. Johnson, and S. A. David, ASM International, Materials Park, OH, June 1995, p. 435.

26. *Principles and Technology of the Fusion Welding of Metal*, Vol. 1., Mechanical Engineering Publishing Company, Peking, 1979 (in Chinese).

27. Flanigan, A. E., *Weld. J.*, **26:** 193s, 1947.

28. Eastwood, L. W., *Gases in Non-ferrous Metals and Alloys*, American Society for Metals, Cleveland, OH, 1953.

29. Devletian, J. H., and Wood, W. E., *Weld. Res. Council Bull.*, **290:** December 1983.

30. Shore, R. J., M.S. Thesis, Ohio State University, Columbus, OH, 1968.

31. Pense, A. W., and Stout, R. D., *Weld. Res. Council Bull.*, **152:** July 1970.

32. Fedoseev, V. A., Ryazantsev, V. I., Shiryaeva, N. V., and Arbuzov, Y. U. P., *Svar. Proiz.*, **6:** 15, 1978.

33. Ishchenko, A. Y., and Chayun, A. G., paper presented at the Second International Conference on Aluminum Weldments, Paper 110, Munich, F.R.G., May 1982.

34. Matsuda, F., Nakata, K., Miyanaga, Y., Kayano, T., and Tsukamoto, K., *Trans. Jpn. Weld. Res. Inst.*, **7:** 181, 1978.

35. Tomsic, M., and Barhorst, S., *Weld. J.*, **63:** 25, 1984.

36. Baeslak, W. A. III, paper presented at the 1982 American Welding Society Convention, Kansas City, MO, 1982.

37. Chai, C. S., and Eagar, T. W., *Weld. J.*, **61:** 229s, 1982.

38. Burck, P. A., Indacochea, J. E., and Olson, D. L., *Weld. J.*, **69:** 115s, 1990.

39. Abraham, K. P., Davies, M. W., and Richardson, F. D., *J. Iron Steel Inst.*, **196:** 82, 1960.

40. Linnert, G. E., in *Welding Metallurgy*, Vol. 1, American Welding Society, Miami, FL, 1965, Chapter 8, p. 367.

41. Gurevich, S. M., *Metallurgy and Technology of Welding Titanium and Its Alloys*, Naukova Dumka Publishing House, Kiev, 1979 (in Russian).

42. Schenck, H., *Physical Chemistry of Steelmaking*, British Iron and Steel Research Association, London, 1945.

43. Tuliani, S. S., Boniszewski, T., and Eaton, N. F., *Weld. Metal Fab.*, **37:** 327, 1969.

44. Eagar, T. W., *Weld. J.*, **57:** 76s, 1978.

45. Chai, C. S., and Eagar, T. W., *Metall. Trans.*, **12B:** 539, 1981.

46. Baune, E., Bonnet, C., and Liu, S., *Weld. J.*, **79:** 66s, 2000.

47. Dallam, C. B., Liu, S., and Olson, D. L., *Weld. J.*, **64:** 140s, 1985.

48. Lathabai, S., and Stout, R. D., *Weld. J.*, **64:** 303s, 1985.

49. Fleck, N. A., Grong, O., Edwards, G. R., and Matlock, D. K., *Weld. J.*, **65:** 113s, 1986.

50. Liu, S., and Olson, D. L., *Weld. J.*, **65:** 139s, 1986.

51. Ito, Y., and Nakanishi, M., *Sumitomo Search*, **15:** 42, 1976.

52. Abson, D. J., Dolby, R. E., and Hart, P. M. H., *Trend in Steel and Consumables fore Welding*, pp. 75–101, 1978.

53. Cochrane, R. C., and Kirkwood, P. R., *Trend in Steel and Consumables fore Welding*, pp. 103–121, 1978.

54. Bose, W. W., Bowen, P., and Strangwood, M., in *Trends in Welding Research*, Eds. H. B. Smartt, J. A. Johnson, and S. A. David, ASM International, Materials Park, OH, June 1995, p. 603.

55. Jackson, C. E., in *Weldability of Steels*, 3rd. ed., Eds. R. D. Stout and W. D. Doty, Welding Research Council, New York, 1978, p. 48.

56. North, T. H., Bell, H. B., Nowicki, A., and Craig, I., *Weld. J.*, **57:** 63s, 1978.

57. Siewert, T. A., and Ferree, S. E., *Metal Prog.*, **119:** 58, 1981.

58. Kim, J. H., Frost, R. H., Olson, D. L., and Blander, M., *Weld. J.*, **69:** 446s, 1990.

59. Kim, J. H., Frost, R. H., and Olson, D. L., *Weld. J.*, **77:** 488s, 1998.

FURTHER READING

1. Olson, D. L., Liu, S., Frost, R. H., Edwards, G. R., and Fleming, D. A., in *ASM Handbook*, Vol. 6: *Welding, Brazing and Soldering*, ASM International, Materials Park, OH, 1993, p. 55.
2. Gaskell, D. R., *Introduction to Metallurgical Thermodynamics*, 2nd ed., McGraw-Hill, New York, 1983.
3. Richardson, F. D., *Physical Chemistry of Melts in Metallurgy*, Vols. 1 and 2. Academic, London, 1974.

PROBLEMS

3.1 Lithium alloys are known to have severe hydrogen porosity problems due to the hydration of oxides and formation of hydrides on the workpiece surface. If you were to do GMAW of these alloys using Al–Li wires, can you avoid porosity? If so, how?

3.2 Do you prefer using an oxidizing or reducing flame in gas welding of high-carbon steels? Explain why or why not.

3.3 (a) Will decreasing welding speed help reduce weld porosity in gas–tungsten arc welds of aluminum if the source of hydrogen is on the workpiece surface?

    (b) What about if the source of hydrogen is in the shielding gas? Explain why or why not.

3.4 When welding rimmed steels or when doing GMAW of carbon steels using CO_2 as the shielding gas, Mn- or Si-containing electrodes are used to prevent gas porosity. Explain why.

3.5 Austenitic stainless steels usually contain very low levels of carbon, around or below 0.05 wt %. When welding stainless steels using CO_2 as the shielding gas or covered electrodes containing abundant $CaCO_3$, the weld metal often tends to carburize. Explain why and indicate how to avoid the problem.

3.6 It has been reported that gas–tungsten arc welds of aluminum made in the overhead position tend to have a significantly higher porosity level than those made in the flat position. Explain why.

3.7 The GTAW of pure iron with Ar–5% H_2 as the shielding gas showed that the weld metal hydrogen content increased with increasing heat input per unit length of the weld. Explain why.

3.8 Electromagnetic stirring has been reported to reduce hydrogen porosity in aluminum welds. Explain why.

3.9 A steel container was welded by SAW with a filler wire containing 1.38% Mn and 0.05% Si and a flux containing 11.22% SiO_2 and 1.15% MnO. Is the electrode tip Mn content expected to be greater or smaller than the wire Mn content and why? What about the Si content?

4 Fluid Flow and Metal Evaporation in Welding

In this chapter fluid flow in both the welding arc and the weld pool will be described. The effect of weld pool fluid flow on the geometry of the resultant weld will be discussed. Evaporation of alloying elements from the weld pool and its effect on the weld metal composition will be presented. Finally, the effect of active fluxes on weld penetration in GTAW will be discussed.

4.1 FLUID FLOW IN ARCS

Figure 4.1 shows a gas–tungsten arc in GTAW with DC electrode negative (1). The shape of the electrode tip is characterized by the tip angle (also called included or vertex angle) and the extent to which the sharp point is removed, that is, the truncation.

4.1.1 Driving Force for Fluid Flow

The welding arc is an ionic gas, that is, a plasma, with an electric current passing through. The driving force for fluid flow in the arc is the *electromagnetic force* or *Lorentz force*. The buoyancy force is negligible. Mathematically, the Lorentz force $\mathbf{F} = \mathbf{J} \times \mathbf{B}$, where $\mathbf{J}$ is the current density vector and $\mathbf{B}$ is the magnetic flux vector. The current density vector $\mathbf{J}$ is in the direction the electric current flows. According to the right-hand rule for the magnetic field, if the thumb points in the direction of the current, the magnetic flux vector $\mathbf{B}$ is in the direction that the fingers curl around the path of the current. Vectors $\mathbf{F}$, $\mathbf{J}$, and $\mathbf{B}$ are perpendicular to each other. According to the right-hand rule for the electromagnetic force, $\mathbf{F}$ is in the direction out of and perpendicular to the palm if the thumb points in the direction of $\mathbf{J}$ and the fingers stretch out and point in the direction of $\mathbf{B}$.

4.1.2 Effect of Electrode Tip Geometry

The welding arc is more or less bell shaped. The tip angle of a tungsten electrode in GTAW is known to have a significant effect on the shape of

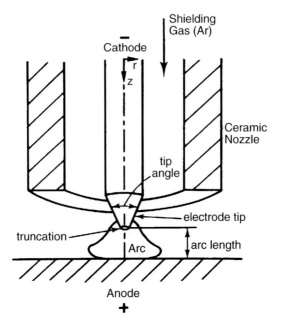

Figure 4.1 Gas–tungsten welding arc. Reprinted from Tsai and Kou (1). Copyright 1990 with permission from Elsevier Science.

the arc (Chapter 2)—it tends to become more constricted as the electrode tip changes from sharp to blunt. The change in the shape of the electrode tip changes fluid flow in the arc plasma, which in turn changes the shape of the arc.

A. Sharp Electrode Consider the case of GTAW with DC electrode negative. The electric current converges from the larger workpiece to the smaller electrode tip. It tends to be perpendicular to the electrode tip surface and the workpiece surface, as illustrated in Figure 4.2a. The electric current induces a magnetic field, and its direction is out of the plane of the paper (as indicated by the front view of an arrow) on the left and into the paper (as indicated by the rear view of an arrow) on the right. The magnetic field and the converging electric current field together produce a downward and inward force **F** to push the ionic gas along the conical surface of the electrode tip. The downward momentum is strong enough to cause the high-temperature ionic gas to impinge on the workpiece surface and turn outward along the workpiece surface, thus producing a bell-shaped arc, as illustrated in Figure 4.2b.

Fluid flow in welding arcs has been studied by computer simulation (1–6). Tsai and Kou (1) investigated the effect of the electrode tip geometry on heat and fluid flow in GTAW arcs. Figure 4.3 (left) shows the current density distribution in a 2-mm-long, 200-A arc produced by a 3.2-mm-diameter electrode

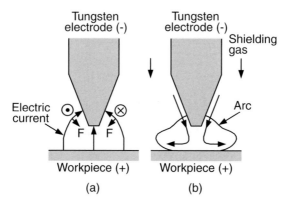

Figure 4.2 Arc produced by a tungsten electrode with a sharp tip: (*a*) Lorentz force (**F**); (*b*) fluid flow.

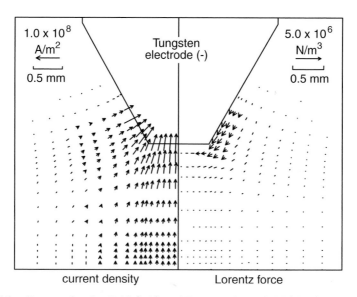

Figure 4.3 Current density field (left) and Lorentz force field (right) in an arc produced by a tungsten electrode with a 60° tip angle. Modified from Tsai and Kou (1).

with a 60° angle tip. The electric current near the electrode tip is essentially perpendicular to the surface. The electromagnetic force, shown in Figure 4.3 (right), is downward and inward along the conical surface of the electrode tip. As shown in Figure 4.4, this force produces a high-velocity jet of more than 200 m/s maximum velocity and the jet is deflected radially outward along the workpiece surface. This deflection of the high-temperature jet causes the isotherms to push outward along the workpiece surface, thus resulting in a bell-shaped arc.

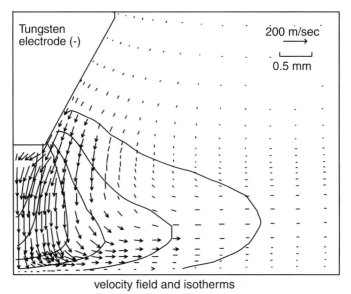

velocity field and isotherms

Figure 4.4   Velocity and temperature fields in an arc produced by a tungsten electrode with a 60° tip angle. The isotherms from right to left are 11,000, 13,000, 15,000, 17,000, 19,000, and 21,000 K. Modified from Tsai and Kou (1).

B. Flat-End Electrode   With a flat-end electrode, on the other hand, there is no longer an electrode tip to act as a fixed cathode spot, where the electric current enters the electrode. Consequently, the cathode spot moves around rather randomly and quickly in the flat electrode end. The time-averaged diameter of the area covered by the moving spot can be considered as the effective cathode spot size. Without a conical surface at the electrode end, the resultant Lorentz force is still inward and downward but the downward component is reduced, as illustrated in Figure 4.5*a*. Consequently, the resultant arc can be expected to be more constricted, as shown in Figure 4.5*b*.

Figure 4.6 (left) shows the current density distribution in a 2-mm-long, 200-A arc produced by a 3.2-mm-diameter electrode with a flat end and a 2.4-mm-diameter cathode spot. The electromagnetic force, shown in Figure 4.6 (right), is less downward pointing than that in the case of a 60° electrode (Figure 4.3). As shown in Figure 4.7, the isotherms are not pushed outward as much and the resultant arc is thus more constricted.

C. Power Density and Current Density Distributions   These distributions at the anode surface can be measured by the split-anode method (7–10). Figure 4.8 shows such distributions for a 100-A, 2.7-mm-long gas–tungsten arc measured by Lu and Kou (10). These distributions are often approximated by the following Gaussian distributions:

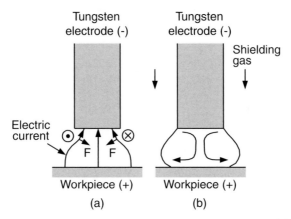

Figure 4.5 Arc produced by a tungsten electrode with a flat end: (*a*) Lorentz force (**F**); (*b*) fluid flow.

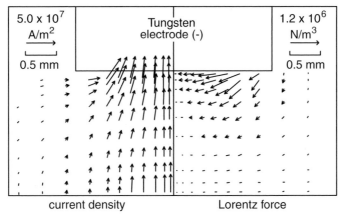

Figure 4.6 Current density field (left) and Lorentz force field (right) in an arc produced by a tungsten electrode with a flat end. Modified from Tsai and Kou (1).

$$q = \frac{3Q}{\pi a^2} \exp\left(\frac{r^2}{-a^2/3}\right) \qquad (4.1)$$

$$j = \frac{3I}{\pi b^2} \exp\left(\frac{r^2}{-b^2/3}\right) \qquad (4.2)$$

where q is the power density, Q the power transfer to the workpiece, a the effective radius of the power density distribution, j the current density, I the welding current, and b the effective radius of the current density distribution.

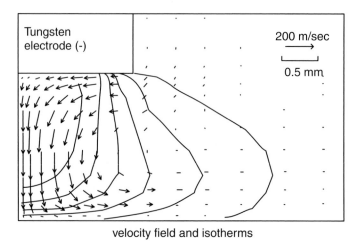

velocity field and isotherms

Figure 4.7   Velocity and temperature fields in an arc produced by a tungsten electrode with a flat end. The isotherms from right to left are 11,000, 13,000, 15,000, 17,000, 19,000, and 21,000 K. Modified from Tsai and Kou (1).

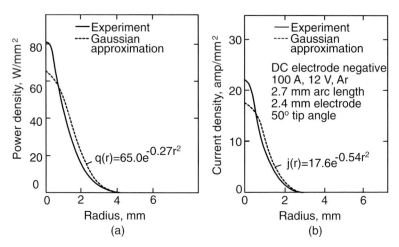

Figure 4.8   Gas–tungsten welding arc: (*a*) power density distribution; (*b*) current density distribution. Reprinted from Lu and Kou (10). Courtesy of American Welding Society.

The effective radius represents the location where q or j drops to 5% of its maximum value. Equation (4.1) is identical to Equation (2.20).

Lee and Na (6) studied, by computer simulation, the effect of the arc length and the electrode tip angle on gas–tungsten arcs. Figure 4.9 shows that both the power and current density distributions at the anode (workpiece) flatten and widen as the arc length increases.

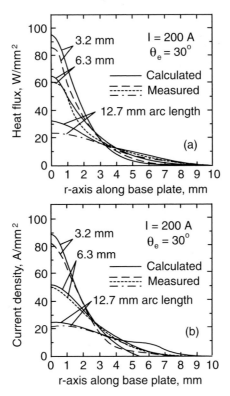

Figure 4.9 Effect of arc length on gas–tungsten welding arcs: (*a*) power density distributions; (*b*) current density distributions. Modified from Lee and Na (6).

4.2 FLUID FLOW IN WELD POOLS

4.2.1 Driving Forces for Fluid Flow

The driving forces for fluid flow in the weld pool include the buoyancy force, the Lorentz force, the shear stress induced by the surface tension gradient at the weld pool surface, and the shear stress acting on the pool surface by the arc plasma. The arc pressure is another force acting on the pool surface, but its effect on fluid flow is small, especially below 200 A (11, 12), which is usually the case for GTAW. The driving forces for fluid flow in the weld pool, shown in Figure 4.10, are explained next.

A. Buoyancy Force The density of the liquid metal (ρ) decreases with increasing temperature (T). Because the heat source is located above the center of the pool surface, the liquid metal is warmer at point *a* and cooler at point *b*. Point *b* is near the pool boundary, where the temperature is lowest at the melting point. As shown in Figure 4.10*a*, gravity causes the heavier liquid

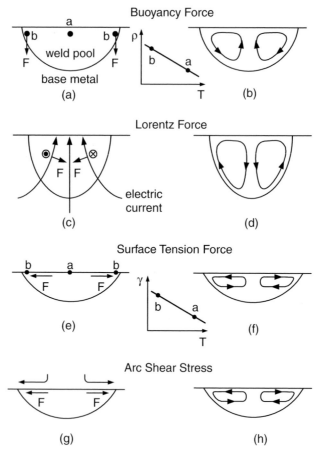

Figure 4.10 Driving forces for weld pool convection: (a, b) buoyancy force; (c, d) Lorentz force; (e, f) shear stress caused by surface tension gradient; (g, h) shear stress caused by arc plasma.

metal at point b to sink. Consequently, the liquid metal falls along the pool boundary and rises along the pool axis, as shown in Figure 4.10b.

B. Lorentz Force Gas–tungsten arc welding with DC electrode negative is used as an example for the purpose of discussion. The electric current in the workpiece converges toward the tungsten electrode (not shown) and hence near the center of the pool surface. This converging current field, together with the magnetic field it induces, causes a downward and inward Lorentz force, as shown in Figure 4.10c. As such, the liquid metal is pushed downward along the pool axis and rises along the pool boundary, as shown in Figure 4.10d. The area on the pool surface where the electric current goes through is called the anode spot (πb^2, where b is the effective radius of the current density distribution).

The smaller the anode spot, the more the current field converges from the workpiece (through the weld pool) to the anode spot, and hence the greater the Lorentz force becomes to push the liquid metal downward.

C. Shear Stress Induced by Surface Tension Gradient   In the absence of a surface-active agent, the surface tension (γ) of the liquid metal decreases with increasing temperature (T), namely, $\partial\gamma/\partial T < 0$. As shown in Figure 4.10e, the warmer liquid metal with a lower surface tension at point a is pulled outward by the cooler liquid metal with a higher surface tension at point b. In other words, an outward shear stress is induced at the pool surface by the surface tension gradient along the pool surface. This causes the liquid metal to flow from the center of the pool surface to the edge and return below the pool surface, as shown in Figure 4.10f. Surface-tension-driven convection is also called thermocapillary convection or Marangoni convection.

D. Shear Stress Induced by Plasma Jet   The plasma moving outward at high speeds along a pool surface (Figure 4.4) can exert an outward shear stress at the pool surface, as shown in Figure 4.10g. This shear stress causes the liquid metal to flow from the center of the pool surface to the pool edge and return below the pool surface, as shown in Figure 4.10h.

These driving forces are included either in the governing equations or as boundary conditions in the computer modeling of fluid flow in the weld pool (13). Oreper et al. (14) developed the first two-dimensional fluid flow model for stationary arc weld pools of known shapes. Kou and Sun (15) developed a similar model but allowed the unknown pool shape to be calculated. Kou and Wang (16–18) developed the first three-dimensional fluid flow model for moving arc and laser weld pools. Numerous computer models have been developed subsequently for fluid flow in weld pools.

4.2.2   Buoyancy Convection

Figure 4.11 shows the buoyancy convection in a stationary weld pool of an aluminum alloy calculated by Tsai and Kou (19). The liquid metal rises along the pool axis and falls along the pool boundary. The maximum velocity is along the pool axis and is only about 2 cm/s. The pool surface is slightly above the workpiece surface because of the expansion of the metal upon heating and melting.

4.2.3   Forced Convection Driven by Lorentz Force

A. Flow Field   Figure 4.12 shows the calculated results of Tsai and Kou (20) for a stationary weld pool in an aluminum alloy. The liquid metal falls along the pool axis and rises along the pool boundary (Figure 4.12a). The electric current converges from the workpiece to the center of the pool surface (Figure 4.12b). The Lorentz force is inward and downward (Figure 4.12c), thus pushing

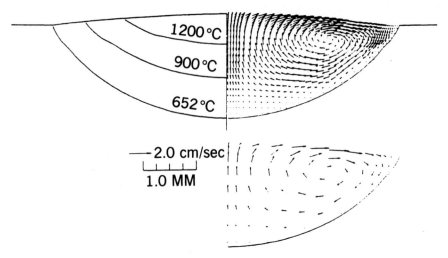

Figure 4.11   Buoyancy convection in aluminum weld pool. From Tsai and Kou (19).

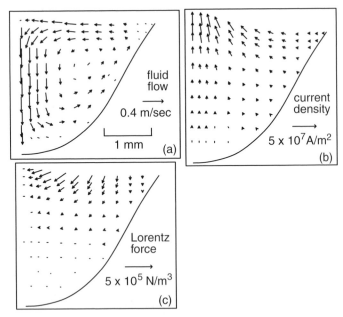

Figure 4.12   Convection in an aluminum weld pool caused by Lorentz force: (*a*) flow field; (*b*) current density field; (*c*) Lorentz force field. Modified from Tsai and Kou (20).

the liquid downward along the pool axis. The maximum velocity, about 40 cm/s, is one order of magnitude greater than that in the case of buoyancy convection (Figure 4.11). The parameters used for the calculation include 150 A for the welding current, 1800 W for the power input, 2 mm for the effective radius of the power density distribution, and 4 mm for the effective radius of the current density distribution.

B. Deep Penetration Caused by Lorentz Force   The Lorentz force (Figure 4.12) makes the weld pool much deeper, as compared to the buoyancy force (Figure 4.11). The liquid pushed downward by the Lorentz force carries heat from the heat source to the pool bottom and causes a deep penetration.

C. Physical Simulation of Effect of Lorentz Force   Kou and Sun (15) welded Woods metal (a low-melting-point alloy) with a heated copper rod in contact with it, as shown in Figure 4.13*a*. The weld became much deeper when a 75-A current was passed through to the weld pool (no arcing and negligible resistance heating), as shown in Figure 4.13*b*. This confirms the effect of the Lorentz force on weld penetration.

4.2.4   Marangoni Convection

A. Heiple's Model   Heiple et al. (21–25) proposed that, when a surface-active agent is present in the liquid metal in a small but significant amount, $\partial\gamma/\partial T$ can be changed from negative to positive, thus reversing Marangoni convection and making the weld pool much deeper. Examples of surface-active

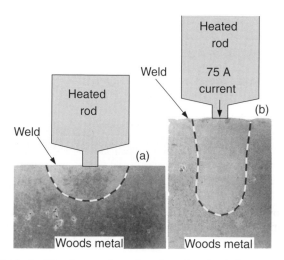

Figure 4.13   Welds in Woods metal produced under the influence of (*a*) buoyancy force and (*b*) Lorentz force. Modified from Kou and Sun (15).

agents in steel and stainless steel are S, O, Se, and Te. Figure 4.14 shows the surface tension data of two different heats of the stainless steel, one with approximately 160 ppm more sulfur than the other (26). Figure 4.15 shows two YAG laser welds made in 6.4-mm- ($\frac{1}{4}$-in.-) thick 304 stainless steel plates at 3000 W and 3.39 mm/s (8 ipm). The plate with the shallower weld contains about 40 ppm sulfur and that with the deeper weld contains about 140 ppm sulfur (27).

Heiple's model is explained in Figure 4.16. In the absence of a surface-active agent (Figures 4.16a–c), the warmer liquid metal of lower surface tension near the center of the pool surface is pulled outward by the cooler liquid metal of higher surface tension at the pool edge. In the presence of a surface-active agent (Figures 4.16d–f), on the other hand, the cooler liquid metal of lower surface tension at the edge of the pool surface is pulled inward by the warmer liquid metal of higher surface tension near the center of the pool surface. The flow pattern in Figure 4.16e favors convective heat transfer from the heat source to the pool bottom. In other words, the liquid metal carries heat from

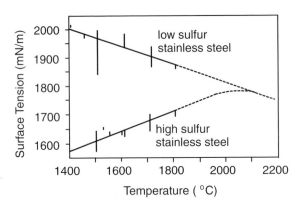

Figure 4.14   Surface tension data of two different heats of 316 stainless steel, one with 160 ppm more sulfur than the other. Modified from Heiple and Burgardt (26).

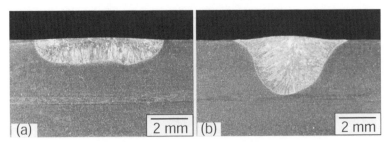

Figure 4.15   YAG laser welds in two 304 stainless steels with (a) 40 ppm sulfur and (b) 140 ppm sulfur. From Limmaneevichitr and Kou (27).

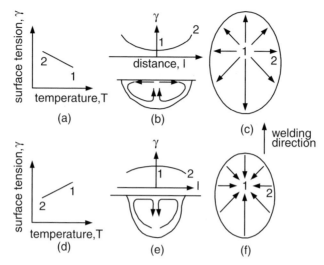

Figure 4.16 Heiple's model for Marangoni convection in a weld pool: (a, b, c) low-sulfur steel; (d, e, f) high-sulfur steel.

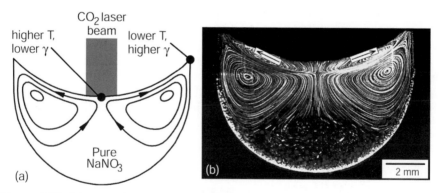

Figure 4.17 Marangoni convection with an outward surface flow in a NaNO$_3$ pool. Reprinted from Limmaneevichitr and Kou (28). Courtesy of American Welding Society.

the heat source to the pool bottom more effectively, thus increasing the weld penetration.

B. Physical Simulation of Marangoni Convection   Limmaneevichitr and Kou (28) induced Marangoni convection in a transparent pool of NaNO$_3$ with a defocused CO$_2$ laser beam. The NaNO$_3$ has a $\partial \gamma / \partial T = -0.056$ dyn/cm/°C. Since its transmission range is from 0.35 to 3 μm, NaNO$_3$ is opaque to CO$_2$ laser (10.6 μm wavelength) just like a metal weld pool is opaque to an arc. Figure 4.17

shows the flow pattern induced by a CO_2 laser beam of 2.5 W power and 3.2 mm diameter (28). The pool surface is just below the two arrows that indicate the directions of flow at the pool surface. The two counterrotating cells are, in fact, the two intersections between the donut-shaped flow pattern and the meridian plane of the pool. The outward surface flow is much faster than the inward return flow, which is typical of Marangoni convection. As the beam diameter is reduced, convection grows stronger and penetrates deeper.

It is worth noting that in conduction-mode (no-keyholing) laser beam welding, the pool surface can be concave due to Marangoni convection and surface tension (29) and, in fact, this has been shown to be the case experimentally (30) and by computer simulation (31). The concave $NaNO_3$ pool surface, however, is just a coincidence; the melt wets the container wall and the meniscus makes the pool surface concave.

Limmaneevichitr and Kou (32) added C_2H_5COOK to the $NaNO_3$ pool and reversed the direction of Marangoni convection. The C_2H_5COOK is a surface-active agent of $NaNO_3$ and, like S reduces the surface tension of liquid steel, it reduces the surface tension of $NaNO_3$, $\partial\gamma/\partial C$ being -22 dyn/(cm/mol %) (33). The $NaNO_3$ pool shown in Figure 4.18 contains 2 mol % of C_2H_5COOK and its surface flow is inward. This flow reversal is because the surface tension is now higher at the center of the pool surface than at the pool edge. At the center of the pool surface, C_2H_5COOK is lower in concentration because it decomposes under the heating of the CO_2 laser beam.

Blocks of solid $NaNO_3$, both pure and with C_2H_5COOK, were welded with a defocused CO_2 laser beam. As shown in Figure 4.19, the weld in pure $NaNO_3$ is shallow and wide (32). This is because the thermal conductivity of $NaNO_3$ is low and heat transfer is dominated by the outward surface flow to the pool

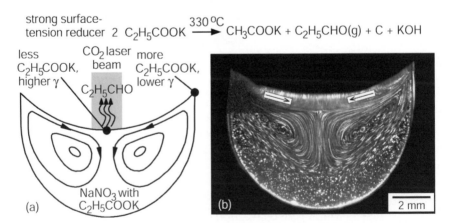

Figure 4.18 Marangoni convection with an inward surface flow in a $NaNO_3$ pool containing 2 mol % C_2H_5COOK as a surface-active agent. Reprinted from Limmaneevitchitr and Kou (32). Courtesy of American Welding Society.

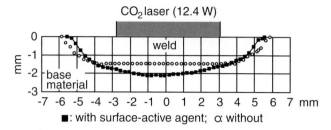

Figure 4.19   Laser welds in solid blocks of pure $NaNO_3$ (open circles) and $NaNO_3$ with 2 mol % C_2H_5COOK (solid squares). Reprinted from Limmaneevichitr and Kou (32). Courtesy of American Welding Society.

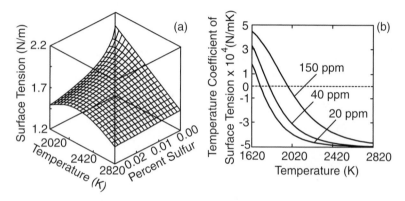

Figure 4.20   Liquid iron with various levels of sulfur: (*a*) surface tension; (*b*) temperature coefficient of surface tension. Reprinted from Pitscheneder et al. (36). Courtesy of American Welding Society.

edge. The weld in $NaNO_3$ with 2 mol % C_2H_5COOK is deeper and slightly narrower, which is consistent with the inward surface flow observed during welding.

C. Thermodynamic Analysis of Surface Tension   Sahoo et al. (34) and McNallan and DebRoy (35) calculated the surface tension of liquid metals based on thermodynamic data. Figure 4.20 shows the surface tension of liquid iron as a function of temperature and the sulfur content (36). For pure Fe, $\partial\gamma/\partial T$ is negative at all temperatures. For sulfur-containing Fe, however, $\partial\gamma/\partial T$ can be positive at lower temperatures, which is consistent with the surface tension measurements by Sundell et al. (37).

Based on the surface tension data in Figure 4.20, Pitscheneder et al. (36) calculated Marangoni convection in stationary steel weld pools. Figure 4.21 shows the results for a laser power of 5200 W and an irradiation time of 5 s.

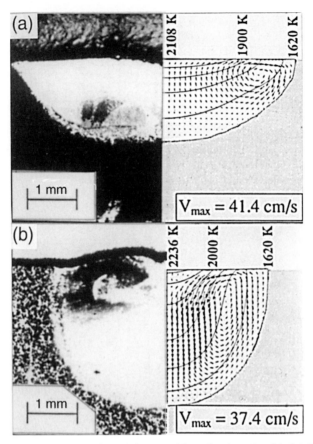

Figure 4.21   Convection in stationary laser weld pools of steels with (*a*) 20 ppm sulfur and (*b*) 150 ppm sulfur. Reprinted from Pitscheneder et al. (36). Courtesy of American Welding Society.

For the steel containing only 20 ppm sulfur, the outward surface flow carries heat from the heat source to the pool edge and results in a shallow and wide pool (Figure 4.21*a*). For the steel containing 150 ppm sulfur, on the other hand, the inward surface flow turns downward to deliver heat to the pool bottom and results in a much deeper pool (Figure 4.21*b*). In the small area near the centerline of the pool surface, the temperature is above 2000 K and the surface flow is outward because of negative $\partial\gamma/\partial T$ (Figure 4.20). It is worth noting that Zacharia et al. (38) showed that computer simulations based on a positive $\partial\gamma/\partial T$ for liquid steel at all temperatures can overpredict the pool depth.

4.2.5   Forced Convection Driven by Plasma Jet

Matsunawa and Shinichiro (39, 40) demonstrated that the plasma shear stress induced by a long arc in GTAW can outweigh both the Lorentz force in the

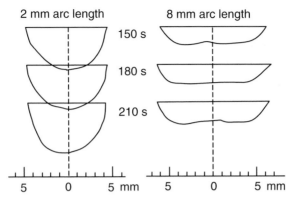

Figure 4.22   Stationary gas–tungsten arc welds in a mild steel made with a 2-mm arc (left) and an 8-mm arc (right) for 150, 180, and 210 s. Modified from Matsunawa et al. (39).

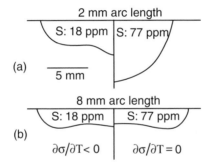

Figure 4.23   Stationary GTA welds in 304 stainless steels with 18 ppm sulfur and 77 ppm sulfur made with an arc length of (*a*) 2 mm and (*b*) 8 mm. Modified from Matsunawa et al. (39).

weld pool and the surface tension gradients along the pool surface. Figure 4.22 shows two series of stationary gas–tungsten arc welds in mild steel with 150, 180, and 210 s of welding times, one with a 2-mm-long arc and the other with a 8-mm-long arc (39, 40). The 8-mm-arc welds are much wider and shallower than the 2-mm-arc welds. For a longer and thus wider arc, the Lorentz force in the weld pool is smaller because of flatter and wider current density distribution at the pool surface (Figure 4.9*b*). The surface tension gradients are also smaller because of the flatter and wider power density distribution (Figure 4.9*a*). However, location of the maximum shear stress shifts outward, thus allowing the shear stress to act on a greater portion of the pool surface.

As shown in Figure 4.23, with a 2-mm-arc length the gas–tungsten arc weld is much deeper in the 304 stainless steel containing 77 ppm sulfur (39, 40). This

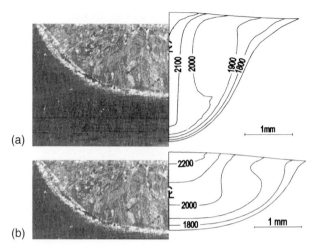

Figure 4.24   Weld pool shapes and isotherms in a 304 stainless steel with 50 ppm sulfur calculated based on (*a*) laminar flow and (*b*) turbulent flow. Reprinted from Hong et al. (42).

is expected because sulfur makes $\partial \gamma / \partial T$ either less negative or even positive. With a longer arc of 8 mm, however, the welds are shallow regardless of the sulfur level. This further demonstrates that forced convection driven by the arc plasma dominates in the welds made with the 8-mm-long arc.

4.2.6   Effect of Turbulence

Choo and Szekely (41) first considered turbulence in gas–tungsten arc weld pools and showed that turbulence can affect the pool depth significantly. Hong et al. (42) demonstrated that a fluid flow model based on laminar can over-predict the pool depth, as shown in Figure 4.24*a* for a GTA weld in a 304 stainless steel. When turbulence is considered, however, the effective viscosity increases ($\mu_{\text{eff}} > \mu$) and convection slows down. Furthermore, the effective thermal conductivity ($k_{\text{eff}} > k$) increases and the effect of convection on the pool shape thus decreases. Consequently, the calculated pool depth decreases and agrees better with the observed one, as shown in Figure 4.24*b*.

4.3   METAL EVAPORATION

4.3.1   Loss of Alloying Elements

Due to the intense heating of pool surface, evaporation from the weld pool can be significant with some alloying elements. Evaporation-induced Mg losses

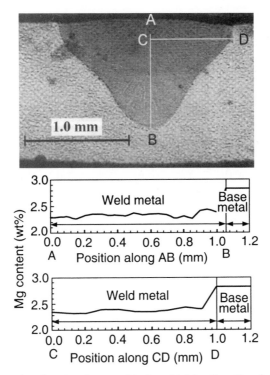

Figure 4.25 Magnesium loss in a laser weld of an Al–Mg alloy. Reprinted from Pastor et al. (45). Courtesy of American Welding Society.

from laser welds of aluminum alloys have been reported (43–45), and Figure 4.25 is an example (45). Since Al–Mg alloys are solution strengthened by alloying with Mg, Mg loss can result in substantial reduction in the tensile strength of the weld metal. Similarly, Figure 4.26 shows evaporation-induced Mn losses in laser welds of stainless steels (46).

Figure 4.27 shows the vapor pressure of several metals (47). It is clear that Mg has a much higher vapor pressure than Al at any temperature; that is, Mg has a greater tendency to evaporate than Al. This explains Mg losses from laser welds of aluminum alloys. It is also clear that Mn has a much higher vapor pressure than Fe, which explains Mn losses from laser welds of stainless steels. The Langmuir equation has been used to predict the evaporation rate of metal from the weld pool surface (47). However, according to DebRoy et al. (48–51), it can overpredict by a factor of 10 or more.

4.3.2   Explosion of Metal Droplets

Evaporation can also occur as metal droplets transfer from the filler wire to the weld pool through the arc, considering the very high temperature of the

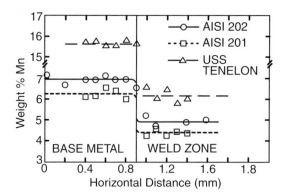

Figure 4.26 Manganese losses in laser welds of stainless steels. Reprinted from DebRoy (46).

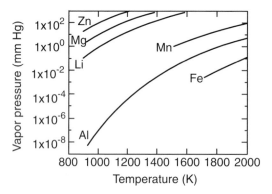

Figure 4.27 Vapor pressure of several metals as a function of temperature.

arc. Unstable metal transfer has been reported in GMAW with Al–Mg and Al–Mg–Zn filler wires (52). In fact, high-speed photography revealed in-flight explosions of metal droplets, resulting in much spattering. Figure 4.27 shows that Zn has an even higher vapor pressure than Mg. Obviously, the high vapor pressures of Mg and Zn are likely to have contributed to the explosions.

4.4   ACTIVE FLUX GTAW

The use of fluxes in GTAW has been found to dramatically increase weld penetration in steels and stainless steels (53–58). The flux usually consists of oxides and halides, and it is mixed with acetone or the like to form a paste and

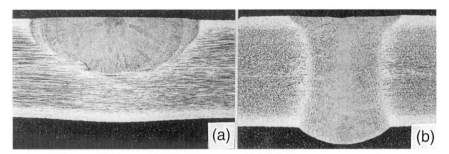

Figure 4.28   Gas–tungsten arc welds of 6-mm-thick 316L stainless steel: (*a*) without a flux; (*b*) with a flux. Reprinted from Howse and Lucas (56).

painted as a thin coating over the area to be welded. Figure 4.28 shows gas–tungsten arc welds made, without and with a flux, in a 6-mm-thick 316L stainless steel containing low S (0.005 wt %) (56).

Howse and Lucus (56) observed that the arc becomes more constricted when a flux is used. Consequently, they proposed that the deeper penetration is caused by arc constriction and the vaporized flux constricts the arc by capturing electrons in the cooler outer region of the arc. For the same welding current, the more the arc is constricted, the smaller the area (πb^2, where b is the effective radius of the current density distribution) on the pool surface to which the current field converges from the workpiece (through the weld pool), and hence the greater the Lorentz force (**F** in Figure 4.10*c*) and deeper penetration.

Tanaka et al. (57), however, proposed that the inward surface flow in the presence of the oxygen from the oxide-containing flux causes the deeper penetration. They gas–tungsten arc welded a 304 stainless steel containing little (0.002 wt %) S, with He for shielding and TiO_2 as the flux (TiO_2 is a main ingredient in several commercial fluxes). They observed both an inward surface flow and a significant decrease in the surface tension when the flux is used. As mentioned previously, $\partial \gamma / \partial T$ can become positive and cause inward surface flow in the presence of a surface-active agent such as oxygen. They observed a steep temperature gradient across the pool surface caused by the inward surface flow. Spectroscopic analysis of the arc showed that the blue luminous plasma appears to be mainly composed of metal vapor (Cr, Fe, etc.) from the weld pool. The steeper temperature gradient across the smaller pool surface suggests a more localized metal evaporation and hence a more constricted arc, which also help increase the penetration.

REFERENCES

1. Tsai, M. C., and Kou, S., *Int. J. Heat Mass Transfer*, **33:** 2089, 1990.
2. Ushio, M., and Matsuda, F., *Trans. JWM*, **11:** 7, 1982.

3. Ushio, M., and Matsuda, F., *J. Jpn. Weld. Soc.*, **6:** 91, 1988.

4. McKelliget, J., and Szekely, J., *Metall. Trans. A*, **17A:** 1139, 1986.

5. Choo, R. T. C., Szekely, J., and Westhoff, R. C., *Weld. J.*, **69:** 346, 1990.

6. Lee, S.-Y., and Na, S.-J., *Weld. J.*, **75:** 269, 1996.

7. Nestor, O. H., *J. Appl. Phys.*, **33:** 1638 (1962).

8. Schoeck, P. A., in *Modern Developments in Heat Transfer*, Ed. W. Ibele, Academic, New York, 1963, p. 353.

9. Tsai, N., Ph.D. Thesis, MIT, Cambridge, MA, 1983.

10. Lu, M., and Kou, S., *Weld. J.*, **67:** 29s, 1988.

11. Lawson, W. H. S., and Kerr, H. W., *Welding Research International*, Vol. 6, p. 1, 1976.

12. Lin, M. L., and Eagar, T. W., *Weld. J.*, **64:** 63s, 985.

13. Kou, S., *Transport Phenomena in Materials Processing*, Wiley, New York, 1996, pp. 3–115.

14. Oreper, G. M., Eagar, T. W., and Szekely, J., *Weld. J.*, **62:** 307s, 1983.

15. Kou, S., and Sun, D. K., *Metall. Trans.*, **16A:** 203, 1985.

16. Kou, S., and Wang, Y. H., *Weld. J.*, **65:** 63s, 1986.

17. Kou, S., and Wang, Y. H., *Metall. Trans.*, **17A:** 2265, 1986.

18. Kou, S., and Wang, Y. H., *Metall. Trans.*, **17A:** 2271, 1986.

19. Tsai, M. C., and Kou, S., *Numerical Heat Transfer*, **17A:** 73, 1990.

20. Tsai, M. C., and Kou, S., *Weld. J.*, **69:** 241s, 1990.

21. Heiple, C. R., and Roper, J. R., *Weld. J.*, **61:** 97s, 1982.

22. Heiple, C. R., and Roper, J. R., in *Trends in Welding Research in the United States*, Ed. S. A. David, American Society for Metals, Metals Park, OH, 1982, pp. 489–520.

23. Heiple, C. R., Roper, J. R., Stagner, R. T., and Aden, R. J., *Weld. J.*, **62:** 72s, 1983.

24. Heiple, C. R., Burgardt, P., and Roper, J. R., in *Modeling of Casting and Welding Processes*, Vol. 2, Eds. J. A. Dantzig and J. T. Berry, TMS-AIME, Warrendale, PA, 1984, pp. 193–205.

25. Heiple, C. R., and Burgardt, P., *Weld. J.*, **64:** 159s, 1985.

26. Keene, B. J., Mills, K. C., and Brooks, R. F., *Mater. Sci. Technol.*, **1:** 568–571, 1985; Heiple, C. R., and Burgardt, P., *ASM Handbook*, Vol. 6: *Welding, Brazing and Soldering*, ASM International, Materials Park, OH, 1993, p. 19.

27. Limmaneevichitr, C., and Kou, S., unpublished research, University of Wisconsin, Madison, WI, 2000.

28. Limmaneevichitr, C., and Kou, S., *Weld. J.*, **79:** 126s, 2000.

29. Duley, W. W., *Laser Welding*, Wiley, New York, 1999, p. 76.

30. Mazumder, J., and Voekel, D., in *Laser Advanced Materials Processing—Science and Applications*, Vol. 1, Eds. A. Matsunawa and S. Katayama, High Temperature Society of Japan, Osaka, Japan, 1992, pp. 373–380.

31. Tsai, M. C., and Kou, S., *Internal J. Numerical Methods Fluids*, **9:** 1503, 1989.

32. Limmaneevichitr, C., and Kou, S., *Weld. J.*, **79:** 324s, 2000.

33. Smechenko, V. K., and Shikobalova, L. P., *Zh. Fiz. Khim.*, **21:** 613, 1947.

34. Sahoo, P., DebRoy, T., and McNallan, M. J., *Metall. Trans. B.*, **19B:** 483–491, 1988.
35. McNallan, M. J., and DebRoy, T., *Metall. Trans. B*, **22B:** 557–560, 1991.
36. Pitscheneder, W., DebRoy, T., Mundra, K., and Ebner, R., *Weld. J.*, **75:** 71s–80s, 1996.
37. Sundell, R. E., Correa, S. M., Harris, L. P., Solomon, H. D., and Wojcik, L. A., Report 86SRD013, General Electric Company, Schenectady, NY, December 1986.
38. Zacharia, T., David, S. A., Vitek, J. M., and DebRoy, T., *Weld. J.*, **68:** 510s–519s, 1989.
39. Matsunawa, A., and Shinichiro, Y., in *Recent Trends in Welding Science and Technology*, Eds. S. A. David and J. M. Vitek, ASM International, Materials Park, OH, 1990, pp. 31–35.
40. Matsunawa, A., in *International Trends in Welding Science and Technology*, Eds. S. A. David and J. M. Vitek, ASM International, Materials Park, OH, 1993, pp. 3–16.
41. Choo, R. T. C., and Szekely, J., *Weld. J.*, **73:** 25s, 1994.
42. Hong, K., Weckman, D. C., Strong, A. B., and Zheng, W., *Science and Technology of Welding and Joining*, in press; Weckman, D. C., in *Trends in Welding Research*, Eds. J. M. Vitek, S. A. David, J. A. Johnson, H. S. Smartt, and T. DebRoy, ASM International, Materials Park, OH, p. 3, 1999.
43. Moon, D. W., and Metzbower, E. A., *Weld. J.*, **62:** 53s, 1983.
44. Cieslak, M. J., and Fuerschbach, P. W., *Metall. Trans.*, **19B:** 319, 1988.
45. Pastor, M., Zhaq, H., Martukanitz, R. P., and Debroy, T., *Weld. J*, **78:** 207s, 1999.
46. DebRoy, T., in *International Trends in Welding Science and Technology*, Eds. S. A. David and J. M. Vitek, ASM International, Materials Park, OH, 1993, p. 18.
47. Block-Bolten, A., and Eagar, T. W., *Metall. Trans.*, **15B:** 461, 1984.
48. Mundra, K., and DebRoy, T., *Weld. J.*, **72:** 1s, 1991.
49. Mundra, K., and DebRoy, T., *Metall. Trans.*, **24B:** 146, 1993.
50. Yang, Z., and DebRoy, T., *Metall. Trans.*, **30B:** 483, 1999.
51. Zhao, H., and DebRoy, T., *Metall. Trans.*, **32B:** 163, 2001.
52. Woods, R. A., *Weld. J.*, **59:** 59s, 1980.
53. Gurevich, S. M., Zamkov, V. N., and Kushnirenko, N. A., Avtomat Svarka, *Automat. Weld.*, **9:** 1, 1965.
54. Gurevich, S. M., and Zamkov, V. N., *Automat. Weld.*, **12:** 13, 1966.
55. Paskell, T., Lundin, C., and Castner, H., *Weld, J.*, **76:** 57, 1997.
56. Howse, D. S., and Lucas, W., *Sci. Technol. Weld. Join.*, **5:** 189, 2000.
57. Tanaka, M., Shimizu, T., Terasaki, H., Ushio, M., Koshi-ishi, F., and Yang, C.-L., *Sci. Technol. Weld. Join.*, **5:** 397, 2000.
58. Kuo, M., Sun, Z., and Pan, D., *Sci. Technol. Weld. Join.*, **6:** 17, 2001.

FURTHER READING

1. Szekely, J., *Fluid Flow Phenomena in Metals Processing*, Academic, New York, 1979.
2. Kou, S., *Transport Phenomena and Materials Processing*, Wiley, New York, 1996.
3. Lancaster, J. F., *The Physics of Welding*, Pergamon, Oxford, 1984.

PROBLEMS

4.1 Experimental results show that the depth–width ratio of stainless steel welds increases with increasing electrode tip angle. How does the angle affect the effective radius of the electric current at the pool surface (the anode spot)? How does this radius in turn affect weld pool convection and the weld depth–width ratio?

4.2 Experimental results show that the depth–width ratio of stainless steel welds decreases with increasing arc length. Explain why.

4.3 It has been suggested that the weldability of stainless steels can be improved by oxidizing the surface by subjecting it to an elevated temperature in an oxidizing environment. From the penetration point of view, do you agree or disagree, and why?

4.4 Two heats of a stainless steel with the same nominal composition but significantly different sulfur contents are butt welded by autogenous GTAW. Which side of the weld is deeper and why?

4.5 Two GTA welds of the same 304 stainless steel were made, one with a shielding gas of Ar and the other with Ar plus 700 ppm SO_2 gas. Which weld was deeper and why?

4.6 Consider Marangoni convection in a simulated weld pool of $NaNO_3$ such as that shown in Figure 4.17. As the laser beam diameter is reduced at the same power from 5.9 to 1.5 mm, does Marangoni convection in the pool become faster or slower and why?

4.7 In electron beam welding with the surface melting mode, which driving force for flow is expected to dominate?

4.8 In conduction-mode laser beam welding of a 201 stainless steel sheet 7 mm thick, the welding speed was 3 mm/s. The power was increased from 400 to 600 W. It was found that: (a) the Mn evaporation rate increased, (b) the weld pool grew significantly larger in volume, and (c) the Mn concentration decrease in the resultant weld metal decreased. The Mn concentration was uniform in the resultant weld metal. Explain (c) based on (a) and (b).

4.9 **(a)** Name the four different driving forces for weld pool convection.

      (b) In GMAW with spray metal transfer, is there an additional driving force for weld pool convection besides the four in (a)? If so, explain what it is and sketch the flow pattern in the pool caused by this force alone.

4.10 Paraffin has been used to study weld pool Marangoni convection. The surface tension of molten paraffin decreases with increasing temperature, and convection is dominated by the surface tension effect. A thin

soldering irons

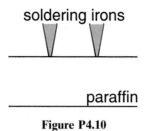

paraffin

Figure P4.10

slice of paraffin is sandwiched between two pieces of glass, and its top surface is in contact with the tips of two hot soldering irons to produce a weld pool that penetrates downward into the thin slice, as shown in Figure P4.10. Sketch and explain the flow pattern and the shape of the pool.

5 Residual Stresses, Distortion, and Fatigue

In this chapter the causes of residual stresses, distortion, and fatigue failure in weldments will be discussed, and the remedies will be described.

5.1 RESIDUAL STRESSES

Residual stresses are stresses that would exist in a body if all external loads were removed. They are sometimes called internal stresses. Residual stresses that exist in a body that has previously been subjected to nonuniform temperature changes, such as those during welding, are often called thermal stresses (1).

5.1.1 Development of Residual Stresses

A. Three-Bar Arrangement The development of residual stresses can be explained by considering heating and cooling under constraint (2). Figure 5.1 shows three identical metal bars connected to two rigid blocks. All three bars are initially at room temperature. The middle bar alone is heated up, but its thermal expansion is restrained by the side bars (Figure 5.1a). Consequently, compressive stresses are produced in the middle bar, and they increase with increasing temperature until the yield stress in compression is reached. The yield stress represents the upper limit of stresses in a material, at which plastic deformation occurs. When heating stops and the middle bar is allowed to cool off, its thermal contraction is restrained by the side bars (Figure 5.1b). Consequently, the compressive stresses in the middle bar drop rapidly, change to tensile stresses, and increase with decreasing temperature until the yield stress in tension is reached. Therefore, a residual tensile stress equal to the yield stress at room temperature is set up in the middle bar when it cools down to room temperature. The residual stresses in the side bars are compressive stresses and equal to one-half of the tensile stress in the middle bar.

B. Welding Roughly speaking, the weld metal and the adjacent base metal are analogous to the middle bar, and the areas farther away from the weld metal are analogous to the two side bars (Figure 5.1c). This is because the

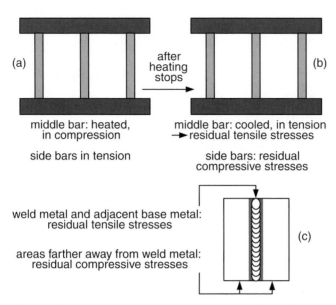

Figure 5.1 Thermally induced stresses: (*a*) during heating; (*b*) during cooling; (*c*) residual stresses in weld.

expansion and contraction of the weld metal and the adjacent base metal are restrained by the areas farther away from the weld metal. Consequently, after cooling to the room temperature, residual tensile stresses exist in the weld metal and the adjacent base metal, while residual compressive stresses exist in the areas farther away from the weld metal. Further explanations are given as follows.

Figure 5.2 is a schematic representation of the temperature change (ΔT) and stress in the welding direction (σ_x) during welding (2). The crosshatched area M–M′ is the region where plastic deformation occurs. Section A–A is ahead of the heat source and is not yet significantly affected by the heat input; the temperature change due to welding, ΔT, is essentially zero. Along section B–B intersecting the heat source, the temperature distribution is rather steep. Along section C–C at some distance behind the heat source, the temperature distribution becomes less steep and is eventually uniform along section D–D far away behind the heat source.

Consider now the thermally induced stress along the longitudinal direction, σ_x. Since section A–A is not affected by the heat input, σ_x is zero. Along section B–B, σ_x is close to zero in the region underneath the heat source, since the weld pool does not have any strength to support any loads. In the regions somewhat away from the heat source, stresses are compressive (σ_x is negative) because the expansion of these areas is restrained by the surrounding metal of lower temperatures. Due to the low yield strength of the high-temperature metal in these areas, σ_x reaches the yield strength of the base metal at

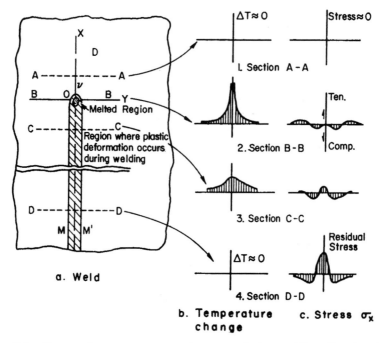

Figure 5.2 Changes in temperature and stresses during welding. Reprinted from *Welding Handbook* (2). Courtesy of American Welding Society.

corresponding temperatures. In the areas farther away from the weld σ_x is tensile, and σ_x is balanced with compressive stresses in areas near the weld.

Along section C–C the weld metal and the adjacent base metal have cooled and hence have a tendency to contract, thus producing tensile stresses (σ_x is positive). In the nearby areas σ_x is compressive. Finally, along section D–D the weld metal and the adjacent base metal have cooled and contracted further, thus producing higher tensile stresses in regions near the weld and compressive stresses in regions away from the weld. Since section D–D is well behind the heat source, the stress distribution does not change significantly beyond it, and this stress distribution is thus the residual stress distribution.

5.1.2 Analysis of Residual Stresses

Figure 5.3 shows typical distributions of residual stresses in a butt weld. According to Masubuchi and Martin (3), the distribution of the longitudinal residual stress σ_x can be approximated by the equation

$$\sigma_x(y) = \sigma_m \left[1 - \left(\frac{y}{b}\right)^2 \right] \exp\left[-\frac{1}{2}\left(\frac{y}{b}\right)^2 \right] \tag{5.1}$$

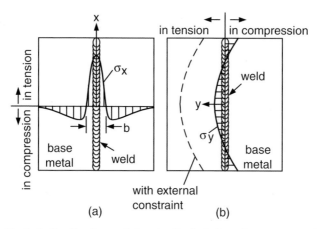

Figure 5.3 Typical distributions of longitudinal (σ_x) and transverse (σ_y) residual stresses in butt weld. Modified from *Welding Handbook* (2).

where σ_m is the maximum residual stress, which is usually as high as the yield strength of the weld metal. The parameter b is the width of the tension zone of σ_x (Figure 5.3*a*).

The distribution of the transverse residual stress σ_y along the length of the weld is shown in Figure 5.3*b*. As shown, tensile stresses of relatively low magnitude are produced in the middle part of the weld, where thermal contraction in the transverse direction is restrained by the much cooler base metal farther away from the weld. The tensile stresses in the middle part of the weld are balanced by compressive stresses at the ends of the weld. If the lateral contraction of the joint is restrained by an external constraint (such as a fixture holding down the two sides of the workpiece), approximately uniform tensile stresses are added along the weld as the reaction stress (1). This external constraint, however, has little effect on σ_x.

Figure 5.4 shows measured and calculated distributions of residual stresses σ_x in a butt joint of two rectangular plates of 5083 aluminum (60 cm long, 27.5 cm wide, and 1 cm thick) welded by GMAW (4). The calculated results are based on the finite-element analysis (FEA), and the measured results are from Satoh and Terasaki (5). The measurement and calculation of weld residual stresses have been described in detail by Masubuchi (1) and will not be repeated here. Residual stresses can cause problems such as hydrogen-induced cracking (Chapter 17) and stress corrosion cracking (Chapter 18). Postweld heat treatment is often used to reduce residual stresses. Figure 5.5 shows the effect of temperature and time on stress relief in steel welds (2). Table 5.1 list the temperature ranges used for postweld heat treatment of various types of materials (2). Other techniques such as preheat, peening, and vibration have also been used for stress relief.

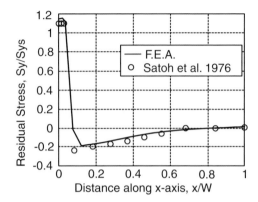

Figure 5.4   Measured and calculated distributions of residual stress in butt joint of 5083 aluminum. Reprinted from Tsai et al. (4). Courtesy of American Welding Society.

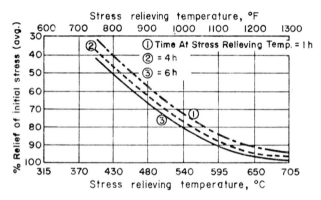

Figure 5.5   Effect of temperature and time on stress relief of steel welds. Reprinted from *Welding Handbook* (2). Courtesy of American Welding Society.

5.2   DISTORTION

5.2.1   Cause

Because of solidification shrinkage and thermal contraction of the weld metal during welding, the workpiece has a tendency to distort. Figure 5.6 illustrates several types of weld distortions (2). The welded workpiece can shrink in the transverse direction (Figure 5.6*a*). It can also shrink in the longitudinal direction along the weld (Figure 5.6*b*). Upward angular distortion usually occurs when the weld is made from the top of the workpiece alone (Figure 5.6*c*). The weld tends to be wider at the top than at the bottom, causing more solidification shrinkage and thermal contraction at the top of the weld than at the bottom. Consequently, the resultant angular distortion is upward. In electron

TABLE 5.1   Typical Thermal Treatments for Stress Relieving Weldments

Material	Soaking Temperature (°C)
Carbon steel	595–680
Carbon–$\frac{1}{2}$% Mo steel	595–720
$\frac{1}{2}$% Cr–$\frac{1}{2}$% Mo steel	595–720
1% Cr–$\frac{1}{2}$% Mo steel	620–730
$1\frac{1}{4}$% Cr–$\frac{1}{2}$% Mo steel	705–760
2% Cr–$\frac{1}{2}$% Mo steel	705–760
$2\frac{1}{4}$% Cr–1% Mo steel	705–770
5% Cr–$\frac{1}{2}$% Mo (Type 502) steel	705–770
7% Cr–$\frac{1}{2}$% Mo steel	705–760
9% Cr–1% Mo steel	705–760
12% Cr (Type 410) steel	760–815
16% Cr (Type 430) steel	760–815
$1\frac{1}{4}$% Mn–$\frac{1}{2}$% Mo steel	605–680
Low-alloy Cr–Ni–Mo steels	595–680
2–5% Ni steels	595–650
9% Ni steels	550–585
Quenched and tempered steels	540–550

Source: *Welding Handbook* (2).

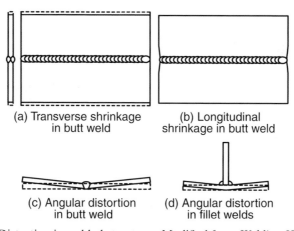

(a) Transverse shrinkage in butt weld

(b) Longitudinal shrinkage in butt weld

(c) Angular distortion in butt weld

(d) Angular distortion in fillet welds

Figure 5.6   Distortion in welded structures. Modified from *Welding Handbook* (2).

beam welding with a deep narrow keyhole, the weld is very narrow both at the top and the bottom, and there is little angular distortion. When fillet welds between a flat sheet at the bottom and a vertical sheet on the top shrink, they pull the flat sheet toward the vertical one and cause upward distortion in the

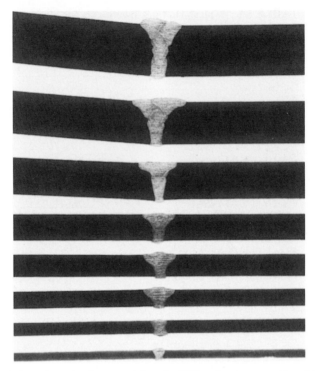

Figure 5.7 Distortion in butt welds of 5083 aluminum with thicknesses of 6.4–38 mm. Reprinted from Gibbs (6). Courtesy of American Welding Society.

flat sheet (Figure 5.6d). Figure 5.7 shows angular distortions in butt welds of 5083 aluminum of various thicknesses (6). As shown, angular distortion increases with workpiece thickness because of increasing amount of the weld metal and hence increasing solidification shrinkage and thermal contraction. The quantitative analysis of weld distortion has also been described in detail by Masubuchi (1) and hence will not be repeated here.

5.2.2 Remedies

Several techniques can be used to reduce weld distortion. Reducing the volume of the weld metal can reduce the amount of angular distortion and lateral shrinkage. Figure 5.8 shows that the joint preparation angle and the root pass should be minimized (7). The use of electron or laser beam welding can minimize angular distortion (Chapter 1). Balancing welding by using a double-V joint in preference to a single-V joint can help reduce angular distortion. Figure 5.9 shows that welding alternately on either side of the double-V joint is preferred (7). Placing welds about the neutral axis also helps reduce distortion. Figure 5.10 shows that the shrinkage forces of an individual weld

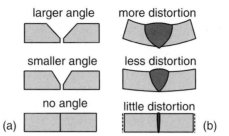

Figure 5.8   Reducing angular distortion by reducing volume of weld metal and by using single-pass deep-penetration welding. Modified from TWI (7).

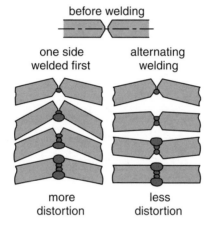

Figure 5.9   Reducing angular distortion by using double-V joint and welding alternately on either side of joint. Modified from TWI (7).

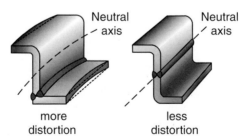

Figure 5.10   Reducing distortion by placing welds around neutral axis. Modified from TWI (7).

can be balanced by placing another weld on the opposite side of the neutral axis (7). Figure 5.11 shows three other techniques to reduce weld distortion (2, 8). Presetting (Figure 5.11a) is achieved by estimating the amount of distortion likely to occur during welding and then assembling the job with members preset to compensate for the distortion. Elastic prespringing (Figure 5.11b) can reduce angular changes after the removal of the restraint. Preheating (Figure 5.11c), thermal management during welding, and postweld heating can also reduce angular distortion (4).

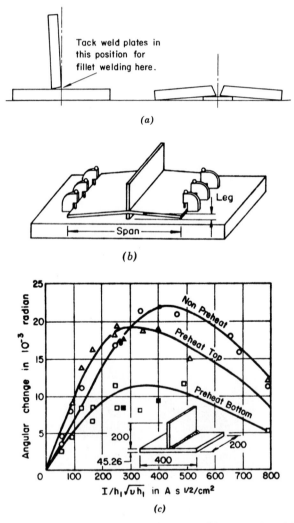

Figure 5.11 Methods for controlling weld distortion: (a) presetting; (b) prespringing; (c) preheating (8). (a), (b) Reprinted from *Welding Handbook* (2). Courtesy of American Welding Society.

5.3   FATIGUE

5.3.1   Mechanism

Failure can occur in welds under repeated loading (9, 10). This type of failure, called fatigue, has three phases: crack initiation, crack propagation, and fracture. Figure 5.12 shows a simple type of fatigue stress cycling and how it can result in the formation of *intrusions* and *extrusions* at the surface of a material along the slip planes (10). A discontinuity point in the material (e.g., inclusions, porosity) can serve as the source for a slip to initiate. Figure 5.13 shows a series of intrusions and extrusions at the free surfaces due to the alternating placement of metal along slip planes (10). Eventually, these intrusions and extrusions become severe enough and initial cracks form along slip planes. The

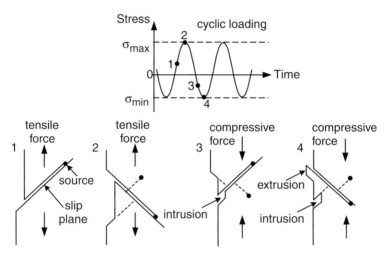

Figure 5.12   Fatigue stress cycling (top) and formation of intrusions and extrusions (bottom).

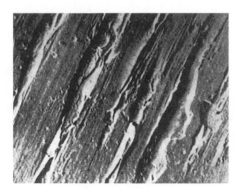

Figure 5.13   Fatigue surface showing extrusions and intrusions. From Hull (10).

direction of crack propagation is along the slip plane at the beginning and then becomes macroscopically normal to the maximum tensile stress (11).

5.3.2 Fractography

As pointed out by Colangelo and Heiser (11), the appearance of fatigue failures is often described as brittle because of the little gross plastic deformation and the fairly smooth fracture surfaces. Fatigue failures are usually easy to distinguish from other brittle failures because they are progressive and they leave characteristic marks. Macroscopically, they appear as "beach," "clam-shell," or "conchoidal" marks, which represent delays in the fatigue loading cycle. Figure 5.14 shows a fatigue fracture surface, where the arrow indicates the origin of fracture (12).

5.3.3 *S–N* Curves

Fatigue data are often presented in the form of *S–N* curves, where the applied stress (*S*) is plotted against number of cycles to failure (*N*). As the applied stress decreases, the number of cycles to failure increases. There are many factors (13) that affect the fatigue behavior, such as material properties, joint configuration, stress ratio, welding procedure, postweld treatment, loading condition, residual stresses, and weld reinforcement geometry. Figures 5.15–5.17 show the effect of some of these factors observed by Sanders and Day (14) in aluminum welds.

Figure 5.14 Fatigue fracture surface showing beach marks and origin of fracture. Reprinted, with permission, from Wulpi (12).

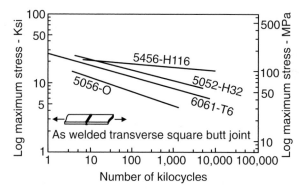

Figure 5.15   Effect of alloy and material properties on fatigue of transverse butt joint. Modified from Sanders and Day (14).

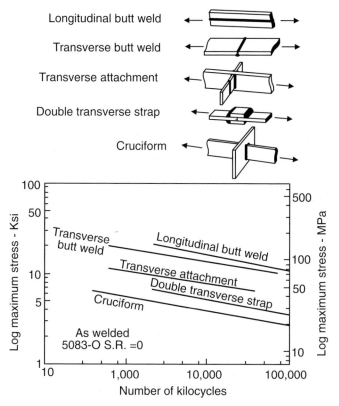

Figure 5.16   Effect of joint configurations on fatigue of 5083–O aluminum. Modified from Sanders and Day (14).

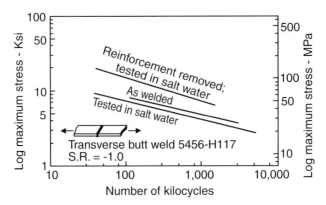

Figure 5.17 Effect of reinforcement removal and saltwater environment on fatigue of 5456-H117 aluminum. Modified from Sanders and Day (14).

5.3.4 Effect of Joint Geometry

As pointed out by Sanders and Day (14), in developing any fatigue behavior criteria for welding, the severity of joint geometry is probably the most critical factor. The more severe the geometry, the lower the fatigue strength, as shown in Figure 5.16. The severity level of the longitudinal butt weld is lowest because both the weld and the base metal carry the load. The severity level of the cruciform, on the other hand, is highest since the welds alone carry the load and the parts are joined perpendicular to each other.

5.3.5 Effect of Stress Raisers

It is well known that stress raisers tend to reduce fatigue life, namely, the so-called *notch effect*. Stress raisers can be mechanical, such as toes with a high reinforcement, lack of penetration, and deep undercuts. They can also be metallurgical, such as microfissures (microcracks), porosity, inclusions, and brittle and sharp intermetallic compounds. Figure 5.18 shows a fatigue crack originating from the toe (Chapter 1) of a gas–metal arc weld of a carbon steel (15). Figure 5.19 shows a fatigue failure originating from an *undercut* at the top of an electron beam weld in a carbon steel and how undercutting can reduce fatigue life (16).

5.3.6 Effect of Corrosion

As also shown in Figure 5.17, a corrosive environment (salt water in this case) can often reduce fatigue life (14). This is called *corrosion fatigue* (17, 18). It has been reported that the damage can be almost always greater than the sum of the damage by corrosion and fatigue acting separately.

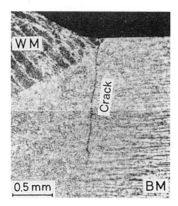

Figure 5.18   Fatigue crack originating from weld toe of gas–metal arc weld of carbon steel. Reprinted from Itoh (15). Courtesy of American Welding Society.

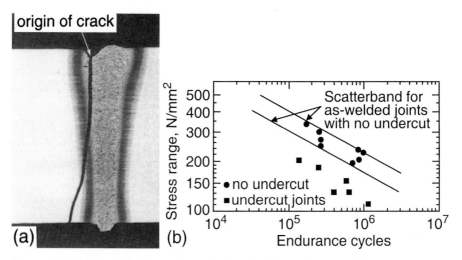

Figure 5.19   Effect of undercutting on fatigue in electron beam welds of carbon steel: (*a*) photograph; (*b*) fatigue life. Reprinted from Elliott (16). Courtesy of American Welding Society.

5.3.7   Remedies

A. Shot Peening   Welding and postweld grinding can create tensile residual stresses at the weld surface and promote fatigue initiation when under cyclic loading. Shot and hammer peening, on the other hand, can introduce surface compressive stresses to suppress the formation of intrusions and extrusions and hence fatigue initiation (13). In shot peening, the metal surface is bombarded with small spherical media called shot (19). Each piece of shot striking the surface acts as a tiny peening hammer, producing a tiny indentation or dimple on the surface. Overlapping indentations or dimples develop a uniform layer of residual compressive stresses. Figure 5.20 shows the residual stresses as a function of depth, namely, the distance below the surface (19). It is clear that tensile residual stresses (>0) in the as-welded condition can be reduced by stress-relieving heat treatment and reversed to become highly compressive residual stresses (<0) by shot peening.

B. Reducing Stress Raisers   Figure 5.21 shows stress raisers caused by improper welding and how they can be reduced or eliminated. Figure 5.17 shows that removing the reinforcement improves the fatigue life (14). This is consistent with the results of Nordmark et al. (20) shown in Figure 5.22 for

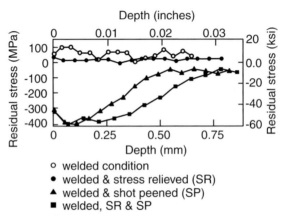

Figure 5.20   Effect of stress relieving and shot peening on residual stresses near the metal surface. Modified from Molzen and Hornbach (19).

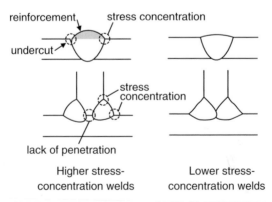

Figure 5.21   Stress raisers in butt and T-welds and their corrections.

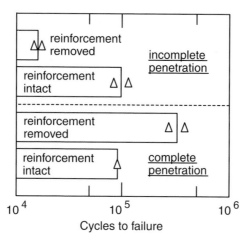

Figure 5.22   Effect of weld reinforcement and penetration on fatigue life of transverse butt welds of 5456 aluminum. Reprinted from Nordmark et al. (20). Courtesy of American Welding Society.

transverse butt welds of 6.4-mm- ($\frac{1}{4}$-in.-) thick 5456-H116 aluminum. However, in the case of incomplete penetration, removing the reinforcement can dramatically reduce the fatigue life.

5.4   CASE STUDIES

5.4.1   Failure of a Steel Pipe Assembly (21)

Figure 5.23a is a sketch showing a steel pipe of 10 mm wall thickness bent to turn in 90° and welded at the ends (b) to connecting flanges. The pipe–flange assembly was connected to a bottle-shaped container (a) and a heavy overhanging valve weighing 95 kg (e). The working pressure was 425 kg/cm² and the operating temperature was 105°C. The assembly was subjected to vibrations from the compressors provided at intervals along the pipe system. Graphical recorders showed evidence of the vibrations. The assembly failed after a service time of about 2500 h.

Examination of the failed assembly showed two fractures: fracture A at position c and fracture B at the weld at position d. Both fractures exhibited radial lines and striations typical of a fatigue fracture. Fracture A (Figure 5.23b) originated in small craters accidentally formed in the flange during arc welding. Fracture B (Figure 5.23c) started at the weld toe, namely, the junction between the weld reinforcement and the base metal surface. The fracture surface (Figure 5.23d) shows the radial lines and striations of fatigue.

The composition of the steel was unavailable, though metallography showed that it had a ferrite–pearlite structure, and the grains are fine and

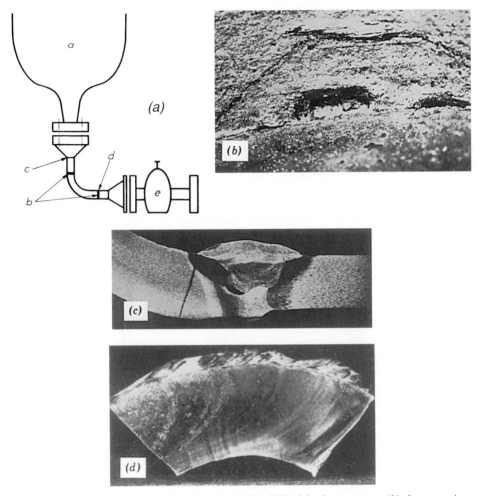

Figure 5.23 Failure of a steel pipe assembly (21): (*a*) pipe system; (*b*) fracture A caused by craters at position *c*; (*c*) fracture B at weld at position *d*; (*d*) surface of fracture B.

equiaxed. Regarding fracture A, many cracks were observed in the fused metal in the craters and in the martensitic structure of the heat-affected zone. Under the cyclic loading from the vibrations, these cracks served as the initiation sites for fatigue. As for fracture B, the notch effect due to reinforcement of the weld promoted the fatigue failure.

5.4.2 Failure of a Ball Mill (22)

Figure 5.24 shows part of a ball mill used for ore crushing. The cylindrical shell of the ball mill was 10.3 m in length, 4.4 m in diameter, and 50 mm in thickness.

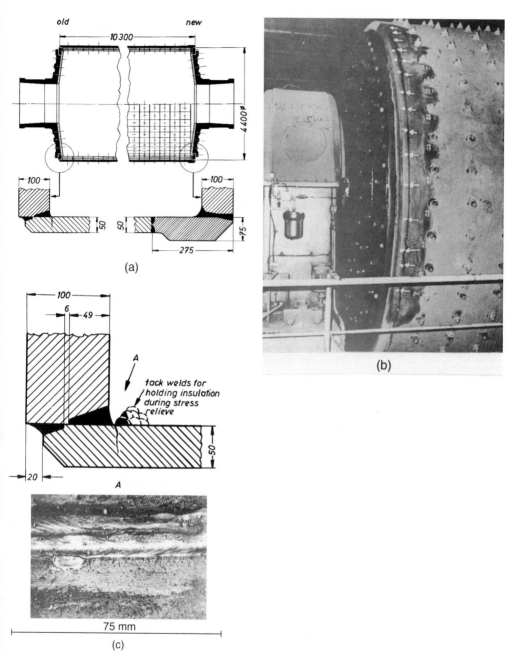

Figure 5.24   Failure of a ball mill: (*a*) design; (*b*) fatigue cracks (indicated by white arrows); (*c*) origin of fatigue cracks. From Wallner (22).

A flange with 100 mm wall thickness was welded to each end of the cylindrical shell (left half of Figure 5.24*a*). Both the cylindrical shell and the flanges were made from a killed steel with the composition of the steel being 0.19% C, 0.25% Si, 0.65% Mn, 0.025% P, and 0.028% S. The entire weight of the mill charge was approximately 320 tons and the drum operated at a rotation speed of 14 rpm. After about 3000 h of operation, long cracks appeared on the outside surface of the drum (Figure 5.24*b*).

The failed drum was emptied, its inside inspected, and cracks ranging from 100 to 1000 mm long were observed. It was found that these cracks had originated from nearby tack welds, which had been made for holding insulation during stress relieving of the drum (Figure 5.24*c*). Apparently, the high notch effect has greatly reduced the fatigue life of the drum. It was subsequently suggested that the new joint design shown in the right half of Figure 5.24*a* be used and a new stress relief method be employed in order to avoid the use of similar tack welds.

REFERENCES

1. Masubuchi, K., *Analysis of Welded Structures*, Pergamon, Elmsford, NY, 1980.
2. *Welding Handbook*, 7th ed., Vol. 1, American Welding Society, Miami, FL, 1976.
3. Masubuchi, K., and Martin, D. C., *Weld. J.*, **40:** 553s, 1961.
4. Tsai, C. L., Park, S. C., and Cheng, W. T., *Weld. J.*, **78:** 156s, 1999.
5. Satoh, K., and Terasaki, T., *J. Jpn. Weld. Soc.*, **45:** 42, 1976.
6. Gibbs, F. E., *Weld. J.*, **59:** 23, 1980.
7. *TWI Job Knowledge for Welders*, Part 34, Welding Institute, Cambridge, UK, March 21, 1998.
8. Watanabe, M., Satoh, K., Morii, H., and Ichikawa, I., *J. Jpn. Weld. Soc.*, **26:** 591, 1957.
9. Hertzberg, R. W., *Deformation and Fracture Mechanics of Engineering Materials*, Wiley, New York, 1976, p. 422.
10. Hull, D., *J. Inst. Met.*, **86:** 425, 1957.
11. Colangelo, V. J., and Heiser, F. A., *Analysis of Metallurgical Failures*, Wiley, New York, 1974.
12. Wulpi, D., *Understanding How Components Fail*, American Society for Metals, Metals Park, OH, 1985, p. 144.
13. Sanders, W. W. Jr., *Weld. Res. Council Bull.*, **171:** 1972.
14. Sanders, W. W. Jr., and Day, R. H., in *Proceedings of the First International Aluminum Welding Conference*, Welding Research Council, New York, 1982.
15. Itoh, Y., *Weld. J.*, **66:** 50s, 1987.
16. Elliott, S., *Weld. J.*, **63:** 8s, 1984.
17. Uhlig, H. H., *Corrosion and Corrosion Control*, 2nd ed., Wiley, New York, 1971.
18. Fontana, M. G., and Greene, N. D., *Corrosion Engineering*, 2nd ed., McGraw-Hill, New York, 1978.
19. Molzen, M. S., and Hornbach, D., *Weld. J.*, **80:** 38, 2001.

20. Nordmark, G. E., Herbein, W. C., Dickerson, P. B., and Montemarano, T. W., *Weld. J.*, **66:** 162s, 1987.

21. *Fatigue Fractures in Welded Constructions*, Vol. II, International Institute of Welding, London, 1979, p. 56.

22. Wallner, F., in *Cracking and Fracture in Welds*, Japan Welding Society, Tokyo, 1972, P. IA3.1.

FURTHER READING

1. Masubuchi, K., *Analysis of Welded Structures*, Pergamon, Elmsford, NY, 1980.

2. *Fatigue Fractures in Welded Constructions*, Vol. II, International Institute of Welding, London, 1979.

3. *Cracking and Fracture in Welds*, Japan Welding Society, Tokyo, 1972.

4. Gurney, T. R., Ed., *Fatigue of Welded Structures*, 2nd ed., Cambridge University Press, Cambridge, 1979.

PROBLEMS

5.1 **(a)** Austenitic stainless steels have a high thermal expansion coefficient and a low thermal conductivity. Do you expect significant distortion in their welds?

   (b) Name a few welding processes that help reduce weld distortion.

5.2 Aluminum alloys have a high thermal expansion coefficient, high thermal conductivity, and high solidification shrinkage. Do you expect significant distortion in their welds?

5.3 The lower the weld reinforcement usually means the less the stress concentration at the weld toe. One quick way to reduce the height of the reinforcement is to grind its upper half flat without changing its contact angle with the workpiece surface. In this a proper treatment of the reinforcement? Explain why or why not.

PART II
The Fusion Zone

6 Basic Solidification Concepts

In order to help explain the weld metal microstructure and chemical inhomogeneities in subsequent chapters, some basic solidification concepts will be presented first in this chapter. These concepts include solute redistribution, solidification modes, constitutional supercooling, microsegregation and banding, the dendrite-arm or cell spacing, and the solidification path.

6.1 SOLUTE REDISTRIBUTION DURING SOLIDIFICATION

When a liquid of uniform composition solidifies, the resultant solid is seldom uniform in composition. The solute atoms in the liquid are redistributed during solidification. The redistribution of the solute depends on both thermodynamics, that is, the phase diagram, and kinetics, that is, diffusion, undercooling, fluid flow, and so on.

6.1.1 Phase Diagram

Figure 6.1a is a portion of a phase diagram near the corner of a pure metal of melting point T_m, with S denoting the solid phase and L the liquid phase. Consider the solidification of alloy C_0, that is, with initial melt composition C_0. A vertical line through C_0 intersects the liquidus line at the liquidus temperature, T_L, and the solidus line at the solidus temperature, T_S. Assume that undercooling is negligible so that the solid begins to form when the liquid cools to T_L. Also assume that equilibrium between the solid and the liquid is maintained at the solid–liquid (S/L) interface throughout solidification. This means at any temperature T the composition of the solid at the interface, C_S, and the composition of the liquid at the interface, C_L, follow the solidus line and the liquidus line, respectively.

A. Equilibrium Partition Ratio   At any temperature T the equilibrium partition ratio, or the *equilibrium segregation coefficient*, k, is defined as

$$k = \frac{C_S}{C_L} \tag{6.1}$$

where C_S and C_L are the compositions of the solid and liquid at the S/L interface, respectively. The value of k depends on temperature T. For simplicity,

145

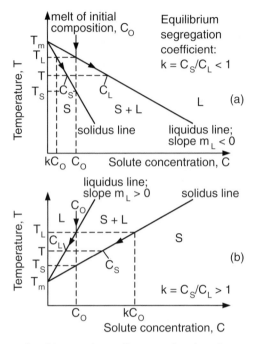

Figure 6.1 Portion of a binary phase diagram showing the equilibrium partition ratio k.

k is assumed constant; that is, the solidus and liquidus lines are both assumed straight lines.

The first solid to form will have the composition kC_0 according to Equation (6.1) and the phase diagram, that is, $C_L = C_0$ at $T = T_L$. Consider the case of $k < 1$ first. As the phase diagram in Figure 6.1a shows, the solid cannot accommodate as much solute as the liquid does, and the solid thus rejects the solute into the liquid during solidification. Consequently, the solute content of the liquid continues to rise during solidification. Since the solid grows from the liquid, its solute content also continues to rise. As indicated by the arrowheads on the solidus and liquidus lines in Figure 6.1a for $k < 1$, C_S and C_L both increase as temperature T of the S/L interface drops during solidification.

Consider now the case of $k > 1$. As the phase diagram in Figure 6.1b shows, the solid can accommodate more solute than the liquid does, and the solid thus absorbs the solute from the liquid during solidification. Consequently, the solute content of the liquid continues to drop during solidification. Since the solid grows from the liquid, its solute content also continues to drop. As indicated by the arrowheads on the solidus and liquidus lines in Figure 6.1b for $k > 1$, C_S and C_L both decrease as temperature T of the S/L interface drops during solidification.

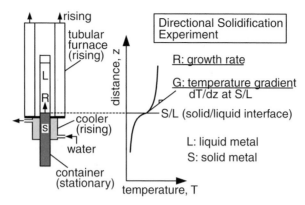

Figure 6.2   Directional solidification experiment.

B. Slope of Liquidus Line   The slope of the liquidus line, m_L, is less than zero when $k < 1$ and vice versa, as shown in Figure 6.1. If the liquidus line is straight, a temperature T on the liquidus line can be expressed as

$$T = T_m + m_L C_L \tag{6.2}$$

6.1.2   Complete Diffusion in Solid and Liquid

To help understand solute redistribution during solidification, consider the one-dimensional solidification experiment shown in Figure 6.2. A metal held in a stationary container of an aluminum oxide tube is heated by a tubular furnace at the top and cooled by a water cooler at the bottom. As the furnace–cooler assembly rises steadily, the metal in the container solidifies upward with a planar S/L interface. The growth rate R of the metal, that is, the travel speed of the S/L interface, can be adjusted by adjusting the rising speed of the furnace–cooler assembly. The temperature gradient G in the liquid metal at the S/L interface can be adjusted by adjusting heating and cooling.

This case of complete diffusion in both the solid and the liquid is shown in Figure 6.3, where a liquid metal of uniform initial composition C_0 is allowed to solidify in one direction with a planar S/L interface, just like in the directional solidification experiment shown in Figure 6.2. This case is also called *equilibrium solidification* because equilibrium exists between the entire solid and the entire liquid, not just at the interface. Diffusion is complete in the solid and the liquid, and the solid and liquid are thus uniform in composition. Uniform composition in the liquid requires either complete mixing by strong convection or complete diffusion in the liquid. Complete diffusion of the solute in the liquid requires $D_L t \gg l^2$, where l is the initial length of the liquid, D_L the diffusion coefficient of the solute in the liquid, and t the time available

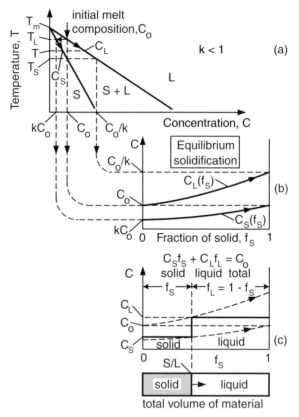

Figure 6.3 Solute redistribution during solidification with complete diffusion in solid and liquid: (*a*) phase diagram; (*b*) $C_L(f_s)$ and $C_S(f_s)$; (*c*) composition profiles in solid and liquid.

for diffusion. This is because the square root of Dt is often considered as an approximation for the diffusion distance, where D is the diffusion coefficient. This requirement can be met when l is very small, when solidification is so slow that the solute has enough time to diffuse across the liquid, or when the diffusion coefficient is very high. Similarly, complete solute diffusion in the solid requires $D_S t \gg l^2$, where D_S is the diffusion coefficient of the solute in the solid. Since D_S is much smaller than D_L, complete diffusion is much more difficult to achieve in solid than in liquid. Interstitial solutes (such as carbon in steel) tend to have a much higher D_S than substitutional solutes (such as Cr in steel) and are more likely to approach complete diffusion in solid.

During solidification the composition of the entire solid follows the solidus line and that of the entire liquid follows the liquidus line (Figure 6.3*a*). The composition of the solid is a function of the fraction of the solid, that is, $C_S(f_s)$. Likewise the composition of the liquid is a function of the fraction of the solid, that is, $C_L(f_S)$. At the onset of solidification at T_L, the fraction of solid $f_s = 0$, and the compositions of the solid and the liquid are kC_0 and C_0, respectively. As solidification continues and temperature drops from T_L to T_S, the composition of the entire solid rises from kC_0 to C_0 and that of the entire liquid from C_0 to C_0/k (Figure 6.3*b*). At any time during solidification, the compositions of the solid and liquid are uniform (as shown by the thick horizontal lines in Figure 6.3*c*). Solidification ends at the solidus temperature T_S and the resultant solid has a uniform composition of C_0; that is, there is *no solute segregation* at all.

In Figure 6.3*c* the hatched areas $C_S f_S$ and $C_L f_L$ represent the amounts of solute in the solid and liquid, respectively, where f_S and f_L are the fractions of solid and liquid, respectively. The area $C_0(f_S + f_L)$ represents the amount of solute in the liquid before solidification. Based on the conservation of solute and the fact that $f_S + f_L = 1$,

$$C_S f_S + C_L f_L = C_0(f_S + f_L) = C_0 \tag{6.3}$$

Substituting $f_S = 1 - f_L$ into the above equation, the following *equilibrium lever rule* is obtained:

$$f_L = \frac{C_0 - C_S}{C_L - C_S} \tag{6.4}$$

$$f_S = \frac{C_L - C_0}{C_L - C_S} \tag{6.5}$$

and the composition of the liquid is

$$C_L = \frac{C_0}{f_L + k(1 - f_L)} \tag{6.6}$$

6.1.3   No Solid Diffusion and Complete Liquid Diffusion

This case is shown in Figure 6.4. Diffusion in the solid is assumed negligible, and the solid is thus not uniform in composition. This requires that $D_S t \ll l^2$. On the other hand, diffusion in the liquid is assumed complete, and the liquid is thus uniform in composition. This requires that $D_L t \gg l^2$. If mixing caused by strong convection is complete in the liquid, the liquid composition can also be uniform. Equilibrium exists between the solid and the liquid only at the interface.

Unlike the case of equilibrium solidification, the solute cannot back diffuse into the solid and all the solute rejected by the growing solid has to go into the liquid. Consequently, C_L rises more rapidly during solidification than in the case of equilibrium solidification. Since the solid grows from the liquid, its composition at the S/L interface C_S also rises more rapidly than in

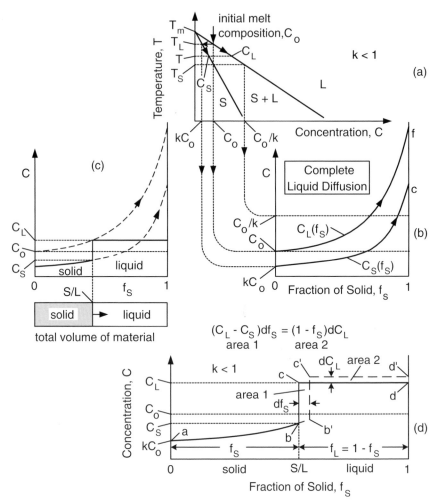

Figure 6.4 Solute redistribution during solidification with complete diffusion in liquid and no diffusion in solid: (*a*) phase diagram; (*b*) $C_L(f_s)$ and $C_S(f_s)$; (*c*) composition profiles in solid and liquid; (*d*) conservation of solute.

the case of equilibrium solidification. When C_L rises beyond C_0/k, C_S rises beyond C_0 (Figure 6.4b).

The two hatched areas in Figure 6.4c represent the amounts of solute in the solid and liquid, and their sum equals the amount of solute in the liquid before solidification, C_0. Consider the conservation of solute shown in Figure 6.4d. Since the total amount of solute in the system (total hatched area) is conserved, the area under $\overline{abcd}$ equals the area $\overline{ab'c'd'}$ under. As such, hatched area 1 equals hatched area 2 and, therefore,

$$(C_L - C_S)\, df_S = (1 - f_S)\, dC_L \tag{6.7}$$

Substituting $C_S = kC_L$ into the above equation and integrating from $C_L = C_0$ at $f_s = 0$,

$$C_S = kC_0(1 - f_S)^{k-1} \tag{6.8}$$

or

$$C_L = C_0 f_L^{k-1} \tag{6.9}$$

The above two equations are the well-known *Scheil equation* for solute segregation (1). Equation (6.9) can be written for the fraction liquid as follows:

$$f_L = \left(\frac{C_L}{C_0}\right)^{1/(k-1)} = \left(\frac{C_0}{C_L}\right)^{1/(1-k)} \tag{6.10}$$

The *nonequilibrium lever rule*, that is, the counterpart of Equation (6.5), can be written as

$$f_S = \frac{C_L - C_0}{C_L - \overline{C_S}} \tag{6.11}$$

From the Scheil equation the *average composition* of the solid, $\overline{C_S}$, can be determined as

$$\overline{C_S} = \frac{\int_0^{fs} kC_0(1 - f_S)^{k-1}\, df_S}{\int_0^{fs} df_S} = \frac{C_0\left[1 - (1 - f_S)^k\right]}{f_S} \tag{6.12}$$

As already mentioned, the solidus and liquidus lines are both assumed straight.

From Figure 6.1a, it can be shown that $(T_L - T_m)/C_0 = m_L$. Furthermore, based on the proportional property between two similar right triangles, it can be shown that $(T_m - T_L)/(T_m - T) = C_0/C_L$. Based on these relationships and Equation (6.10), the Scheil equation can be rewritten as follows:

$$f_L = 1 - f_S = \left(\frac{T_m - T_L}{T_m - T}\right)^{1/1-k} = \left(\frac{(-m_L)C_o}{T_m - T}\right)^{1/1-k} \tag{6.13}$$

This equation can be used to determine the fraction of solid f_s at a temperature T below the liquidus temperature T_L.

6.1.4 No Solid Diffusion and Limited Liquid Diffusion

This case is shown in Figure 6.5. In the solid diffusion is assumed negligible. In the liquid diffusion is assumed limited, and convection is assumed negligible. Consequently, neither the solid nor the liquid is uniform in composition during solidification. Because of limited liquid diffusion, the solute rejected by the solid piles up and forms a *solute-rich boundary layer* ahead of the growth front, as shown in Figure 6.5c. Again, equilibrium exists between the solid and the liquid only at the interface.

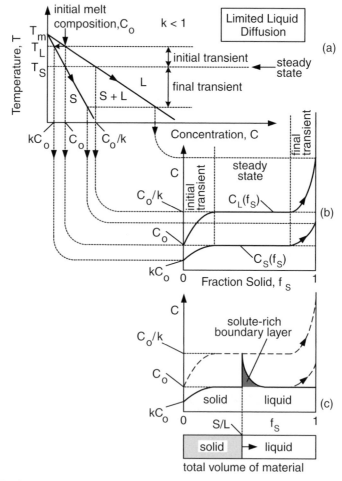

Figure 6.5 Solute redistribution during solidification with limited diffusion in liquid and no diffusion in solid: (*a*) phase diagram; (*b*) $C_L(f_s)$ and $C_S(f_s)$; (*c*) composition profiles in solid and liquid.

Unlike the previous case of complete liquid diffusion and no solid diffusion, the solute rejected by the growing solid forms a solute-rich boundary layer ahead of the growth front, rather than spreading uniformly in the entire liquid. Consequently, C_L and hence C_S rise more rapidly than those in the case of complete liquid diffusion and no solid diffusion. The period of rising C_L and C_S is called the *initial transient* (Figure 6.5b). When C_L and C_S reach C_0/k and C_0, respectively, a *steady-state* period is reached, within which C_L, C_S, and the boundary layer (Figure 6.5c) remain unchanged. As the boundary layer moves forward, it takes in a liquid of composition C_0 and it leaves behind a solid of the same composition C_0. Since the input equals the output, the boundary layer remains unchanged. This steady-state condition continues until the boundary layer touches the end of the liquid, that is, when the thickness of the remaining liquid equals that of the steady-state boundary layer. Here the volume of the remaining liquid is already rather small, and any further decrease in volume represents a very significant percentage drop in volume. As such, the remaining liquid quickly becomes much more concentrated in solute as solidification continues, and C_L and hence C_S rise sharply. This final period of rapidly rising C_L and C_S is called the *final transient*, and its length equals the thickness of the steady-state boundary layer.

The solute-rich boundary layer is further examined in Figure 6.6. It has been shown mathematically (2) that the steady-state composition profile in the boundary layer is

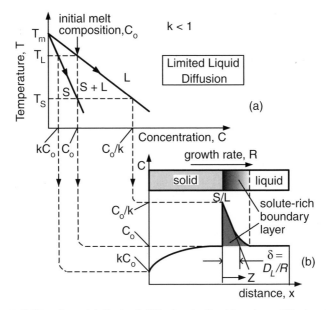

Figure 6.6   Solidification with limited diffusion in liquid and no diffusion in solid: (a) phase diagram; (b) solute-rich boundary layer.

$$\boxed{\frac{C_\mathrm{L}-C_0}{C_0/k-C_0}=e^{-(R/D_L)Z}} \tag{6.14}$$

where R is the growth rate and Z the distance from the S/L interface pointing into the liquid. At $Z = 0$, $C_\mathrm{L} - C_0$ has a maximum value of $C_0/k - C_0$ and at $Z = D_\mathrm{L}/R$ it drops to $1/e$ (about one-third) of the maximum value. As such, the thickness of the boundary layer at the steady state can be taken as $\delta \approx D_\mathrm{L}/R$. As such, the *characteristic length of the final transient* is D_L/R.

It also has been shown mathematically (3, 4) that, for $k \ll 1$, the composition profile in the initial transient can be expressed as

$$\boxed{\frac{C_0-C_\mathrm{S}}{C_0-kC_0}=e^{-k(R/D_L)x}} \tag{6.15}$$

where x is the distance from the starting point of solidification. At $x = 0$, $C_0 - C_\mathrm{S}$ has a maximum value of $C_0 - kC_0$ and at $x = D_\mathrm{L}/kR$ it drops to $1/e$ of the maximum value. As such, the *characteristic length of the initial transient* is D_L/kR, which is much greater than that of the final transient because k is significantly less than 1.

Equation (6.15) can be rearranged as follows:

$$1-\frac{C_\mathrm{S}}{C_0}=(1-k)e^{-k(R/D_L)x} \tag{6.16}$$

Upon taking the logarithm on both sides, the above equation becomes

$$\log\!\left(1-\frac{C_\mathrm{S}}{C_0}\right)=\log(1-k)-\left(\frac{\log e}{D_\mathrm{L}/kR}\right)x \tag{6.17}$$

The left-hand-side of Equation (6.17) can be plotted against the distance x. From the intercept the value of k can be checked, and from the slope the length of the initial transient D_L/kR can be found. From this the growth rate R can be found if the diffusion coefficient D_L is known or vice versa.

Figure 6.7 shows three different types of solute of redistributions that can occur when diffusion in the solid is negligible during solidification. In type 1 either liquid diffusion or convection-induced mixing in the liquid is complete, and the resultant solute segregation is most severe (Figure 6.7d). In type 3, on the other hand, liquid diffusion is limited and there is no convection-induced mixing in the liquid, and the resultant solute segregation is least severe. Type 2 is intermediate, so is the resultant solute segregation.

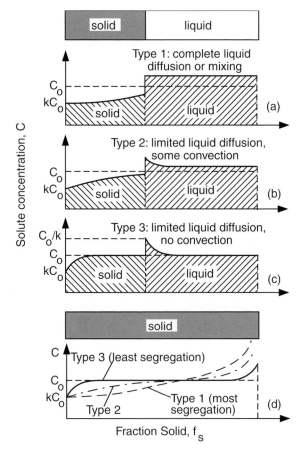

Figure 6.7 Solute redistributions in the absence of solid diffusion: (*a*) type 1; (*b*) type 2; (*c*) type 3; (*d*) resultant segregation profiles.

6.2 SOLIDIFICATION MODES AND CONSTITUTIONAL SUPERCOOLING

6.2.1 Solidification Modes

During the solidification of a pure metal the S/L interface is usually planar, unless severe thermal undercooling is imposed. During the solidification of an alloy, however, the S/L interface and hence the mode of solidification can be planar, cellular, or dendritic depending on the solidification condition and the material system involved. In order to directly observe the S/L interface during solidification, transparent organic materials that solidify like metals have been used. Shown in Figure 6.8 are the four basic types of the S/L interface morphology observed during the solidification of such transparent materials: planar, cellular, columnar dendritic, and equiaxed dendritic (5, 6). Typical

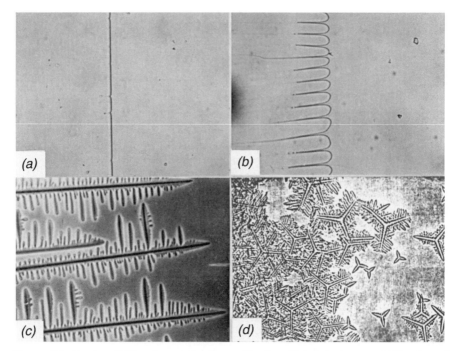

Figure 6.8   Basic solidification modes (magnification 67×): (*a*) planar solidification of carbon tetrabromide (5); (*b*) cellular solidification of carbon tetrabromide with a small amount of impurity (5); (*c*) columnar dendritic solidification of carbon tetrabromide with several percent impurity (5); (*d*) equiaxed dendritic solidification of cyclohexanol with impurity (6). From *Solidification* (5), pp. 132–134, with permission.

microstructures resulting from the cellular, columnar dendritic, and equiaxed dendritic modes of solidification in alloys are shown in Figures 6.9*a*, *b*, and *c*, respectively (7–9). A three-dimensional view of dendrites is shown in Figure 6.9*d* (10).

6.2.2   Constitutional Supercooling

Two major theories have been proposed to quantitatively describe the breakdown of a planar S/L interface during solidification: the constitutional supercooling theory by Chalmer and co-workers (11, 12) and the interface stability theory by Mullins and Sekerka (13–15). The former theory considers only the thermodynamic aspect of the problem while the latter incorporates the interface kinetic and heat transfer aspects. For simplicity, however, only the constitutional supercooling theory will be described here.

Consider the solidification of alloy C_0 at the steady state with a planar S/L interface, as shown in Figure 6.10. As shown previously in Figure 6.6, the composition distribution in the solute-rich boundary layer is shown in Figure 6.10*b*.

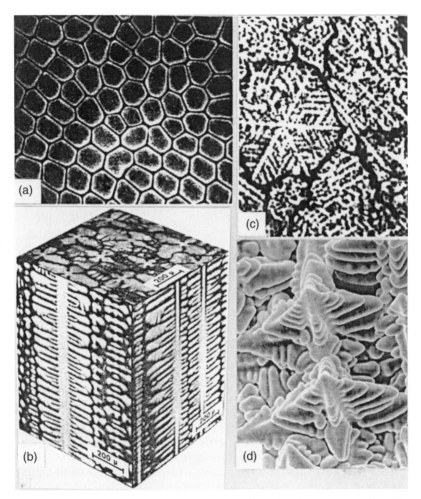

Figure 6.9 Nonplanar solidification structure in alloys. (*a*) Transverse section of a cellularly solidified Pb–Sn alloy from Journal of Crystal Growth (7) (magnification 48×). (*b*) Columnar dendrites in a Ni alloy. From *New Trends in Materials Processing* (8), with permission. (*c*) Equiaxed dendrites of a Mg–Zn alloy from Journal of Inst. of Metals (9) (magnification 55×). (*d*) Three-dimensional view of dendrites in a Ni-base superalloy. Reprinted from *International Trends in Welding Science and Technology* (10).

The liquidus temperature distribution corresponding to this composition distribution can be constructed point by point from the liquidus line of the phase diagram and is shown in Figure 6.10c. A boundary layer consisting of the liquid phase alone is thermodynamically stable only if its temperature is above the liquidus temperature. If its temperature is below the liquidus temperature, solid and liquid should coexist. This means that the planar S/L interface should

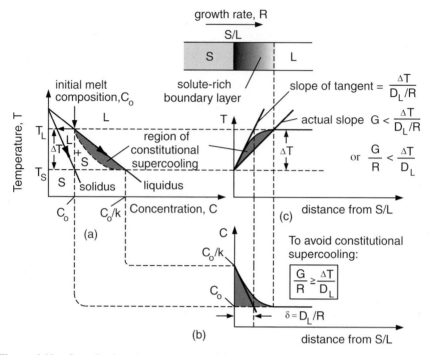

Figure 6.10 Constitutional supercooling: (*a*) phase diagram; (*b*) composition profile in liquid; (*c*) liquidus temperature profile in liquid.

break down to a cellular or dendritic one so that solid cells or dendrites can coexist with the intercellular or interdendritic liquid (Figures 6.8*b–d*). The shaded area under the liquidus temperature distribution in Figure 6.10*c* indicates the region where the actual liquid temperature is below the liquidus temperature, that is, the region of constitutional supercooling. The shaded area below the liquidus line in Figure 6.10*a*, which depends on D_L and R, indicates the corresponding region of constitutional supercooling in the phase diagram. Obviously, this area lies within the solid–liquid region of the phase diagram, that is, liquid alone is unstable and solid and liquid should coexist.

The temperature difference across the boundary layer is the equilibrium freezing range $\Delta T = T_L - T_S$. As shown previously, the thickness of the boundary layer at the steady state is D_L/R. As such, the slope of the tangent to the liquidus temperature distribution at the S/L interface is $\Delta T/(D_L/R)$ or $R\Delta T/D_L$. To ensure that a planar S/L interface is stable, the actual temperature gradient G at the S/L interface must be at least $R\Delta T/D_L$. Therefore, for a planar S/L interface to be stable at the steady state, the following criterion must be met:

$$\boxed{\frac{G}{R} \geq \frac{\Delta T}{D_L}} \tag{6.18}$$

This is the steady-state form of the *criterion for planar growth*. It says that for an alloy to be able to grow with a planar solidification mode, the ratio G/R must be no less than $\Delta T/D_L$. The constitutional supercooling theory has been verified experimentally by many investigators (16–23). In general, this theory predicts fairly closely the conditions required to initiate the breakdown of a planar S/L interface in alloys with isotropic surface energy.

According to Equation (6.18), the higher the temperature gradient G and the lower the growth rate, the easier for a planar S/L interface to be stable. On the other hand, the higher the freezing range ΔT and the lower the diffusion coefficient D_L, the more difficult for a planar S/L interface to be stable. For an Al–4% Cu alloy, for example, T_L, T_S and D_L, are 650°C, 580°C, and 3×10^{-5} cm²/s, respectively. If the temperature gradient G is 700°C/cm, the growth rate R has to be less than or equal to 3×10^{-4} cm/sec in order to have planar solidification. If the growth rate is higher than this, the planar S/L interface will break down and cellular or dendritic solidification will take place.

Figure 6.11 shows that the solidification mode changes from planar to cellular, to columnar dendritic, and finally to equiaxed dendritic as the degree of constitutional supercooling continues to increase. The region where dendrites (columnar or equiaxed) and the liquid phase coexist is often called the *mushy zone* (2). It is interesting to note that at a very high degree of constitutional supercooling (Figure 6.11d) the mushy zone can become so wide that it is easier for equiaxed dendrites to nucleate than for columnar dendrites to stretch all the way across the mushy zone. Unfortunately, simple theories

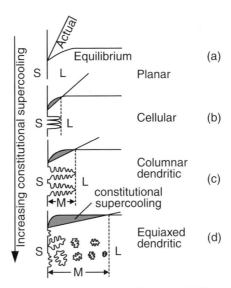

Figure 6.11   Effect of constitutional supercooling on solidification mode: (*a*) planar; (*b*) cellular; (*c*) columnar dendritic; (*d*) equiaxed dendritic (S, L, and M denote solid, liquid, and mushy zone, respectively).

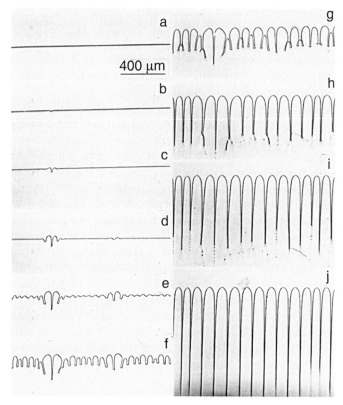

Figure 6.12 Breakdown of a planar S/L interface during the solidification of pivalic acid–ethanol: (*a*) 0 s, (*b*) 120 s, (*c*) 210 s, (*d*) 270 s, (*e*) 294 s, (*f*) 338 s, (*g*) 378 s, (*h*) 456 s, (*i*) 576 s, and (*j*) 656 s. Reprinted from Liu and Kirkaldy (24). Copyright 1994 with permission from Elsevier Science.

similar to the constitutional supercooling theory are not available for predicting the transitions from the cellular mode to the columnar dendritic mode and from the columnar dendritic mode to the equiaxed dendritic mode.

Figure 6.12 is a series of photographs showing the breakdown of a planar S/L interface into a cellular one during the solidification of a pivalic acid alloyed with 0.32 mol % ethanol (24). The temperature gradient G was 15°C/mm, and the growth rate R was suddenly raised to a higher level of 5.7 μm/s, to suddenly lower the G/R ratio and trigger the breakdown by constitutional supercooling.

6.3 MICROSEGREGATION AND BANDING

6.3.1 Microsegregation

Solute redistribution during solidification results in microsegregation across cells or dendrite arms. The analysis of solute redistribution during the directional solidification of a liquid metal (Section 6.1) can be applied to solute redistribution during the solidification of an intercellular or interdendritic liquid during welding (or casting). The total volume of material in directional solidification (Figures 6.3–6.5) is now a volume element in a cell or a dendrite arm, as shown in Figure 6.13. Within the volume element the S/L interface is still planar even though the overall structure is cellular or dendritic. The volume element covers the region from the centerline of the cell or dendrite arm to the boundary between cells or dendrite arms. Solidification begins in the volume element when the tip of the cell or dendrite arm reaches the volume element.

The case of the equilibrium partition ratio $k < 1$ is shown in Figure 6.14a. No segregation occurs when diffusion is complete in both the liquid and solid.

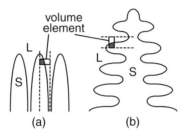

Figure 6.13   Volume elements for microsegregation analysis: (*a*) cellular solidification; (*b*) dendritic solidification.

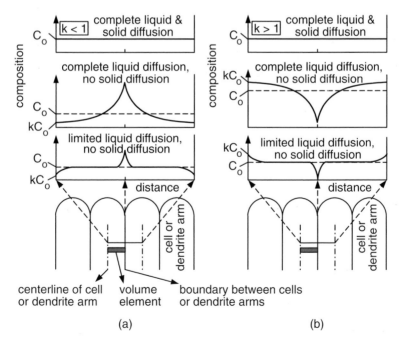

Figure 6.14   Microsegregation profiles across cells or dendrite arms: (*a*) $k < 1$; (*b*) $k > 1$.

This requires that $D_L t \gg l^2$ and that $D_S t \gg l^2$, where l is now half the cell or dendrite arm spacing (the length of the volume element). Segregation is worst with complete diffusion in the liquid but no diffusion in the solid. This requires that $D_L t \gg l^2$ and that $D_S t \ll l^2$. Segregation is intermediate with limited diffusion in the liquid and no diffusion in the solid. When this occurs, there is a clear concentration minimum at the centerline of the cell or dendrite arm. Usually, there is some diffusion in the solid and the concentration minimum may not always be clear. The case of $k > 1$ is shown in Figure 6.14b. The segregation profiles are opposite to those of $k < 1$.

Consider the case of a eutectic phase diagram, which is common among aluminum alloys. Assume complete liquid diffusion and no solid diffusion. As shown in Figure 6.15, the solid composition changes from kC_0 to C_{SM}, the maximum possible solute content in the solid, when the eutectic temperature T_E is reached. The remaining liquid at this point has the eutectic composition

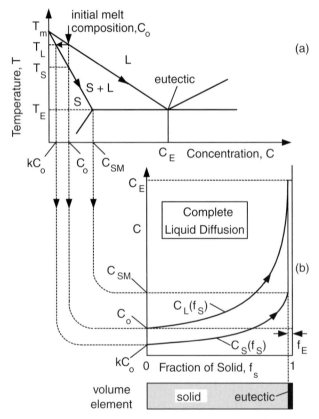

Figure 6.15 Solute redistribution during solidification with complete diffusion in liquid and no diffusion in solid: (a) eutectic phase diagram; (b) composition profiles in liquid and solid.

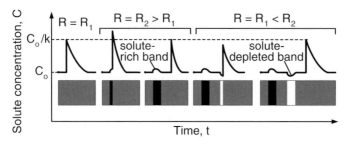

Figure 6.16   Formation of banding due to changes in the solidification rate R.

C_E and, therefore, solidifies as solid eutectic at T_E. The *fraction of eutectic*, f_E, can be calculated from Equation (6.13) with $f_L = f_E$ and $T = T_E$, that is,

$$f_E = \left(\frac{T_m - T_L}{T_m - T_E} \right)^{1/(1-k)} \tag{6.19}$$

6.3.2   Banding

In addition to microsegregation, solute segregation can also occur as a result of growth rate fluctuations caused by thermal fluctuations. This phenomenon, shown in Figure 6.16, is known as banding (2). As shown, steady-state solidification occurs at the growth rate R_1. When the growth rate is suddenly increased from R_1 to R_2, an extra amount of solute is rejected into the liquid at the S/L interface, causing its solute content to rise. As a result, the material solidifies right after the increase in the growth rate has a higher solute concentration than that before the increase. The boundary layer at R_2 is thinner than that at R_1 because, as mentioned previously, the boundary layer thickness is about D_L/R. The solute concentration eventually resumes its steady value if no further changes take place. If, however, the growth rate is then decreased suddenly from R_2 back to R_1, a smaller amount of solute is rejected into the liquid at the S/L interface, causing its solute content to drop. As a result, the material solidifies right after the decrease in the growth rate has a lower solute concentration than that just before the decrease. In practice, the growth rate can vary as a result of thermal fluctuations during solidification caused by unstable fluid flow in the weld pool, and solute-rich and solute-depleted bands form side by side along the solidification path. When the flow carrying heat from the heat source and impinging on the growth front suddenly speeds up, the growth rate is reduced and vice versa.

6.4   EFFECT OF COOLING RATE

It has been observed that the higher the cooling rate, that is, the shorter the solidification time, the finer the cellular or dendritic structure becomes (2).

Figure 6.17 shows the dendrite arm spacing as a function of the cooling rate or solidification time in three different materials (25–27). The relationship can be expressed as (2)

$$d = at_f^n = b(\varepsilon)^{-n} \qquad (6.20)$$

where d is the secondary dendrite arm spacing, t_f is the local solidification time, ε is the cooling rate, and a and b are proportional constants.

Figure 6.18 depicts the growth of a dendrite during solidification (2). The window for viewing the growing dendrite remains stationary as the dendrite tip advances. As can be seen through the window, large dendrite arms grow at the expense of smaller ones as solidification proceeds. Since smaller dendrite arms have more surface area per unit volume, the total surface energy of the solidifying material in the window can be reduced if larger dendrite arms grow at the expense of the smaller ones. The slower the cooling rate during solidi-

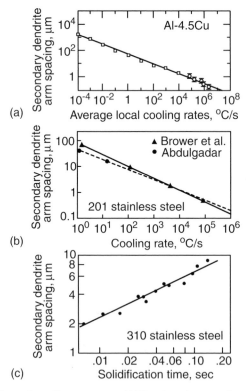

Figure 6.17 Effect of cooling rate or solidification time on dendrite arm spacing: (*a*) For Al–4.5Cu. From Munitz (25), (*b*) For 201 stainless steel. From Paul and DebRoy (26). (*c*) For 310 stainless steel. From Kou and Le (27).

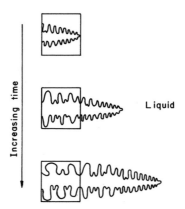

Figure 6.18 Schematic showing dendritic growth of an alloy at a fixed position at various stages of solidification. Note that several smaller arms disappear while larger ones grow. Reprinted from Flemings (2).

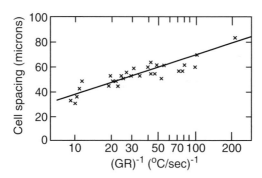

Figure 6.19 Effect of *GR* on cell spacing in Sn–Pb alloys. Modified from Flemings (2) based on data from Plaskett and Winegard (28).

fication, the longer the time available for coarsening and the larger the dendrite arm spacing (Figure 6.17).

Figure 6.19 shows how *GR* affects the cell spacing in Sn–Pb alloys solidified in the cellular mode (2, 28). The product *GR* (°C/cm times cm/s) is, in fact, the cooling rate, judging from its unit of °C/s. As shown, the higher the cooling rate, that is, the lower $(GR)^{-1}$, the finer the cells become.

The effect of the temperature gradient *G* and the growth rate *R* on the solidification microstructure of alloys is summarized in Figure 6.20. Together, *G* and *R* dominate the solidification microstructure. The ratio *G/R* determines the mode of solidification while the product *GR* governs the size of the solidification structure.

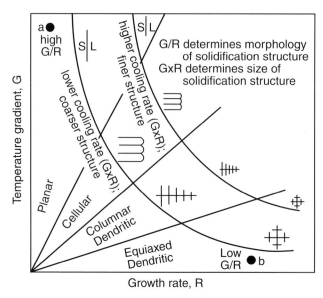

Figure 6.20   Effect of temperature gradient G and growth rate R on the morphology and size of solidification microstructure.

6.5   SOLIDIFICATION PATH

Consider the eutectic phase diagram of a binary system A–B shown by the broken lines in Figure 6.21a. To the left of the eutectic point, the primary solidification phase is γ, namely, γ is the first thing to solidify. To the right, the primary solidification phase is α. The solid line represents the solidification path of alloy C_0. The arrow in the solidification path indicates the direction in which the liquid composition changes as temperature decreases. At step 1, where the fraction of the liquid phase L is $f_L = 1$, γ begins to form from the liquid phase, assuming negligible undercooling before solidification. The composition of the liquid phase follows the solid line representing $L \rightarrow \gamma$. At step 2, solidification is complete and $f_L = 0$.

In the case of a ternary system A–B–C, the phase diagram is three dimensional, with the base plane showing the composition and the vertical direction showing the temperature. The liquidus is a surface (curved) instead of a line as in a binary phase diagram (Figure 6.21a). The intersection between two liquidus surfaces is a line called the *line of twofold saturation*, instead of a point intersection between two liquidus lines in a binary phase diagram. As shown in Figure 6.21b, when projected vertically downward to the base plane, the liquidus surfaces show areas of primary solidification phases with the lines of twofold saturation separating them.

The solid line shows the solidification path of alloy C_0. At step 1, where the fraction of the liquid phase L is $f_L = 1$, the solid phase γ begins to form from

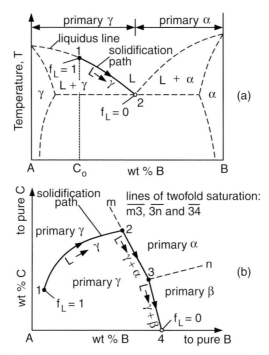

Figure 6.21   Solidification paths: (*a*) a binary A–B system; (*b*) a ternary A–B–C system.

the liquid phase. As temperature decreases, the composition of the liquid phase follows the solid line representing $L \rightarrow \gamma$ until the line of twofold saturation $\overline{m3}$ is reached. At step 2, the α phase begins to form from the liquid, too. As temperature decreases further, the composition of the liquid phase follows the solid line representing the ternary eutectic reaction $L \rightarrow \gamma + \alpha$ until the line of twofold saturation $\overline{3n}$ is reached. At step 3, the formation of α from the liquid stops and is taken over by the formation of β. As temperature decreases still further, the composition of the liquid phase follows the solid line representing the ternary eutectic reaction $L \rightarrow \gamma + \beta$. This can go on until the line representing binary A–B (the abscissa) is reached, if there is still liquid available to get there. The fraction of liquid $f_L = 0$ when solidification is complete.

REFERENCES

1. Scheil, E., *Z. Metallk.*, **34:** 70, 1942.
2. Flemings, M. C., *Solidification Processing*, McGraw-Hill, New York, 1974.
3. Pohl, R. G., *J. Appl. Phys.*, **25:** 668, 1954.

4. Kou, S., *Transport Phenomena and Materials Processing*, Wiley, New York, 1996.

5. Jackson, K. A., in *Solidification*, American Society for Metals, Metals Park, OH, 1971, p. 121.

6. Jackson, K. A., Hunt, J. D., Uhlmann, D. R., Sewand III, T. P., *Trans. AIME*, **236:** 149, 1966.

7. Morris, L. R., and Winegard, W. C., *J. Crystal Growth*, **5:** 361, 1969.

8. Giamei, A. F., Kraft, E. H., and Lernkey, F. D., in *New Trends in Materials Processing*, American Society for Metals, Metals Park, OH, 1976, p. 48.

9. Kattamis, T. Z., Holmber, U. T., and Flemings, M. C., *J. Inst. Metals*, **55:** 343, 1967.

10. David, S. A., and Vitek, J. M., in *International Trends in Welding Science and Technology*, Eds. by S. A. David and J. M. Vitek, ASM International, Materials Park, OH, 1993, p. 147.

11. Rutter, J. W., and Chalmer, B., *Can. J. Physiol.*, **31:** 15, 1953.

12. Tiller, W. A., Jackson, K. A., Rutter, J. W., and Chalmer, B., *Acta Met.*, **1:** 428, 1953.

13. Mullins, W. W., and Sekerka, R. F., *J. Appl. Physiol.*, **34:** 323, 1963; **35:** 444, 1964.

14. Sekerka, R. F., *J. Appl. Physiol.*, **36:** 264, 1965.

15. Sekerka, R. F., *J. Phys. Chem. Solids*, **28**: 983, 1967.

16. Flemings, M. C., *Solidification Processing*, McGraw-Hill, New York, 1974.

17. Jackson, K. A., and Hunt, J. S., *Acta Met.*, **13:** 1212, 1965.

18. Walton, D., Tiller, W. A., Rutter, J. W., and Winegard, W. C., *Trans. AIME*, **203:** 1023, 1955.

19. Cole, G. S., and Winegard, W. C., *J. Inst. Metals*, **92:** 323, 1963.

20. Hunt, M. D., Spittle, J. A., and Smith, R. W., *The Solidification of Metals*, Publication No. 110, Iron and Steel Institute, London, 1968, p. 57.

21. Plaskett, T. S., and Winegard, W. C., *Can. J. Physiol.*, **37:** 1555, 1959.

22. Bardsley, W., Callan, J. M., Chedzey, H. A., and Hurle, D. T. J., *Solid-State Electron*, **3:** 142, 1961.

23. Coulthard, J. O, and Elliott, R., *The Solidification of Metals*, Publication No. 110, Iron and Steel Institute, 1968, p. 61.

24. Liu, L. X., and Kirkaldy, J. S., *J. Crystal Growth*, 144: 335, 1994.

25. Munitz, A., *Metall. Trans.*, **16B:** 149, 1985.

26. Paul, A. J., and DebRoy, T., *Rev. Modern Phys.*, **67**: 85, 1995.

27. Kou, S., and Le, Y., *Metall. Trans.*, **13A:** 1141, 1982.

28. Plaskett, T. S., and Winegard, W. C., *Can. J. Physiol.*, **38:** 1077, 1960.

FURTHER READING

1. Chalmers, B., *Principles of Solidification*, Wiley, New York, 1964.

2. Davies, G. J., *Solidification and Casting*, Wiley, New York, 1973.

3. Davies, G. J., and Garland, J. G., *Int. Metall. Rev.*, **20:** 83, 1975.

4. Flemings, M. C., *Solidification Processing*, McGraw-Hill, New York, 1974.

5. Savage, W. F., *Weld. World*, **18:** 89, 1980.

PROBLEMS

6.1 From the Al–Mg phase diagram, the equilibrium freezing range of 5052 aluminum (essentially Al–2.5Mg) is about 40°C. Suppose the welding speed is 4 mm/s and the diffusion coefficient D_L is 3×10^{-5} cm^2/s. Calculate the minimum temperature gradient required for planar solidification at the weld centerline. What is the corresponding cooling rate? Can this level of cooling rate be achieved in arc welding.

6.2 Let C_E and C_{SM} be respectively 35% and 15% Mg, and both the solidus and liquidus lines are essentially straight in the Al–Mg system. The melting point of pure Al is 660°C, and the eutectic temperature is 451°C. What is the approximate volume fraction of the aluminum-rich dendrites in the fusion zone of autogenous 5052 aluminum weld?

6.3 It has been observed that aluminum alloys welded with the electron beam welding process show much finer secondary dendrite arm spacing in the weld metal than those welded with GMAW. Explain why.

6.4 Which alloy has a greater tendency for planar solidification to break down, Al-0.01Cu or Al-6.3Cu and why?

6.5 How would preheating of the workpiece affect the secondary dendrite arm spacing in welds of aluminum alloys and why?

7 Weld Metal Solidification I: Grain Structure

In this chapter we shall discuss the development of the grain structure in the fusion zone and the effect of welding parameters on the grain structure. Then, we shall discuss various mechanisms of and techniques for grain refining. The grain structure of the fusion zone can significantly affect its susceptibility to solidification cracking during welding and its mechanical properties after welding.

7.1 EPITAXIAL GROWTH AT FUSION BOUNDARY

7.1.1 Nucleation Theory

Figure 7.1 shows the nucleation of a crystal from a liquid on a flat substrate with which the liquid is in contact (1, 2). The parameters γ_{LC}, γ_{LS}, and γ_{CS} are the surface energies of the liquid–crystal interface, liquid–substrate interface, and crystal–substrate interface, respectively. According to Turnbull (2), the energy barrier ΔG for the crystal to nucleate on the substrate is

$$\Delta G = \frac{4\pi\gamma_{LC}^3 T_m^2}{3(\Delta H_m \Delta T)^2}(2 - 3\cos\theta + \cos^3\theta) \tag{7.1}$$

where T_m is the equilibrium melting temperature, ΔH_m the latent heat of melting, ΔT the undercooling below T_m, and θ the contact angle. If the liquid wets the substrate completely, the contact angle θ is zero and so is ΔG. This means that the crystal can nucleate on the substrate without having to overcome any energy barrier required for nucleation. The energy barrier can be significant if no substrate is available or if the liquid does not wet the substrate completely.

In fusion welding the existing base-metal grains at the fusion line act as the substrate for nucleation. Since the liquid metal of the weld pool is in intimate contact with these substrate grains and wets them completely ($\theta = 0$), crystals nucleate from the liquid metal upon the substrate grains without difficulties. When welding without a filler metal (autogenous welding), nucleation occurs by arranging atoms from the liquid metal upon the substrate grains without

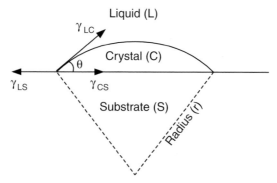

Figure 7.1 Spherical cap of a crystal nucleated on a planar substrate from a liquid.

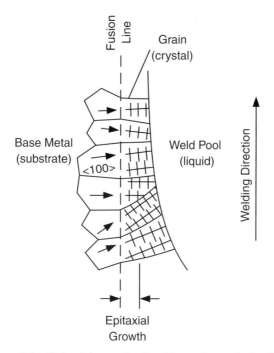

Figure 7.2 Epitaxial growth of weld metal near fusion line.

altering their existing crystallographic orientations. Such a growth initiation process, shown schematically in Figure 7.2, is called *epitaxial growth* or epitaxial nucleation. The arrow in each grain indicates its <100> direction. For materials with a face-centered-cubic (fcc) or body-centered-cubic (bcc) crystal structure, the trunks of columnar dendrites (or cells) grow in the <100> direction. As shown, each grain grows without changing its <100> direction.

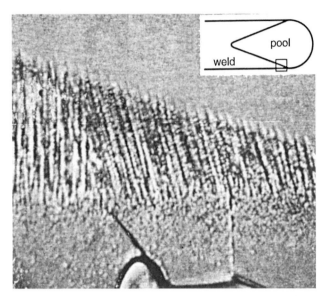

Figure 7.3 Epitaxial growth during the "welding" of camphene in the area indicated by the square (magnification 75×). Modified from Savage (6).

7.1.2   Epitaxial Growth in Welding

Savage et al. (3–8) first discovered epitaxial growth in fusion welding. By using the Laue x-ray back-reflection technique, they confirmed the continuity of crystallographic orientation across the fusion boundary. Savage and Hrubec (6) also studied epitaxial growth by using a transparent organic material (camphene) as the workpiece, as shown in Figure 7.3. Three grains are visible in the base metal at the fusion line. All dendrites growing from each grain point in one direction, and this direction varies from one grain to another. From the welds shown in Figure 7.4 epitaxial growth at the fusion line is evident (9).

Epitaxial growth can also occur when the workpiece is a material of more than one phase. Elmer et al. (10) observed epitaxial growth in electron beam welding of an austenitic stainless steel consisting of both austenite and ferrite. As shown in Figure 7.5, both austenite (A) and ferrite (F) grow epitaxially at the fusion line (dotted line) from the base metal to the weld metal (resolidified zone).

7.2   NONEPITAXIAL GROWTH AT FUSION BOUNDARY

When welding with a filler metal (or joining two different materials), the weld metal composition is different from the base metal composition, and the weld metal crystal structure can differ from the base metal crystal structure. When

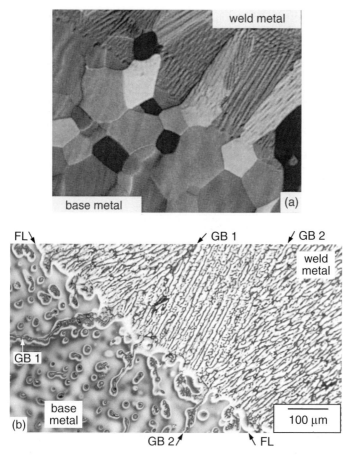

Figure 7.4 Epitaxial growth. (*a*) Near the fusion boundary of electron beam weld of C103 alloy (magnification 400×). Reprinted from O'Brien (9). Courtesy of American Welding Society. (*b*) Near the fusion boundary of as-cast Al–4.5Cu welded with 4043 filler (Al–5Si).

this occurs, epitaxial growth is no longer possible and new grains will have to nucleate at the fusion boundary.

Nelson et al. (11) welded a type 409 ferritic stainless steel of the bcc structure with a Monel (70Ni–30Cu) filler metal of the fcc structure and produced a fcc weld metal. Figure 7.6 shows the fusion boundary microstructure. They proposed that, when the base metal and the weld metal exhibit two different crystal structures at the solidification temperature, nucleation of solid weld metal occurs on heterogeneous sites on the partially melted base metal at the fusion boundary. The fusion boundary exhibits random misorientations between base metal grains and weld metal grains as a result of heterogeneous nucleation at the pool boundary. The weld metal grains may or may not follow

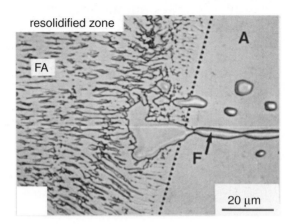

Figure 7.5   Epitaxial growth of austenite (A) and ferrite (F) from the fusion line of an austenitic stainless steel containing both phases. From Elmer et al. (10).

special orientation relationships with the base metal grains they are in contact with, namely, orient themselves so that certain atomic planes are parallel to specific planes and directions in the base-metal grains.

7.3   COMPETITIVE GROWTH IN BULK FUSION ZONE

As described in the previous section, the grain structure near the fusion line of a weld is dominated either by epitaxial growth when the base metal and the weld metal have the same crystal structure or by nucleation of new grains when they have different crystal structures. Away from the fusion line, however, the grain structure is dominated by a different mechanism known as competitive growth.

During weld metal solidification grains tend to grow in the direction perpendicular to pool boundary because this is the direction of the maximum temperature gradient and hence maximum heat extraction. However, columnar dendrites or cells within each grain tend to grow in the easy-growth direction. Table 7.1 shows the easy-growth directions in several materials (12) and, as shown, it is <100> for both fcc and bcc materials. Therefore, during solidification grains with their easy-growth direction essentially perpendicular to the pool boundary will grow more easily and crowd out those less favorably oriented grains, as shown schematically in Figure 7.7. This mechanism of competitive growth dominates the grain structure of the bulk weld metal.

7.4   EFFECT OF WELDING PARAMETERS ON GRAIN STRUCTURE

The weld pool becomes teardrop shaped at high welding speeds and elliptical at low welding speeds (Chapter 2). Since the trailing pool boundary of a

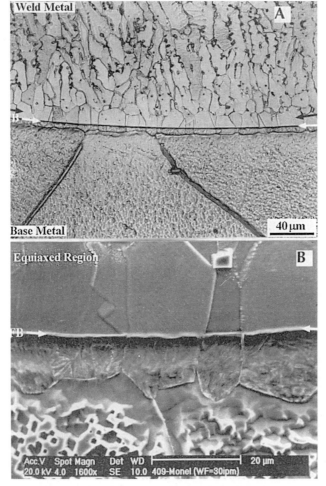

Figure 7.6 Fusion boundary microstructure in 409 ferritic stainless steel (bcc) welded with Monel filler wire (fcc): (*a*) optical micrograph; (*b*) scanning electron micrograph. White arrows: fusion boundary; dark arrows: new grains nucleated along fusion boundary. Reprinted from Nelson et al. (11). Courtesy of American Welding Society.

TABLE 7.1 Easy-Growth Directions

Crystal Structure	Easy-Growth Direction	Examples
Face-centered-cubic (fcc)	<100>	Aluminum alloys, austenitic stainless steels
Body-centered-cubic (bcc)	<100>	Carbon steels, ferritic stainless steels
Hexagonal-close-packed (hcp)	<10$\bar{1}$0>	Titanium, magnesium
Body-centered-tetragonal (bct)	<110>	Tin

Source: From Chalmers (12).

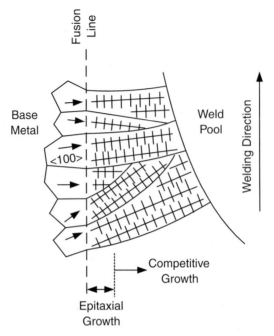

Figure 7.7   Competitive growth in bulk fusion zone.

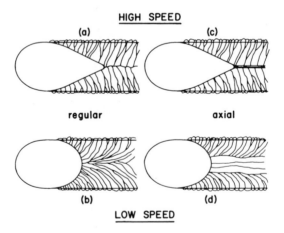

Figure 7.8   Effect of welding speed on columnar-grain structure in weld metal: (*a*, *b*) regular structure; (*c*, *d*) with axial grains.

teardrop-shaped weld pool is essentially straight, the columnar grains are also essentially straight in order to grow perpendicular to the pool boundary, as shown schematically in Figure 7.8*a*. On the other hand, since the trailing boundary of an elliptical weld pool is curved, the columnar grains are also

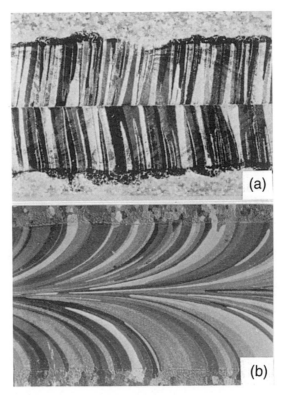

Figure 7.9 Gas–tungsten arc welds of 99.96% aluminum: (*a*) 1000 mm/min welding speed; (*b*) 250 mm/min welding speed. From Arata et al. (13).

curved in order to grow perpendicular to the pool boundary, as shown in Figure 7.8*b*. Figure 7.9 shows the gas–tungsten arc welds of high-purity (99.96%) aluminum made by Arata et al. (13). At the welding speed of 1000 mm/min straight columnar grains point toward the centerline, while at 250 mm/min curved columnar grains point in the welding direction.

Axial grains can also exist in the fusion zone. Axial grains can initiate from the fusion boundary at the starting point of the weld and continue along the length of the weld, blocking the columnar grains growing inward from the fusion lines. Like other columnar grains, these axial grains also tend to grow perpendicular to the weld pool boundary. With a teardrop-shaped pool, only a short section of the trailing pool boundary can be perpendicular to the axial direction, and the region of axial grains is thus rather narrow, as shown in Figure 7.8*c*. With an elliptical weld pool, however, a significantly longer section of the trailing pool boundary can be perpendicular to the axial direction, and the region of axial grains can thus be significantly wider, as shown in Figure 7.8*d*. Figure 7.10 shows aluminum welds with these two types of grain

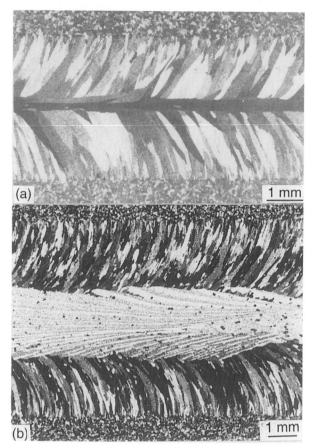

Figure 7.10 Axial grains in GTAW: (*a*) 1100 aluminum at 12.7 mm/s welding speed; (*b*) 2014 aluminum at 3.6 mm/s welding speed.

structure. Axial grains have been reported in aluminum alloys (14–16), austenitic stainless steels (17), and iridium alloys (18).

7.5 WELD METAL NUCLEATION MECHANISMS

The mechanisms of nucleation of grains in the weld metal will be discussed in the present section. In order to help understand these mechanisms, the microstructure of the material around the weld pool will be discussed first.

Figure 7.11 shows a 2219 aluminum (essentially Al–6.3% Cu) weld pool quenched with ice water during GTAW (19). The S + L region around the weld pool consists of two parts (Figure 7.11*a*): the partially melted material

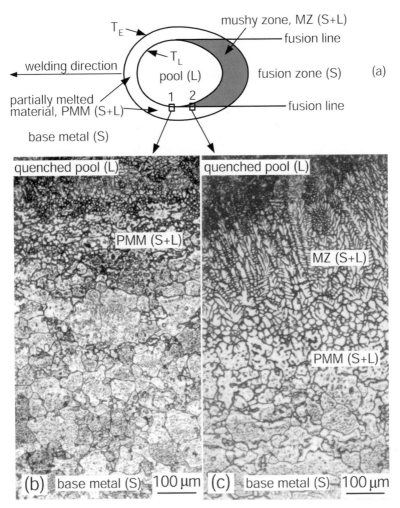

Figure 7.11 Weld pool of 2219 aluminum quenched during GTAW: (*a*) overall view; (*b*) microstructure at position 1; (*c*) microstructure at position 2. Modified from Kou and Le (19). Courtesy of American Welding Society.

(clear) associated with the leading portion of the pool boundary and the mushy zone (shaded) associated with the trailing portion. Area 1 (Figure 7.11*b*) covers a small portion of the partially melted material. Area 2 (Figure 7.11*c*) covers a small portion of the mushy zone as well as the partially melted material.

Based on Figure 7.11, the microstructure around the weld pool boundary of an alloy is shown schematically in Figure 7.12, along with thermal cycles at the weld centerline and at the fusion line and a phase diagram. The eutectic-type phase diagram is common among aluminum alloys. As shown, the mushy

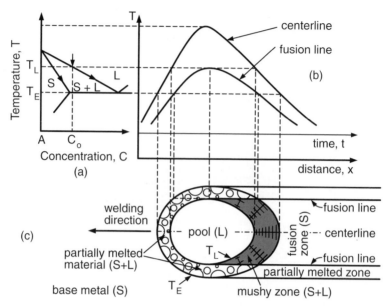

Figure 7.12 Microstructure around the weld pool boundary: (*a*) phase diagram; (*b*) thermal cycles; (*c*) microstructure of solid plus liquid around weld pool.

zone behind the trailing portion of the pool boundary consists of solid dendrites (S) and the interdendritic liquid (L). The partially melted material around the leading portion of the pool boundary, on the other hand, consists of solid grains (S) that are partially melted and the intergranular liquid (L). In summary, there is a region of solid–liquid mixture surrounding the weld pool of an alloy.

Figure 7.13*a* shows three possible mechanisms for new grains to nucleate during welding: dendrite fragmentation, grain detachment, and heterogeneous nucleation (19). Figure 7.13*b* shows the fourth nucleation mechanism, surface nucleation. These mechanisms, which have been well documented in metal casting, will be described briefly below. The techniques for producing new grains in the weld metal by these mechanisms will be discussed in a subsequent section.

7.5.1 Dendrite Fragmentation

Weld pool convection (Chapter 4) can in principle cause fragmentation of dendrite tips in the mushy zone, as illustrated in Figure 7.13*a*. These dendrite fragments are carried into the bulk weld pool and act as nuclei for new grains to form if they survive the weld pool temperature. It is interesting to note that this mechanism has been referred to frequently as the grain refining mechanism for weld metals without proof.

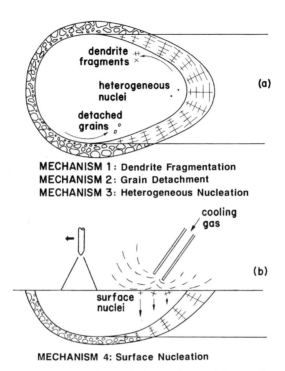

MECHANISM 1: Dendrite Fragmentation
MECHANISM 2: Grain Detachment
MECHANISM 3: Heterogeneous Nucleation

MECHANISM 4: Surface Nucleation

Figure 7.13 Nucleation mechanisms during welding: (*a*) top view; (*b*) side view. Reprinted from Kou and Le (19). Courtesy of American Welding Society.

7.5.2 Grain Detachment

Weld pool convection can also cause partially melted grains to detach themselves from the solid–liquid mixture surrounding the weld pool, as shown in Figure 7.13*a*. Like dendrite fragments, these partially melted grains, if they survive in the weld pool, can act as nuclei for the formation of new grains in the weld metal.

7.5.3 Heterogeneous Nucleation

Foreign particles present in the weld pool upon which atoms in the liquid metal can be arranged in a crystalline form can act as heterogeneous nuclei. Figure 7.14 depicts heterogeneous nucleation and the growth of new grains in the weld metal. Figure 7.15*a* shows two (dark) heterogeneous nuclei at the centers of two equiaxed grains in an autogenous gas–tungsten arc weld of a 6061 aluminum containing 0.043% titanium (20). Energy dispersive spectrometry (EDS) analysis, shown in Figure 7.15*b*, indicates that these nuclei are rich in titanium (boron is too light to be detected by EDS). A scanning electron microscopy (SEM) image of a nucleus is shown in Figure 7.15*c*. The

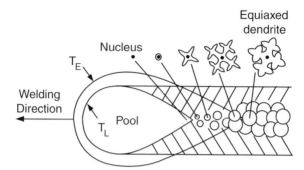

Figure 7.14 Heterogeneous nucleation and formation of equiaxed grains in weld metal.

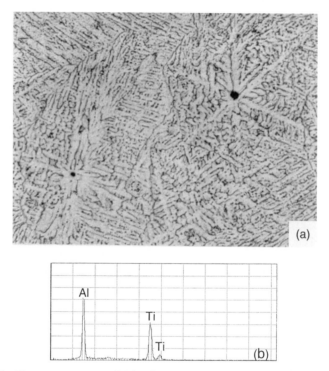

Figure 7.15 Heterogeneous nuclei in GTAW of 6061 aluminum: (*a*) optical micrograph; (*b*) EDS analysis; (*c*) SEM image; (*d*) SEM image of TiB_2 particles in a grain refiner for aluminum casting. (*a, b*) From Kou and Le (20). (*d*) Courtesy of Granger (21).

Figure 7.15 *Continued*

morphology of the nucleus is similar to that of the agglomerated TiB$_2$ particles, shown in Figure 7.15*d*, in an Al–5% Ti–0.2% B grain refiner for ingot
casting of aluminum alloys (21). This suggests that the TiB$_2$ particles in the
weld metal are likely to be from the Al–Ti–B grain refiner in aluminum ingot
casting. Figure 7.16 shows a heterogeneous nucleus of TiN at the center of an
equiaxed grain in a GTA weld of a ferritic stainless steel (22).

As mentioned previously, Nelson et al. (11) have observed nucleation of
solid weld metal on heterogeneous sites on the partially melted base metal at
the fusion boundary when the weld metal and the base metal differ in crystal
structure. Gutierrez and Lippold (23) have also studied the formation of the
nondendritic equiaxed zone in a narrow region of the weld metal adjacent to
the fusion boundary of 2195 aluminum (essentially Al–4Cu–1Li). Figure 7.17
shows an example of the equiaxed zone in a 2195 weld made with a 2319
(essentially Al–6.3Cu) filler metal. The width of the equiaxed zone was found
to increase with increasing Zr and Li contents in the alloy. They proposed that
near the pool boundary the cooler liquid is not mixed with the warmer bulk
weld pool. Consequently, near the pool boundary heterogeneous nuclei such
as Al$_3$Zr and Al$_3$(Li$_x$Zr$_{1-x}$), which are originally present as dispersoids in the

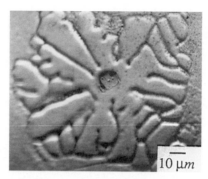

Figure 7.16   TiN particle as heterogeneous nucleus in GTAW of ferritic stainless steel. Reprinted from Villafuerte and Kerr (22).

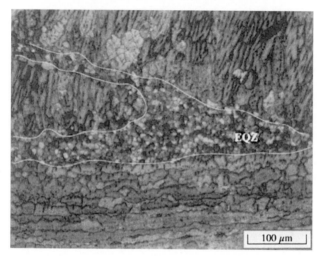

Figure 7.17   Nondendritic equiaxed zone in narrow region adjacent to fusion boundary of 2195 Al–Cu–Li alloy. Reprinted from Gutierrez and Lippold (23). Courtesy of American Welding Society.

base metal, are able to survive and form the nondendritic equiaxed zone, as illustrated in Figure 7.18. By using a Gleeble thermal simulator, Kostrivas and Lippold (24) found that the equiaxed zone could be formed by heating in the temperature range of approximately 630–640°C and at temperatures above 640°C the normal epitaxial growth occurred. This is consistent with the proposed mechanism in the sense that the heterogeneous nuclei can only survive near the cooler pool boundary and not in the warmer bulk weld pool.

In the metal casting process the metal is superheated during melting before it is cast. Since the nuclei are unstable at higher temperatures, they dissolve in the superheated liquid in the casting process. As such, there is no equiaxed

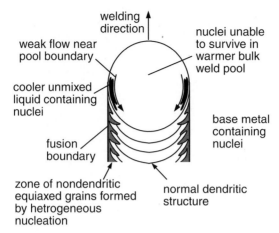

Figure 7.18 Mechanism for formation of nondendritic equiaxed zone in Al–Cu–Li weld according to Gutierrez and Lippold (23).

zone in the weld of an as-cast alloy 2195. However, the equiaxed zone occurs again if the as-cast alloy 2195 is solution heat treated first and then welded. Gutierrez and Lippold (23) proposed that because of the relatively low solubility of Zr in solid aluminum, Al_3Zr and $Al_3(Li_xZr_{1-x})$ particles precipitate out of the solid solution during heat treating.

7.5.4   Surface Nucleation

The weld pool surface can be undercooled thermally to induce surface nucleation by exposure to a stream of cooling gas or by instantaneous reduction or removal of the heat input. When this occurs, solid nuclei can form at the weld pool surface, as illustrated in Figure 7.13*b*. These solid nuclei then grow into new grains as they shower down from the weld pool surface due to their higher density than the surrounding liquid metal.

7.5.5   Effect of Welding Parameters on Heterogeneous Nucleation

Before leaving this section, an important point should be made about the effect of welding parameters on heterogeneous nucleation. Kato et al. (25), Arata et al. (26), Ganaha et al. (16), and Kou and Le (20) observed in commercial aluminum alloys that the formation of equiaxed grains is enhanced by higher heat inputs and welding speeds. As shown in Figure 7.19, equiaxed grains can form a band along the centerline of the weld and block off columnar grains as the heat input and welding speed are increased (20). Kou and Le (27) showed in Figure 7.20 that, as the heat input and the welding speed are increased, the temperature gradient (G) at the end of the weld pool is reduced. Furthermore, as the welding speed is increased, the solidification rate of the

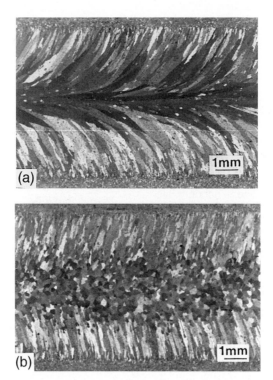

Figure 7.19 Effect of welding parameters on grain structure in GTAW of 6061 aluminum: (*a*) 70 A × 11 V heat input and 5.1 mm/s welding speed; (*b*) 120 A × 11 V heat input and 12.7 mm/s welding speed. From Kou and Le (20).

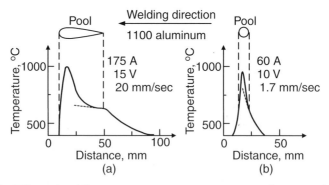

Figure 7.20 Effect of welding parameters on temperature gradient at weld pool end of 1100 aluminum: (*a*) higher welding speed and heat input; (*b*) lower welding speed and heat input. From Kou and Le (27).

weld metal (R) is also increased. As illustrated in Figure 7.21, the ratio G/R should be decreased and the constitutional supercooling (28) in front of the advancing solid–liquid interface should, therefore, be increased. Kato et al. (25) and Arata et al. (26) proposed that the transition to an equiaxed grain

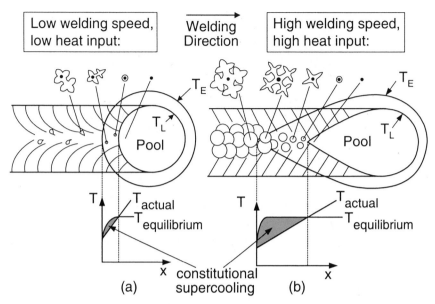

Figure 7.21 Effect of welding parameters on heterogeneous nucleation: (*a*) low constitutional supercooling at low welding speed and heat input; (*b*) heterogeneous nucleation aided by high constitutional supercooling at high welding speed and heat input.

structure is due to the existence of a sufficiently long constitutionally under-cooled zone in the weld pool. Ganaha et al. (16), however, indicated that the transition is not due to constitutional supercooling alone. In fact, it was observed that significant amounts of equiaxed grains formed only in those alloys containing around or more than 0.01 wt % Ti or 0.10 wt % Zr. Further-more, tiny second-phase particles rich in titanium and/or zirconium (possibly Ti_xZr_yC compounds) existed at the dendrite centers of the equiaxed grains. Consequently, it was proposed that equiaxed grains in the fusion zone form by heterogeneous nucleation aided by constitutional supercooling. The obser-vation of Ganaha et al. (16) was confirmed by Kou and Le (19, 20, 29). The effect of welding parameters on grain refining in aluminum welds was dis-cussed by Kou and Le (20).

7.6 GRAIN STRUCTURE CONTROL

The weld metal grain structure can affect its mechanical properties signifi-cantly. Arata et al. (13) tensile tested aluminum welds in the welding direc-tion. The weld metal ductility of 99.96% aluminum dropped greatly when the columnar grains pointed to the weld centerline (Figure 7.9*a*), that is, when the grains became nearly normal to the tensile axis. Also, the weld metal tensile strength of 5052 aluminum increased as the amount of equiaxed grains increased.

The formation of fine equiaxed grains in the fusion zone has two main advantages. First, fine grains help reduce the susceptibility of the weld metal to solidification cracking during welding (Chapter 11). Second, fine grains can improve the mechanical properties of the weld, such as the ductility and fracture toughness in the case of steels and stainless steels. Therefore, much effort has been made to try to grain refine the weld fusion zone. This includes the application of grain refining techniques that were originally developed for casting. Described below are several techniques that have been used to control the weld metal grain structure.

7.6.1 Inoculation

This technique has been used extensively in metal casting. It involves the addition of nucleating agents or inoculants to the liquid metal to be solidified. As a result of inoculation, heterogeneous nucleation is promoted and the liquid metal solidifies with very fine equiaxed grains. In the work by Davies and Garland (1), inoculant powders of titanium carbide and ferrotitanium–titanium carbide mixtures were fed into the weld pool during the submerged arc welding of a mild steel and very fine grains were obtained. Similarly, Heintze and McPherson (30) grain refined submerged arc welds of C-Mn and stainless steels with titanium. Figure 7.22 shows the effect of inoculation on the grain structure of the weld fusion zone of the C-Mn steel. It is interesting to note that Petersen (31) has grain refined Cr–Ni iron base alloys with aluminum nitride and reported a significant increase in the ductility of the resultant welds, as shown in Figure 7.23.

Pearce and Kerr (32), Matsuda et al. (33), Yunjia et al. (34), and Sundaresan et al. (35) grain refined aluminum welds by using Ti and Zr as inoculants. Yunjia et al. (34) showed the presence of $TiAl_3$ particles at the origins of equiaxed grains in Ti microalloyed 1100 aluminum welds. Figure 7.24 shows grain refining in a 2090 Al–Li–Cu alloy gas–tungsten arc welded with a 2319 Al–Cu filler inoculated with 0.38% Ti (35).

7.6.2 External Excitation

Different dynamic grain refining techniques, such as liquid pool stirring, mold oscillation, and ultrasonic vibration of the liquid metal, have been employed in metal casting, and recently similar techniques, including weld pool stirring, arc oscillation, and arc pulsation, have been applied to fusion welding.

A. Weld Pool Stirring Weld pool stirring can be achieved by electromagnetic stirring, as shown in Figure 7.25, by applying an alternating magnetic field parallel to the welding electrode (36). Matsuda et al. (37, 38) and Pearce and Kerr (32) increased the degree of grain refinement in aluminum alloys containing small amounts of titanium by electromagnetic pool stirring. Pearce and Kerr (32) suggested that the increased grain refinement was due to heteroge-

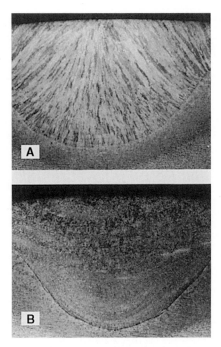

Figure 7.22 Effect of inoculation on grain structure in submerged arc welds of C–Mn steel (magnification 6×): (*a*) without inoculation; (*b*) inoculation with titanium. Reprinted from Heintze and McPherson (30). Courtesy of American Welding Society.

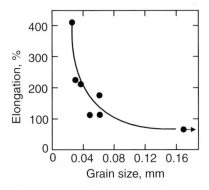

Figure 7.23 Effect of grain size on weld metal ductility of a Cr–Ni iron base alloy at 925°C. Modified from Petersen (31).

neous nucleation, rather than to dendrite fragmentation. This is because in GTAW, unlike ingot casting, the liquid pool and the mushy zone are rather small, and it is therefore difficult for the liquid metal in the pool to penetrate and break away dendrites, which are so short and so densely packed together.

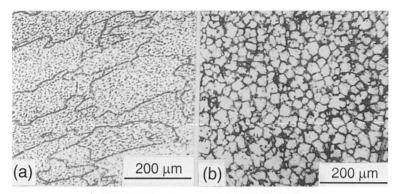

Figure 7.24   Effect of inoculation on grain structure in GTAW of 2090 Al–Li–Cu alloy: (*a*) 2319 Al–Cu filler metal; (*b*) 2319 Al–Cu filler metal inoculated with 0.38% Ti. Reprinted from Sundaresan et al. (35).

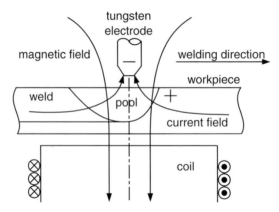

Figure 7.25   Schematic sketch showing application of external magnetic field during autogenous GTAW. Modified from Matsuda et al. (36).

Stirring of the weld pool tends to lower the weld pool temperature, thus helping heterogeneous nuclei survive. Figure 7.26 shows the increased grain refinement by weld pool stirring in an Al–2.5 wt % Mg alloy containing 0.11 wt % Ti (32). As shown in Figure 7.27, Villafuerte and Kerr (39) grain refined a weld of 409 ferritic stainless steel containing 0.32% Ti by the same technique. The higher the Ti content, the more effective grain refining was. Titanium-rich particles were found at the origin of equiaxed grains, suggesting heterogeneous nucleation as the grain refining mechanism.

Pearce and Kerr (32) also found that by applying weld pool stirring, the partially melted grains along the leading portion of the pool boundary, if they are only loosely held together, could be swept by the liquid metal into the weld

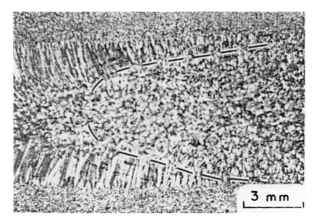

Figure 7.26 Widening of equiaxed zone in GTAW of alloy Al–2.5Mg–0.011Ti by magnetic stirring (at dotted line). Reprinted from Pearce and Kerr (32).

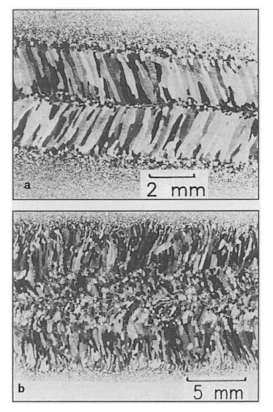

Figure 7.27 Effect of electromagnetic pool stirring on grain structure in GTAW of 409 ferritic stainless steel: (*a*) without stirring; (*b*) with stirring. Reprinted from Villafuerte and Kerr (39). Courtesy of American Welding Society.

pool, where they served as nuclei. Significant amounts of equiaxed grains were so produced in gas–tungsten arc welds of a 7004 aluminum alloy containing very little Ti.

B. Arc Oscillation   Arc oscillation, on the other hand, can be produced by magnetically oscillating the arc column using a single- or multipole magnetic probe or by mechanically vibrating the welding torch. Davies and Garland (1) produced grain refining in gas–tungsten arc welds of Al–2.5 wt % Mg alloy by torch vibration. Resistance to weld solidification cracking was improved in these welds. Figure 7.28 shows the effect of the vibration amplitude on the grain size (1). Dendrite fragmentation was proposed as the grain refining mechanism. It is, however, suspected that heterogeneous nucleation could have been the real mechanism, judging from the fact that the Al–2.5 wt % Mg used actually contained about 0.15 wt %Ti. Venkataraman et al. (40, 41) obtained grain refining in the electroslag welds of steels by electrode vibration and enhanced electromagnetic stirring of the weld pool and improved their toughness and resistance to centerline cracking. Dendrite fragmentation was also considered as the mechanism for grain refining.

Sharir et al. (42) obtained grain refinement in gas–tungsten arc welds of pure tantalum sheets by arc oscillation. Due to the high melting point of pure tantalum (about 3000°C), the surface heat loss due to radiation was rather significant. As a result, the liquid metal was cooled down rapidly and was in fact undercooled below its melting point when the heat source was deflected away during oscillated arc welding. This caused surface nucleation and resulted in grain refinement.

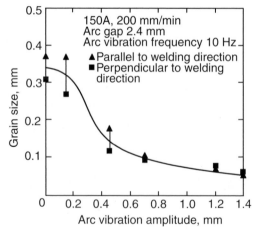

Figure 7.28   Effect of arc vibration amplitude on grain size in Al–2.5Mg welds. Modified from Davies and Garland (1).

Sundaresan and Janaki Ram (43) grain refined Ti alloys by magnetic arc oscillation and improved the weld metal tensile ductility. No specific grain refining mechanism was identified.

C. Arc Pulsation   Arc pulsation can be obtained by pulsating the welding current. Sharir et al. (42) also obtained grain refinement in gas–tungsten arc welds of pure tantalum sheets by arc pulsation. The liquid metal was undercooled when the heat input was suddenly reduced during the low-current cycle of pulsed arc welding. This caused surface nucleation and resulted in grain refinement.

Figure 7.29 shows grain refining in a pulsed arc weld of a 6061 aluminum alloy containing 0.04 wt % Ti (44). Heterogeneous nucleation, aided by thermal undercooling resulting from the high cooling rate produced by the relatively high welding speed used and arc pulsation, could have been responsible for the grain refinement in the weld.

7.6.3   Stimulated Surface Nucleation

Stimulated surface nucleation was originally used by Southin (45) to obtain grain refinement in ingot casting. A stream of cool argon gas was directed on the free surface of molten metal to cause thermal undercooling and induce surface nucleation. Small solidification nuclei formed at the free surface and showered down into the bulk liquid metal. These nuclei then grew and became small equiaxed grains. This technique was used to produce grain refining in Al–2.5Mg welds by Davies and Garland (1) and in Ti alloys by Wells (46).

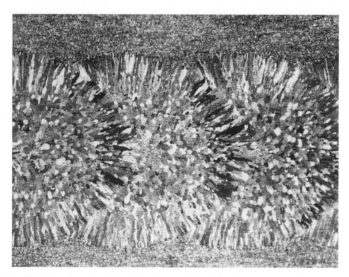

Figure 7.29   Equiaxed grains in pulsed arc weld of 6061 aluminum (magnification 9×). From Kou and Le (44).

7.6.4 Manipulation of Columnar Grains

Kou and Le (14, 15, 47) manipulated the orientation of columnar grains in aluminum welds by low-frequency arc oscillation. Figure 7.30*a* shows a 2014 aluminum weld made with 1 Hz transverse arc oscillation, that is, with arc oscillating normal to the welding direction (14). Similarly, Figure 7.30*b* shows a 5052 aluminum weld made with 1 Hz circular arc oscillation (15). In both

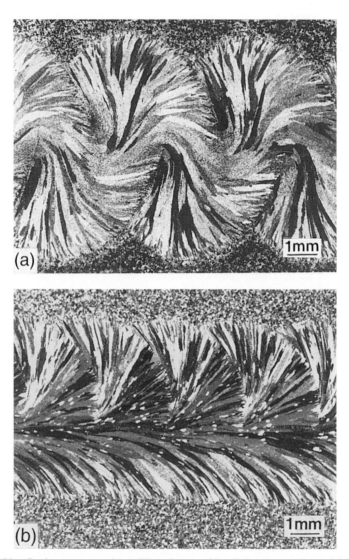

Figure 7.30   Grain structures in oscillated arc welds of aluminum alloys. (*a*) For alloy 2014 with transverse arc oscillation. From Kou and Le (14). (*b*) For alloy 5052 with circular arc oscillation. From Kou and Le (15).

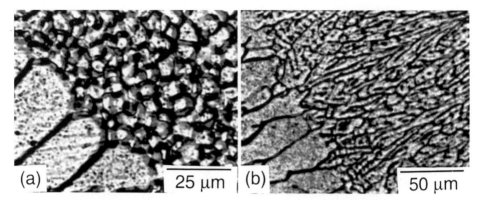

Figure 7.31   Effect of gravity on grain structure in GTAW of 2090 Al–Li–Cu alloy: (*a*) 1*g*; (*b*) 10*g* and with equiaxed zone (EQZ) near the fusion boundary eliminated. Reprinted from Aidun and Dean (48). Courtesy of American Welding Society.

cases, columnar grains grew perpendicular to the trailing portion of the weld pool, and the weld pool in turn followed the path of the moving oscillating arc.

As will be discussed later in Chapter 11, periodic changes in grain orientation, especially that produced by transverse arc oscillation at low frequencies, can reduce solidification cracking and improve both the strength and ductility of the weld metal (47).

7.6.5   Gravity

Aidun and Dean (48) gas–tungsten arc welded 2195 aluminum under the high gravity produced by a centrifuge welding system and eliminated the narrow band of nondendritic equiaxed grains along the fusion boundary. As shown in Figure 7.31, the band disappeared when gravity was increased from 1*g* to 10*g*. It was suggested that buoyancy convection enhanced by high gravity caused Al_3Zr and $Al_3(Li_xZr_{1-x})$ nuclei near the pool boundary to be swept into the bulk pool and completely dissolved, thus eliminating formation of equiaxed grains by heterogeneous nucleation.

REFERENCES

1. Davies, G. J., and Garland, J. G., *Int. Metall. Rev.*, **20(196):** 83, 1975.
2. Turnbull, D., *J. Chem. Physiol.*, **18:** 198, 1950.
3. Savage, W. F., Lundin, C. D., and Aronson, A. H., *Weld. J.*, **44:** 175s, 1965.
4. Savage, W. F., and Aronson, A. H., *Weld. J.*, **45:** 85s, 1966.
5. Savage, W. F., Lundin, C. D., and Chase, T. F., *Weld. J.*, **47:** 522s, 1968.
6. Savage, W. F., and Hrubec, R. J., *Weld. J.*, **51:** 260s, 1972.

7. Savage, W. F., *Weld. World*, **18:** 89, 1980.

8. Savage, W. F., in *Weld Imperfections*, Eds. A. R. Hunger and R. E. Lewis, Addison-Wesley, Reading, MA, 1968, p. 13.

9. 18th ed., O'brien R. L., Ed., *Jefferson's Welding Encyclopedia*, American Welding Society, Miami, FL, 1997, p. 316.

10. Elmer, J. W., Allen, S. M., and Eagar, T. W., *Metall. Trans. A*, **20A:** 2117, 1989.

11. Nelson, T. W., Lippold, J. C., and Mills, M. J., *Weld. J.*, **78:** 329s, 1999.

12. Chalmers, B., *Principles of Solidification*, Wiley, New York, 1964, p. 117.

13. Arata, Y., Matsuda, F., Mukae, S., and Katoh, M., *Trans. JWRI*, **2:** 55, 1973.

14. Kou, S., and Le, Y., *Metall. Trans.*, **16A:** 1887, 1985.

15. Kou, S., and Le, Y., *Metall. Trans.*, **16A:** 1345, 1985.

16. Ganaha, T., Pearce, B. P., and Kerr, H. W., *Metall. Trans.*, **11A:** 1351, 1980.

17. Kou, S., and Le, Y., *Metall. Trans.*, **13A:** 1141, 1982.

18. David, S. A., and Liu, C. T., *Metals Technol.*, **7:** 102, 1980.

19. Kou, S., and Le, Y., *Weld. J.*, **65:** 305s, 1986.

20. Kou, S., and Le, Y., *Metall. Trans.*, **19A:** 1075, 1988.

21. Granger, D. A., Practical Aspects of Grain Refining Aluminum Alloy Melts, paper presented at International Seminar on Refining and Alloying of Liquid Aluminum and Ferro-Alloys, Trondheim, Norway, August 26–28, 1985.

22. Villafuerte, J. C., and Kerr, H. W., in *International Trends in Welding Science and Technology*, Eds. S. A. David and J. M. Vitek, ASM International, Materials Park, OH, March 1993, p. 189.

23. Gutierrez, A., and Lippold, J. C., *Weld. J.*, **77:** 123s, 1998.

24. Kostrivas, A., and Lippold, J. C., *Weld. J.*, **79:** 1s, 2000.

25. Kato, M., Matsuda, F., and Senda, T., *Weld. Res. Abroad*, **19:** 26, 1973.

26. Arata, Y., Matsuda, F., and Matsui, A., *Trans. Jpn. Weld. Res. Inst.*, **3:** 89, 1974.

27. Kou, S., and Le, Y., unpublished research, University of Wisconsin, Madison, 1985.

28. Rutter, J. W., and Chalmer, B., *Can. J. Physiol.*, **31:** 15, 1953.

29. Le, Y., and Kou, S., in *Advances in Welding Science and Technology*, Ed. S. A. David, ASM International, Metals Park, OH, 1986, p. 139.

30. Heintze, G. N., and McPherson, R., *Weld. J.*, **65:** 71s, 1986.

31. Petersen, W. A., *Weld. J.*, **53:** 74s, 1973.

32. Pearce, B. P., and Kerr, H. W., *Metall. Trans.*, **12B:** 479, 1981.

33. Matsuda, F., Nakata, K., Tsukamoto, K., and Arai, K., *Trans. JWRI*, **12:** 93, 1983.

34. Yunjia, H., Frost, R. H., Olson, D. L., and Edwards, G. R., *Weld. J.*, **68:** 280s, 1983.

35. Sundaresan, S., Janaki Ram, G. D., Murugesan, R., and Viswanathan, N., *Sci. Technol. Weld. Join.*, **5:** 257, 2000.

36. Matsuda, F., Ushio, M., Nakagawa, H., and Nakata, K., in *Proceedings of the Conference on Arc Physics and Weld Pool Behavior*, Vol. 1, Welding Institute, Cambridge, 1980, p. 337.

37. Matsuda, F., Nakagawa, H., Nakata, K., and Ayani, R., *Trans. JWRI*, **7:** 111, 1978.

38. Matsuda, F., Nakata, K., Miyawaga, Y., Kayano, T., and Tsukarnoto, K., *Trans. JWRI*, **7:** 181, 1978.

39. Villafuerte, J. C., and Kerr, H. W., *Weld. J.*, **69:** 1s, 1990.

40. Venkataraman, S., Devietian, J. H., Wood, W. E., and Atteridge, D. G., in *Grain Refinement in Castings and Welds*, Eds. G. J. Abbaschian and S. A. David, Metallurgical Society of AIME, Warrendale, PA, 1983, p. 275.

41. Venkataraman, S., Wood, W. E., Atteridge, D. G., and Devletian, J. H., in *Trends in Welding Research in the United States*, Ed. S. A. David, American Society for Metals, Metals Park, OH, 1982.

42. Sharir, Y., Pelleg, J., and Grill, A., *Metals Technol.*, **5:** 190, 1978.

43. Sundaresan, S., and Janaki Ram, G. D., *Sci. Technol. Weld. Join.*, **4:** 151, 1999.

44. Kou, S., and Le, Y., unpublished research, University of Wisconsin, Madison, 1986.

45. Southin, R. T., *Trans. AIME*, **239:** 220, 1967.

46. Wells, M. E., and Lukens, W. E., *Weld. J.*, **65:** 314s, 1986.

47. Kou, S., and Le, Y., *Weld. J.*, **64:** 51, 1985.

48. Aidun, D. K., and Dean, J. P., *Weld. J.*, **78:** 349s, 1999.

FURTHER READING

1. Chalmers, B., *Principles of Solidification*, Wiley, New York, 1964.

2. Davies, G. J., *Solidification and Casting*, Wiley, New York, 1973.

3. Davies, G. J., and Garland, J. G., *Int. Metall. Rev.*, **20(196):** 83, 1975.

4. Flemings, M. C., *Solidification Processing*, McGraw-Hill, New York, 1974.

5. Savage, W. F., *Weld. World*, **18:** 89, 1980.

6. Abbaschian, G. J., and David, S. A., Eds., *Grain Refinement in Castings and Welds*, Metallurgical Society of AIME, Warrendale, PA, 1983.

PROBLEMS

7.1 In aluminum alloys such as 6061 and 5052, which often contain small amounts of Ti (say about 0.02 wt %), the Ti-rich particles in the workpiece can be dissolved with a gas–tungsten arc by multipass melting. If the preweld is a multipass weld intended to dissolve such particles and the grain structure is shown in Figure P7.1, what is the grain refining mechanism in the test weld and why?

7.2 Equiaxed grains can often be found in the crater of a weld that exhibits an essentially purely columnar grain structure, as shown in Figure P7.2. Explain why.

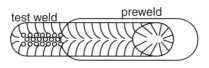

Figure P7.1

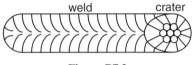

Figure P7.2

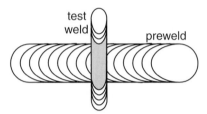

Figure P7.3

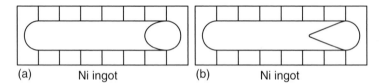

Figure P7.4

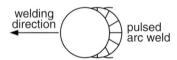

Figure P7.5

7.3 Gutierrez and Lippold (23) made a preweld in aluminum alloy 2195 and then a test weld perpendicular to it, as shown in Figure P7.3. (a) Do you expect to see a nondendritic equiaxed zone near the fusion boundary of the test weld in the overlap region and why or why not? (b) Same as (a) but with the workpiece and the preweld solution heat treated before making the test weld.

7.4 Part of a pure Ni ingot with large columnar grains is welded perpendicular to the grains, as shown in Figure P7.4. Sketch the grain structure in the weld.

7.5 A pulsed arc weld is shown in Figure P7.5. Sketch the grain structure in the area produced by the last pulse.

8 Weld Metal Solidification II: Microstructure within Grains

In this chapter we shall discuss the microstructure within the grains in the fusion zone, focusing on the solidification mode, dendrite spacing and cell spacing, how they vary across the weld metal, and how they are affected by welding parameters. The advantages of fine microstructure and techniques for microstructural refining will also be discussed.

8.1 SOLIDIFICATION MODES

As constitutional supercooling increases, the solidification mode changes from planar to cellular and from cellular to dendritic (Chapter 6). Figure 8.1 shows schematically the effect of constitutional supercooling on the microstructure within the grains in the weld metal. The solidification mode changes from planar to cellular, columnar dendritic, and equiaxed dendritic as the degree of constitutional supercooling at the pool boundary increases. Heterogeneous nucleation aided by constitutional supercooling promotes the formation of equiaxed grains in the weld metal (Chapter 7).

8.1.1 Temperature Gradient and Growth Rate

While the solidification mode can vary from one weld to another (Figure 8.1), it can also vary within a single weld from the fusion line to the centerline. This will be explained following the discussion on the growth rate R and the temperature gradient G.

Figure 8.2 shows the relationship between the growth rate R and the welding speed V. The distance a given point on the pool boundary travels in the normal direction n during a very small time interval dt is

$$R_n \, dt = (V \, dt)\cos\alpha = (R \, dt)\cos(\alpha - \beta) \qquad (8.1)$$

Dividing the above equation by $dt\cos(\alpha - \beta)$ yields

$$R = \frac{V \cos\alpha}{\cos(\alpha - \beta)} \qquad (8.2)$$

199

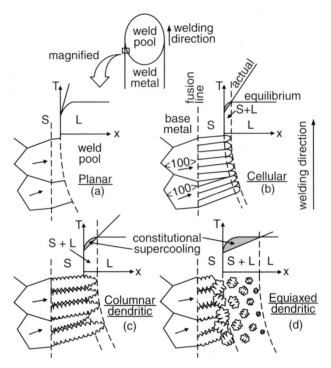

Figure 8.1 Effect of constitutional supercooling on solidification mode during welding: (*a*) planar; (*b*) cellular; (*c*) columnar dendritic; (*d*) equiaxed dendritic. Constitutional supercooling increases from (*a*) through (*d*).

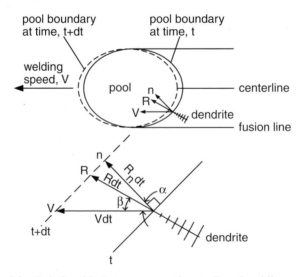

Figure 8.2 Relationship between growth rate R and welding speed V.

where α is the angle between the welding direction and the normal to the pool boundary and β is the angle between the welding direction and the growth direction of a dendrite at the point (<100> in fcc and bcc materials). This relationship has been shown by Nakagawa et al. (1). If the difference between the two angles is neglected ($\cos 0° = 1$) as an approximation, Equation (8.2) becomes

$$\boxed{R = V \cos \alpha} \tag{8.3}$$

As shown in Figure 8.3, $\alpha = 0°$ and $90°$ at the weld centerline and the fusion line, respectively. Therefore, the solidification rate at the centerline $R_{CL} = V$ (maximum) while that at the fusion line $R_{FL} = 0$ (minimum). As shown in Figure 8.4, the distance between the maximum pool temperature (T_{max}) and the pool boundary (T_L) is greater at the centerline than at the fusion line

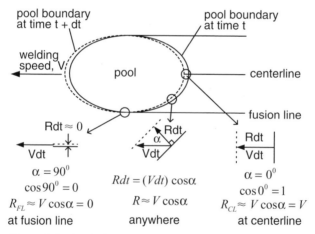

Figure 8.3 Variation in growth rate along pool boundary.

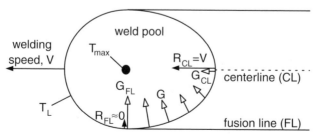

Figure 8.4 Variations in temperature gradient G and growth rate R along pool boundary.

because the weld pool is elongated. Consequently, the temperature gradient normal to the pool boundary at the centerline, G_{CL}, is less than that at the fusion line, G_{FL}. Since $G_{CL} < G_{FL}$ and $R_{CL} \gg R_{FL}$,

$$\left(\frac{G}{R}\right)_{CL} \ll \left(\frac{G}{R}\right)_{FL} \tag{8.4}$$

8.1.2   Variations in Growth Mode across Weld

According to Equation (8.4), the ratio G/R decreases from the fusion line toward the centerline. This suggests that the solidification mode may change from planar to cellular, columnar dendritic, and equiaxed dendritic across the fusion zone, as depicted in Figure 8.5. Three grains are shown to grow epitaxially from the fusion line. Consider the one on the right. It grows with the planar mode along the easy-growth direction <100> of the base-metal grain. A short distance away from the fusion line, solidification changes to the cellular mode. Further away from the fusion line, solidification changes to the columnar dendritic mode. Some of the cells evolve into dendrites and their side arms block off the neighboring cells. Near the weld centerline equiaxed dendrites nucleate and grow, blocking off the columnar dendrites. The solidification-mode transitions have been observed in several different materials (2–4).

Figure 8.6 shows the planar-to-cellular transition near the weld fusion line of an autogenous gas–tungsten weld of Fe–49Ni (2). Figure 8.7 shows the planar-to-cellular transition and the cellular-to-dendritic transition in 1100 aluminum (essentially pure Al) welded with a 4047 (Al–12Si) filler metal. Figure 8.8 shows the transition from columnar to equiaxed dendrites in an electron beam weld of a Fe–15Cr–15Ni single crystal containing some sulfur (4). These columnar dendrites, with hardly visible side arms, follow the easy-growth

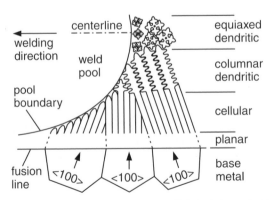

Figure 8.5   Variation in solidification mode across the fusion zone.

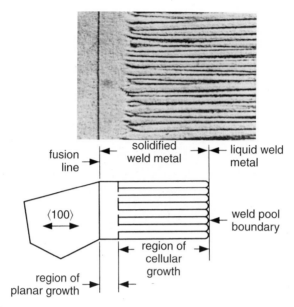

fusion line | solidified weld metal | liquid weld metal

⟨100⟩

weld pool boundary

region of cellular growth

region of planar growth

Figure 8.6 Planar-to-cellular transition in an autogenous weld of Fe–49Ni. Modified from Savage et al. (2).

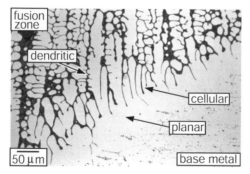

fusion zone

dendritic

cellular

planar

50 μm

base metal

Figure 8.7 Planar-to-cellular and cellular-to-dendritic transitions in 1100 Al welded with 4047 filler.

direction <100> of the single crystal, which happens to be normal to the fusion line in this case.

Before leaving the subject of the solidification mode, it is desirable to further consider the weld metal microstructure of a workpiece with only one and very large grain, that is, a single-crystal workpiece. Figure 8.9 shows the columnar dendritic structure in a Fe–15Cr–15Ni single crystal of high purity electron beam welded along a [110] direction on a (001) surface (5). The weld is still a single crystal because of epitaxial growth. However, it can have

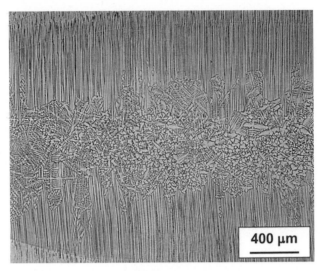

Figure 8.8   Electron beam weld of single crystal of Fe–15Cr–15Ni with sulfur showing transition from columnar to equiaxed dendrites. Reprinted from David and Vitek (4).

colonies of columnar dendrites of different orientations, as shown by the top weld pass in Figure 8.9b (ϕ is one of the angles characterizing the normal to the pool boundary). This is because of competitive growth between columnar dendrites along the three <100> easy-growth directions [100], [010], and [001]. Dendrites with an easy-growth direction closest to the heat flow direction (normal to the pool boundary) compete better.

8.2   DENDRITE AND CELL SPACING

The spacing between dendrite arms or cells, just as the solidification mode, can also vary across the fusion zone. As already mentioned in the previous section, $G_{CL} < G_{FL}$ and $R_{CL} \gg R_{FL}$. Consequently,

$$\boxed{(G \times R)_{CL} > (G \times R)_{FL}} \tag{8.5}$$

where $G \times R$ is the *cooling rate*, as explained previously in Chapter 6. According to Equation (8.5), the cooling rate ($G \times R$) is higher at the weld centerline and lower at the fusion line. This suggests that the dendrite arm spacing decreases from the fusion line to the centerline because the dendrite arm spacing decreases with increasing cooling rate (Chapter 6).

The variation in the dendrite arm spacing across the fusion zone can be further explained with the help of thermal cycles (Chapter 2). Figure 8.10

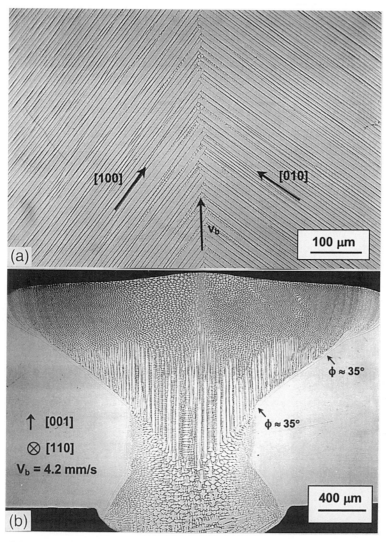

Figure 8.9 Electron beam weld of single crystal of pure Fe–15Cr–15Ni made in a [110] direction on a (001) surface: (*a*) top cross section; (*b*) transverse cross section. Reprinted from Rappaz et al. (5).

shows a eutectic-type phase diagram and the thermal cycles at the weld centerline and fusion line of alloy C_0. As shown, the cooling time through the solidification temperature range is shorter at the weld centerline ($\overline{24}/V$) and longer at the fusion line ($\overline{13}/V$). As such, the cooling rate through the solidification temperature range increases and the dendrite arm spacing decreases from the fusion line to the centerline. As shown by the aluminum weld in

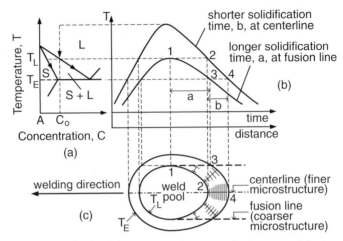

Figure 8.10   Variation in dendrite arm spacing across fusion zone: (*a*) phase diagram; (*b*) thermal cycles; (*c*) top view of weld pool.

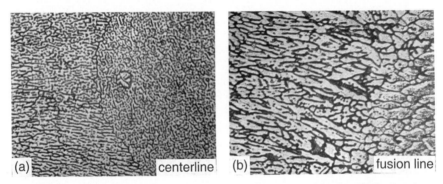

Figure 8.11   Transverse cross-section of gas–tungsten arc weld in 6061 aluminum: (*a*) finer microstructure near centerline; (*b*) coarser microstructure near fusion line. Magnification 115×. Reprinted from Kou et al. (6). Courtesy of American Welding Society.

Figure 8.11, the solidification microstructure gets finer from the fusion line to the centerline (6). The same trend was observed in other aluminum welds by Kou et al. (6, 7) and Lanzafame and Kattamis (8).

8.3   EFFECT OF WELDING PARAMETERS

8.3.1   Solidification Mode

The heat input and the welding speed can affect the solidification mode of the weld metal significantly. The solidification mode changes from planar to cel-

TABLE 8.1   Effect of Welding Parameters on Weld Metal Microstructure

Travel speed	150 A	300 A	450 A
0.85 mm/s (2 ipm)	Cellular	Cellular dendritic	Coarse cellular dendritic
1.69 mm/s (4 ipm)	Cellular	Fine cellular dendritic	Coarse cellular dendritic
3.39 mm/s (8 ipm)	Fine cellular	Cellular, slight undercutting	Severe undercutting
6.77 mm/s (16 ipm)	Very fine cellular	Cellular, undercutting	Severe undercutting

Source: From Savage et al. (9).

lular and dendritic as the ratio G/R decreases (Chapter 6). Table 8.1 summarizes the observations of Savage et al. (9) in HY-80 steel. At the welding speed of 0.85 mm/s (2 ipm), the weld microstructure changes from cellular to dendritic when the welding current increases from 150 to 450 A. According to Equation (2.15), the higher the heat input (Q) under the same welding speed (V), the lower the temperature gradient G and hence the lower the ratio G/R. Therefore, at higher heat inputs G/R is lower and dendritic solidification prevails, while at lower heat inputs G/R is higher and cellular solidification prevails. Although analytical equations such as Equations (2.15) and (2.17) are oversimplified, they can still qualitatively tell the effect of welding parameters.

8.3.2   Dendrite and Cell Spacing

The heat input and the welding speed can also affect the spacing between dendrite arms and cells. The dendrite arm spacing or cell spacing decreases with increasing cooling rate (Chapter 6). As compared to arc welding, the cooling rate in laser or electron beam welding is higher and the weld metal microstructure is finer. The 6061 aluminum welds in Figure 8.12 confirm that this is the case (10).

As shown in Table 8.1, at the welding current of 150 A, the cells become finer as the welding speed increases. From Equation (2.17), under the same heat input (Q), the cooling rate increases with increasing welding speed V. Therefore, at higher welding speeds the cooling rate is higher and the cells are finer, while at lower welding speeds the cooling rate is lower and the cells are coarser. Elmer et al. (11) also observed this trend in EBW of two austenitic stainless steels of similar compositions, as shown in Figure 8.13. The difference in the cell spacing can be seen even though the magnifications of the micrographs are different.

According to Equation (2.17), the cooling rate increases with decreasing heat input–welding speed ratio Q/V. This ratio also represents the amount

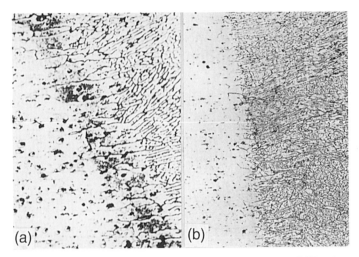

Figure 8.12   Autogenous welds of 6061 aluminum: (*a*) coarser solidification structure in gas–tungsten arc weld: (*b*) finer solidification structure in electron beam weld. Reprinted from *Metals Handbook* (10).

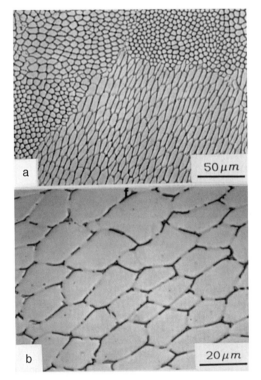

Figure 8.13   Effect of welding speed on cell spacing in EBW of austenitic stainless steels: (*a*) 100 mm/s; (*b*) 25 mm/s. From Elmer et al. (11).

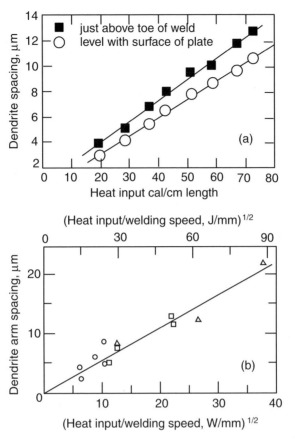

Figure 8.14   Effect of heat input per unit length of weld on dendrite arm spacing. (*a*) For Al–Mg–Mn alloy. From Jordan and Coleman (13). (*b*) For 2014 Al–Cu alloy. Modified from Lanzafame and Kattamis (14). Courtesy of American Welding Society.

of heat per unit length of the weld (J/cm or cal/cm). Therefore, the dendrite arm spacing or cell spacing can be expected to increase with increasing Q/V or amount of heat per unit length of the weld. This has been observed in several aluminum alloys (7, 8, 12–14), including those shown in Figure 8.14.

8.4   REFINING MICROSTRUCTURE WITHIN GRAINS

It has been shown in aluminum alloys that the finer the dendrite arm spacing, the higher the ductility (15) and yield strength (8, 15) of the weld metal and the more effective the postweld heat treatment (8, 12, 15), due to the finer distribution of interdendritic eutectics.

8.4.1 Arc Oscillation

Kou and Le (16) studied the microstructure in oscillated arc welds of 2014 aluminum alloy. It was observed that the dendrite arm spacing was reduced significantly by transverse arc oscillation at low frequencies, as shown in Figure 8.15. This reduction in the dendrite arm spacing has contributed to the significant improvement in both the strength and ductility of the weld, as shown in Figure 8.16.

As illustrated in Figure 8.17, when arc oscillation is applied, the weld pool gains a lateral velocity, v, in addition to its original velocity u in the welding direction (16). The magnitude of v can be comparable with that of u depending on the amplitude and frequency of arc oscillation.

As shown, the resultant velocity of the weld pool, w, is greater than that of the unoscillated weld pool, u. Furthermore, the temperature gradient ahead of the solid–liquid interface, G, could also be increased due to the smaller distance between the heat source and the pool boundary. Consequently, the product GR or the cooling rate is increased significantly by the action of arc oscillation. This explains why the microstructure is finer in the oscillated arc weld.

Example: Suppose the welding speed of a regular weld is 4.2 mm/s (10 ipm). Calculate the increase in the velocity of the weld pool if the arc is oscillated

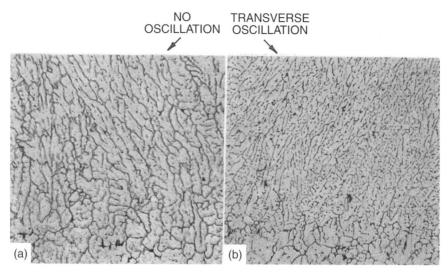

Figure 8.15 Microstructures near fusion line of gas-tungsten arc welds of 2014 aluminum: (*a*) coarser dendrites in weld made without arc oscillation; (*b*) finer dendrites in weld made with transverse arc oscillation. Magnification 200×. Reprinted from Kou and Le (16). Courtesy of American Welding Society.

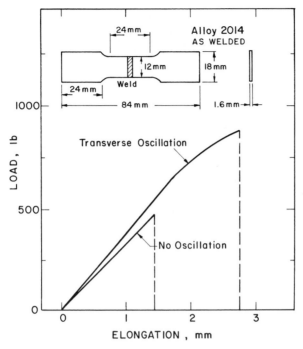

Figure 8.16   Tensile testing of two gas–tungsten arc welds of 2014 aluminum made without arc oscillation and with transverse arc oscillation. From Kou and Le (16).

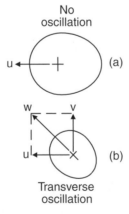

Figure 8.17   Increase in weld pool travel speed due to transverse arc oscillation. Modified from Kou and Le (16). Courtesy of American Welding Society.

transversely at a frequency of 1 Hz and an amplitude (the maximum deflection of the arc from the weld centerline) of 1.9 mm. Do you expect the dendrite arm spacing to keep on decreasing if the frequency keeps on increasing to, say, 100 Hz?

Since the arc travels a distance equal to four oscillation amplitudes per second, v = 4 × 1.9 mm/s = 7.6 mm/s. Therefore, the resultant velocity w = $(u^2 + v^2)^{1/2} = (4.2^2 + 7.6^2)^{1/2} = 8.7$ mm/s. The increase in the weld pool velocity is 8.7 − 4.2 = 4.5 mm/s. The increase in the weld pool velocity and hence the decrease in the dendrite arm spacing cannot keep on going with increasing oscillation frequency. This is because when the arc oscillates too fast, say at 100 Hz, the weld pool cannot catch up with it because there is not enough time for melting and solidification to occur.

It is interesting to point out that Kou and Le (16) also observed that the microstructure in oscillated arc welds is much more uniform than that in welds without oscillation. In oscillated arc welds the weld centerline is no longer a location where the cooling rate (or GR) is clearly at its maximum.

Tseng and Savage (17) studied the microstructure in gas–tungsten arc welds of HY-80 steel made with transverse and longitudinal arc oscillation. Refining of the grain structure was not obtained, but refining of the dendritic structure within grains (subgrain structure) was observed and solidification cracking was reduced, as shown in Figure 8.18. It is possible that solidification cracking was reduced because the crack-causing constituents were diluted to a greater extent by the larger interdendritic area in the welds with a finer dendritic structure.

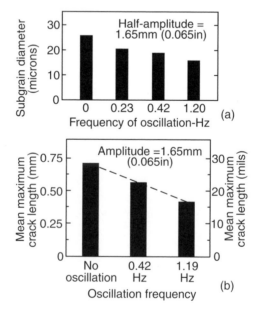

Figure 8.18   Effect of magnetic arc oscillation on gas–tungsten arc welds of HY-80 steel: (*a*) refining of microstructure within grains; (*b*) solidification cracking. Modified from Tseng and Savage (17).

8.4.2   Arc Pulsation

Becker and Adams (18) studied the microstructure in pulsed arc welds of titanium alloys. It was observed that the cell spacing varied periodically along the weld, larger where solidification took place during the high-current portion of the cycle. Obviously, the cooling rate was significantly lower during the high-current portion of the cycle. This is because, from Equation (2.17), the cooling rate can be expected to decrease with increasing heat input, though, strictly speaking, this equation is for steady-state conditions only.

REFERENCES

1. Nakagawa, H., et al., *J. Jpn. Weld. Soc.*, **39:** 94, 1970.
2. Savage, W. F., Nippes, E. F., and Erickson, J. S., *Weld. J.*, **55:** 213s, 1976.
3. Kou, S., and Le, Y., unpublished research, University of Wisconsin, Madison, 1983.
4. David, S. A., and Vitek, J. M., *Int. Mater. Rev.*, **34:** 213, 1989.
5. Rappaz, M., David, S. A., Vitek, J. M., and Boatner, L. A., in *Recent Trends in Welding Science and Technology*, Eds. S. A. David and J. M. Vitek, ASM International, Materials Park, OH, May 1989, p. 147.
6. Kou, S., Kanevsky, T., and Fyfitch, S., *Weld. J.*, **61:** 175s, 1982.
7. Kou, S., and Le, Y., *Metall. Trans. A,* **14A:** 2245, 1983.
8. Lanzafame, J. N., and Kattamis, T. Z., *Weld. J.*, **52:** 226s, 1973.
9. Savage, W. F., Lundin, C. D., and Hrubec, R. J., *Weld. J.*, **47:** 420s, 1968.
10. *Metals Handbook*, Vol. 7, 8th ed., American Society for Metals, Metals Park, OH, 1972, pp. 266, 269.
11. Elmer, J. W., Allen, S. M., and Eagar, T. W., *Metall. Trans.*, **20A:** 2117, 1989.
12. Brown, P. E., and Adams, C. M. Jr., *Weld. J.*, **39:** 520s, 1960.
13. Jordan, M. F., and Coleman, M. C., *Br. Weld. J.*, **15:** 552, 1968.
14. Lanzafame, J. N., and Kattamis, T. Z., *Weld. J.*, **52:** 226s, 1973.
15. Fukui, T., and Namba, K., *Trans. Jpn. Weld. Soc.*, **4:** 49, 1973.
16. Kou, S., and Le, Y., *Weld. J.*, **64:** 51, 1985.
17. Tseng, C., and Savage, W. F., *Weld. J.*, **50:** 777, 1971.
18. Becker, D. W., and Adams, C. M. Jr., *Weld. J.*, **58:** 143s, 1979.
19. Savage, W. F., in *Weldments: Physical Metallurgy and Failure Phenomena*, Eds. R. J. Christoffel, E. F. Nippes, and H. D. Solomon, General Electric Co., Schenectady, NY, 1979, p. 1.

FURTHER READING

1. Davies, G. J., and Garland, J. G., *Int. Metall. Rev.*, **20:** 83, 1975.
2. Savage, W. F., *Weld. World*, **18:** 89, 1980.

3. David, S. A., and Vitek, J. M., *Int. Mater. Rev.*, **34:** 213, 1989.

4. Flemings, M. C., *Solidification Processing*, McGraw-Hill, New York, 1974.

PROBLEMS

8.1 It has been suggested that the secondary dendrite arm spacing d along the weld centerline can be related quantitatively to the heat input per unit length of weld, Q/V. Based on the data of the dendrite arm spacing d as a function of cooling rate ε, similar to those shown in Figure 6.17a, it can be shown that $d = a\varepsilon^{-1/b}$, where a and b are constant with b being in the range of 2–3. (a) Express the dendrite arm spacing in terms of Q/V for bead-on-plate welds in thick-section aluminum alloys. (b) How do the preheat temperature and thermal conductivity affect the dendrite arm spacing? (c) Do you expect the relationship obtained to be very accurate?

8.2 The size of the mushy zone is often an interesting piece of information for studying weld metal solidification. Let $d = a\varepsilon^{-1/b}$, where d is the dendrite arm spacing and ε the cooling rate. Consider how measurements of the dendrite arm spacing across the weld metal can help determine the size of the mushy zone. Express the width of the mushy zone in the welding direction Δx, as shown in Figure P8.2, in terms of the dendrite arm spacing d, the welding speed V, and the freezing temperature range $\Delta T \; (= T_L - T_E)$.

8.3 It has been observed that the greater the heat input per unit length of weld (Q/V), the longer it takes to homogenize the microsegregation in the weld metal of aluminum alloys for improving its mechanical properties. Let $d = a\varepsilon^{-1/b}$, where d is the dendrite arm spacing and ε the cooling rate. Express the time required for homogenization (t) in terms of Q/V.

8.4 An Al–1% Cu alloy is welded autogenously by GTAW, and an Al–5% Cu alloy is welded under identical condition. Which alloy is expected to develop more constitutional supercooling and why? Which alloy is likely to have more equiaxed dendrites in the weld metal and why?

8.5 An Al–5% Cu alloy is welded autogenously by GTAW and by EBW under the same welding speed but different heat inputs (much less in the

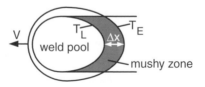

Figure P8.2

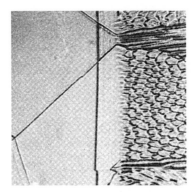

Figure P8.8

case of EBW). Which weld is expected to experience more constitutional supercooling and why? Which weld is likely to have more equiaxed dendrites and why?

8.6 In autogenous GTAW of aluminum alloys, how do you expect the amount of equiaxed grains in the weld metal to be affected by preheating and why?

8.7 In autogenous GTAW of aluminum alloys, how do you expect the dendrite arm spacing of the weld metal to be affected by preheating and why?

8.8 Figure P8.8 is a micrograph near the fusion line of an autogenous gas–tungsten arc weld in a Fe–49% Ni alloy sheet (19). Explain the solidification microstructure, which is to the right of the fusion line (dark vertical line).

9 Post-Solidification Phase Transformations

Post-solidification phase transformations, when they occur, can change the solidification microstructure and properties of the weld metal. It is, therefore, essential that post-solidification phase transformations be understood in order to understand the weld metal microstructure and properties. In this chapter two major types of post-solidification phase transformations in the weld metal will be discussed. The first involves the ferrite-to-austenite transformation in welds of austenitic stainless steels, and the second involves the austenite-to-ferrite transformation in welds of low-carbon, low-alloy steels.

9.1 FERRITE-TO-AUSTENITE TRANSFORMATION IN AUSTENITIC STAINLESS STEEL WELDS

9.1.1 Primary Solidification Modes

The welds of austenitic stainless steels normally have an austenite (fcc) matrix with varying amounts of δ-ferrite (bcc) (1–7). A proper amount of δ-ferrite in austenitic stainless steel welds is essential—too much δ-ferrite ($\gtrsim 10$ vol %) tends to reduce the ductility, toughness, and corrosion resistance, while too little δ-ferrite (≤ 5 vol %) can result in solidification cracking.

A. Phase Diagram Figure 9.1 shows the ternary phase diagram of the Fe–Cr–Ni system (8). The heavy curved line in Figure 9.1a represents the trough on the liquidus surface, which is called the *line of twofold saturation*. The line declines from the binary Fe–Ni peritectic reaction temperature to the ternary eutectic point at 49Cr–43Ni–8Fe. Alloys with a composition on the Cr-rich (upper) side of this line have δ-ferrite as the primary solidification phase, that is, the first solid phase to form from the liquid. On the other hand, alloys with a composition on the Ni-rich (lower) side have austenite as the primary solidification phase. The heavy curved lines on the solidus surface in Figure 9.1b more or less follow the trend of the liquidus trough and converge at the ternary eutectic temperature.

The development of weld metal microstructure in austenitic stainless steels is explained in Figure 9.2. The weld metal ferrite can have three different types

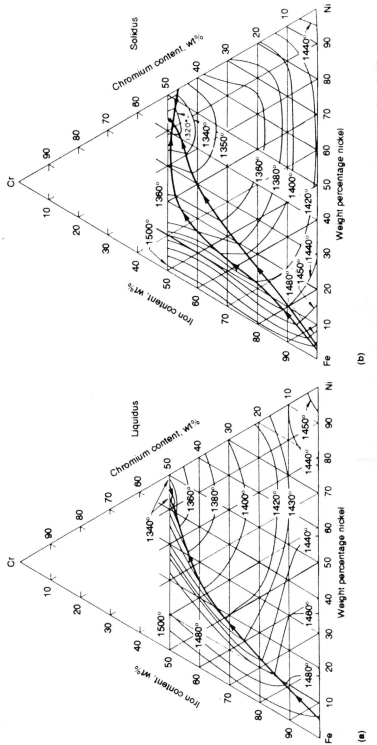

Figure 9.1 The Fe–Cr–Ni ternary system: (*a*) liquidus surface; (*b*) solidus surface. Reprinted from *Metals Handbook* (8).

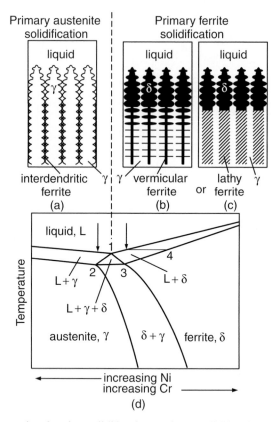

Figure 9.2   Schematics showing solidification and postsolidification transformation in Fe–Cr–Ni welds: (*a*) interdendritic ferrite; (*b*) vermicular ferrite; (*c*) lathy ferrite; (*d*) vertical section of ternary-phase diagram at approximately 70% Fe.

of morphology: interdendritic (Figure 9.2*a*), vermicular (Figure 9.2*b*), and lathy (Figure 9.2*c*). Figure 9.2*d* shows a schematic vertical (isoplethal) section of the ternary phase diagram in Figure 9.1, for instance, at 70 wt % Fe and above 1200°C. This has also been called a *pseudo-binary phase diagram*. The apex (point 1) of the three-phase eutectic triangle (L + γ + δ) corresponds to the intersection between the vertical section and the heavy curved line in Figure 9.1*a*. The two lower corners (points 2 and 3) of the triangle, on the other hand, correspond to the intersections between the vertical section and the two heavy curved lines in Figure 9.1*b*.

B. Primary Austenite   For an alloy on the Ni-rich (left-hand) side of the apex of the three-phase eutectic triangle, austenite (γ) is the primary solidification phase. The light dendrites shown in Figure 9.2*a* are austenite, while the dark particles between the primary dendrite arms are the δ-ferrite that forms when the three-phase triangle is reached during the terminal stage of solidification. These are called the *interdendritic ferrite*. For dendrites with long secondary arms, interdendritic ferrite particles can also form between secondary dendrite arms.

C. Primary Ferrite   For an alloy on the Cr-rich (right-hand) side of the apex of the three-phase eutectic triangle, δ-ferrite is the primary solidification phase. The dark dendrites shown in Figure 9.2*b* are δ-ferrite. The core of the δ-ferrite dendrites, which forms at the beginning of solidification, is richer in Cr (point 4), while the outer portions, which form as temperature decreases, have lower chromium contents. Upon cooling into the $(\delta + \gamma)$ two-phase region, the outer portions of the dendrites having less Cr transform to austenite, thus leaving behind Cr-rich "skeletons" of δ-ferrite at the dendrite cores. This skeletal ferrite is called *vermicular ferrite*. In addition to vermicular ferrite, primary δ-ferrite dendrites can also transform to *lathy* or *lacy ferrite* upon cooling into the $(\delta + \gamma)$ two-phase region, as shown in Figure 9.2*c*.

D. Weld Microstructure   Figure 9.3*a* shows the solidification structure at the centerline of an autogenous gas–tungsten arc weld of a 310 stainless steel sheet, which contains approximately 25% Cr, 20% Ni, and 55% Fe by weight (9). The composition is on the Ni-rich (left) side of the apex of the three-phase eutectic triangle, as shown in Figure 9.4*a*, and solidification occurs as primary austenite. The microstructure consists of austenite dendrites (light etching; mixed-acids etchant) and interdendritic δ-ferrite (dark etching; mixed-acids etchant) between the primary and secondary dendrite arms, similar to those shown in Figure 9.2*a*.

Figure 9.3*b*, on the other hand, shows the solidification structure at the centerline of an autogenous gas–tungsten arc weld of a 309 stainless steel sheet, which contains approximately 23 wt % Cr, 14 wt % Ni, and 63 wt % Fe. The composition lies just to the Cr-rich side of the apex of the three-phase eutectic triangle, as shown in Figure 9.4*b*, and solidifies as primary δ-ferrite. The microstructure consists of vermicular ferrite (dark etching; mixed-acids etchant) in an austenite matrix (light etching; mixed-acids etchant) similar to those shown in Figure 9.2*b*. In both welds columnar dendrites grow essentially perpendicular to the teardrop-shaped pool boundary as revealed by the columnar dendrites.

Kou and Le (9) quenched welds during welding in order to preserve the as-solidified microstructure, that is, the microstructure before post-solidification phase transformations. For stainless steels liquid-tin quenching is more effective than water quenching because steam and bubbles reduce heat transfer. With the help of quenching, the evolution of microstructure during welding can be better studied. Figure 9.5 shows the δ-ferrite dendrites (light etching; mixed-chloride etchant) near the weld pool of an autogenous gas–tungsten arc weld of 309 stainless steel, quenched in during welding with liquid tin before the $\delta \rightarrow \gamma$ transformation changed it to vermicular ferrite like that

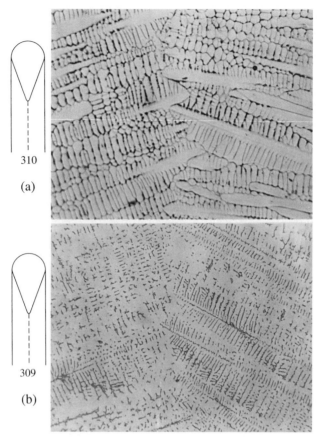

Figure 9.3   Solidification structure at the weld centerline: (*a*) 310 stainless steel; (*b*) 309 stainless steel. Magnification 190×. Reprinted from Kou and Le (9).

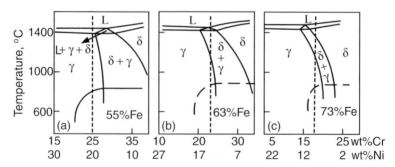

Figure 9.4   The Fe–Cr–Ni pseudo-binary phase diagrams: (*a*) at 55 wt % Fe; (*b*) at 63 wt % Fe; (*c*) at 73 wt % Fe. Reprinted from Kou and Le (9).

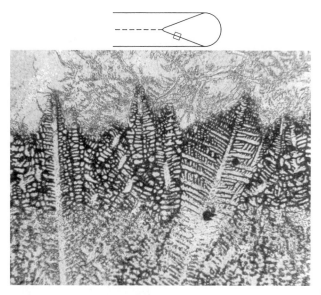

Figure 9.5 Liquid-tin quenched solidification structure near the pool of an auto-genous gas–tungsten arc weld of 309 stainless steel. Magnification 70×. Mixed-chloride etchant. Reprinted from Kou and Le (9).

shown in Figure 9.3*b*. Liquid-tin quenching was subsequently used by other investigators to study stainless steel welds (10, 11).

9.1.2 Mechanisms of Ferrite Formation

Inoue et al. (11) studied vermicular and lathy ferrite in autogenous GTAW of austenitic stainless steels of 70% Fe with three different Cr–Ni ratios. It was found that, as the Cr–Ni ratio increases, the ratio of lathy ferrite to total ferrite does not change significantly even though both increase. A schematic of the proposed formation mechanism of vermicular and lathy ferrite is shown in Figure 9.6. Austenite first grows epitaxially from the unmelted austenite grains at the fusion boundary, and δ-ferrite soon nucleates at the solidification front. The crystallographic orientation relationship between the δ-ferrite and the austenite determines the ferrite morphology after the postsolidification trans-formation. If the closed-packed planes of the δ-ferrite are parallel to those of the austenite, the $\delta \rightarrow \gamma$ transformation occurs with a planar δ/γ interface, resulting in vermicular ferrite. However, if the so-called Kurdjumov–Sachs ori-entation relationships, namely, $(\bar{1}10)_\delta$ //$(\bar{1}11)_\gamma$ and $[\bar{1}\bar{1}1]_\delta$ //$[\bar{1}\bar{1}0]_\gamma$, exist between the δ-ferrite and the austenite, the transformation occurs along the austenite habit plane into the δ-ferrite dendrites. The resultant ferrite morphology is lathy, as shown in Figure 9.7. For the lathy ferrite to continue to grow, the

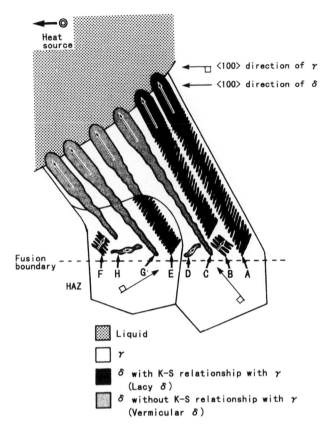

Figure 9.6 Mechanism for the formation of vermicular and lathy ferrite. Reprinted from Inoue et al. (11).

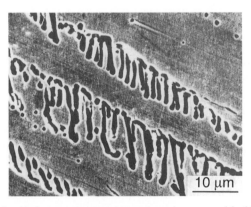

Figure 9.7 Lathy ferrite in an autogenous gas–tungsten arc weld of Fe–18.8Cr–11.2Ni. Reprinted from Inoue et al. (11).

preferred growth direction <100> of both δ-ferrite and austenite must be aligned with the heat flow direction.

9.1.3   Prediction of Ferrite Content

Schaeffler (12) first proposed the quantitative relationship between the composition and ferrite content of the weld metal. As shown by the constitution diagram in Figure 9.8, the chromium equivalent of a given alloy is determined from the concentrations of ferrite formers Cr, Mo, Si, and Cb, and the austenite equivalent is determined from the concentrations of austenite formers Ni, C, and Mn. DeLong (13) refined Schaeffler's diagram to include nitrogen, a strong austenite former, as shown in Figure 9.9. Also, the ferrite content is expressed in terms of the ferrite number, which is more reproducible than the ferrite percentage and can be determined nondestructively by magnetic means. Figure 9.10 shows that nitrogen, introduced into the weld metal by adding various amounts of N_2 to the Ar shielding gas, can reduce the weld ferrite content significantly (14). Cieslak et al. (6), Okagawa et al. (7), and Lundin et al. (15) have reported similar results previously.

The WRC-1992 diagram of Kotecki and Siewert (16), shown in Figure 9.11, was from the Welding Research Council in 1992. It was modified from the WRC-1988 diagram of McCowan et al. (17) by adding to the nickel equivalent the coefficient for copper (18) and showing how the axes could be extended to make Schaeffler-like calculations for dissimilar metal joining.

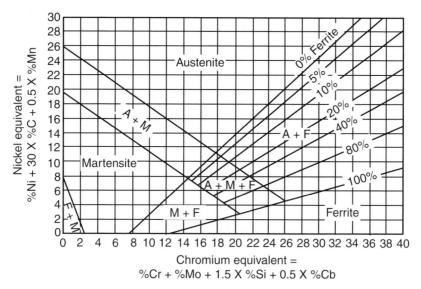

Figure 9.8   Schaeffler diagram for predicting weld ferrite content and solidification mode. From Schaeffler (12).

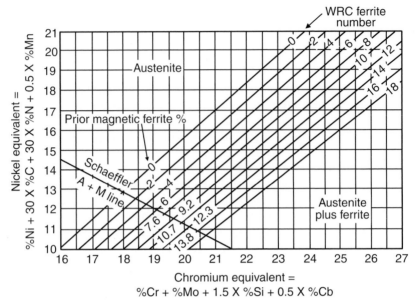

Figure 9.9   DeLong diagram for predicting weld ferrite content and solidification mode. Reprinted from DeLong (13). Courtesy of American Welding Society.

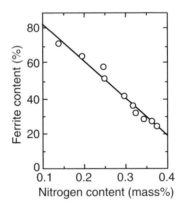

Figure 9.10   Effect of nitrogen on ferrite content in gas–tungsten arc welds of duplex stainless steel. Reprinted from Sato et al. (14).

Kotecki (19, 20) added the martensite boundaries to the WRC-1992 diagram, as shown in Figure 9.12. More recent investigations of Kotecki (21, 22) have revealed that the boundaries hold up well with Mo and N variation but not as well with C variation. Balmforth and Lippold (23) proposed the ferritic–martensitic constitution diagram shown in Figure 9.13. Vitek et al. (24,

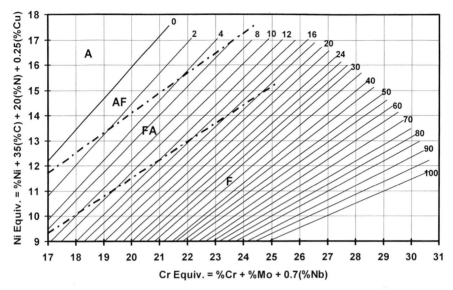

Figure 9.11 WRC-1992 diagram for predicting weld ferrite content and solidification mode. Reprinted from Kotecki and Siewert (16). Courtesy of American Welding Society.

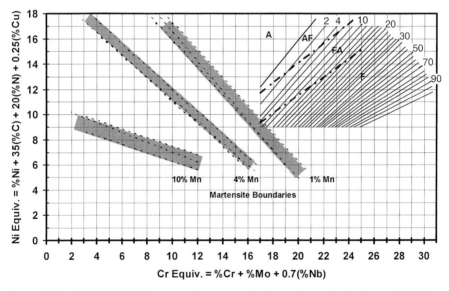

Figure 9.12 WRC-1992 diagram with martensite boundaries for 1, 4, and 10% Mn. Reprinted from Kotecki (20). Courtesy of American Welding Society.

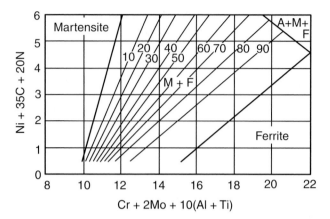

Figure 9.13 Ferritic–martensitic stainless steel constitution diagram containing a boundary for austenite formation and with iso-ferrite lines in volume percent of ferrite. Reprinted from Balmforth and Lippold (23).

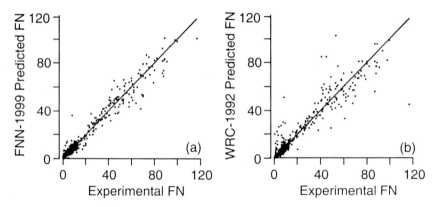

Figure 9.14 Experimentally measured ferrite number (FN) versus predicted FN: (a) FNN-1999; (b) WRC-1992. Reprinted from Vitek et al. (25). Courtesy of American Welding Society.

25) developed a model FNN-1999 using artificial neural networks to improve ferrite number prediction, as shown in Figure 9.14. Twelve alloying elements besides Fe were considered: C, Cr, Ni, Mo, N, Mn, Si, Cu, Ti, Cb, V, and Co. The model is not in a simple pictorial form, such as the WRC-1992 diagram, because it allows nonlinear effects and element interactions.

9.1.4   Effect of Cooling Rate

A. Changes in Solidification Mode   The prediction of the weld metal ferrite content based on the aforementioned constitution diagrams can be inaccurate when the cooling rate is high, especially in laser and electron beam welding (3, 26–37). Katayama and Matsunawa (28, 29), David et al. (31), and Brooks and Thompson (37) have compared microstructures that form in slow-cooling-rate arc welds with those that form in high-cooling-rate, high-energy-beam welds. Their studies show two interesting trends. For low Cr–Ni ratio alloys the ferrite content decreases with increasing cooling rate, and for high Cr–Ni ratio alloys the ferrite content increases with increasing cooling rate. Elmer et al. (33) pointed out that in general low Cr–Ni ratio alloys solidify with austenite as the primary phase, and their ferrite content decreases with increasing cooling rate because solute redistribution during solidification is reduced at high cooling rates. On the other hand, high Cr–Ni ratio alloys solidify with ferrite as the primary phase, and their ferrite content increases with increasing cooling rate because the $\delta \rightarrow \gamma$ transformation has less time to occur at high cooling rates.

Elmer et al. (33, 34) studied a series of Fe–Ni–Cr alloys with 59% Fe and the Cr–Ni ratio ranging from 1.15 to 2.18, as shown in Figure 9.15. The apex of the three-phase triangle is at about Fe–25Cr–16Ni. Figure 9.16 summarizes the microstructural morphologies of small welds made by scanning an electron beam over a wide range of travel speeds and hence cooling rates (33). At low travel speeds such as 0.1–1 mm/s, the cooling rates are low and the alloys with a low Cr–Ni ratio (especially alloys 1 and 2) solidify as primary austenite. The solidification mode is either single-phase austenite (A), that is, no ferrite between austenite dendrites or cells (cellular–dendritic A), or primary austenite with second-phase ferrite (AF), that is, only a small amount of ferrite between austenite dendrites (interdendritic F). The alloys with a high Cr–Ni

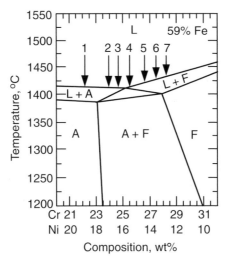

Figure 9.15 Vertical section of Fe–Ni–Cr phase diagram at 59% Fe showing seven alloys with Cr–Ni ratio ranging from 1.15 to 2.18. Modified from Elmer et al. (33).

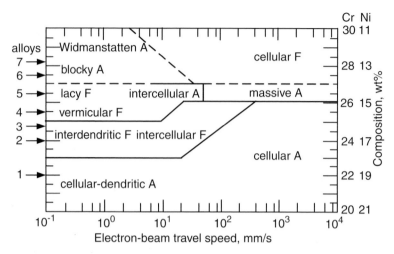

Figure 9.16 Electron beam travel speed (cooling rate) versus composition map of microstructural morphologies of the seven alloys in Figure 9.15 (A and F denote austenite and ferrite, respectively). The solid lines indicate the regions of the four primary solidification modes, while the dashed lines represent the different morphologies resulting from postsolidification transformation from ferrite to austenite. Modified from Elmer et al. (33).

ratio (especially alloys 5–7), on the other hand, solidify as primary ferrite. The solidification mode is primary ferrite with second-phase austenite (FA), that is, vermicular ferrite, lacy ferrite, small blocks of austenite in a ferrite matrix (blocky A), or Widmanstatten austenite platelets originating from ferrite grain boundaries (Widmanstatten A).

At very high welding speeds such as 2000 mm/s, however, the cooling rates are high and the alloys solidify in only the single-phase austenite mode (A) or the single-phase ferrite mode (F). An example of the former is the alloy 3 (about Fe–24.75Cr–16.25Ni) shown in Figure 9.17a. At the travel speed of 25 mm/s (2×10^3 °C/s cooling rate) the substrate solidifies as primary austenite in the AF mode, with austenite cells and intercellular ferrite. At the much higher travel speed of 2000 mm/s (1.5×10^6 °C/s cooling rate) the weld at the top solidifies as primary austenite in the A mode, with much smaller austenite cells and no intercellular ferrite (cellular A). An example of the latter is alloy 6 (about Fe–27.5Cr–13.5Ni) shown in Figure 9.17b. At 25 mm/s the substrate solidifies as primary ferrite in the FA mode, with blocky austenite in a ferrite matrix. At 2000 mm/s the weld at the top solidifies as primary ferrite in the F mode, with ferrite cells alone and no austenite (cellular F).

Figure 9.16 also demonstrates that under high cooling rates an alloy that solidifies as primary ferrite at low cooling rates can change to primary austenite solidification. For instance, alloy 4 (about Fe–25.5Cr–15.5Ni) can solidify as primary ferrite at low cooling rates (vermicular F) but solidifies as primary

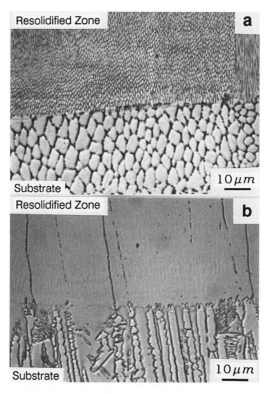

Figure 9.17   Microstructure of the low-cooling-rate substrate (2×10^3 °C/s) and the high-cooling-rate electron beam weld at the top: (*a*) alloy 3 in Figure 9.15; (*b*) alloy 6. Reprinted from Elmer et al. (34).

austenite at higher cooling rates (intercellular F or cellular A). Another interesting point seen in the same figure is that at high cooling rates alloy 5 can solidify in the fully ferritic mode and undergoes a massive (diffusionless) transformation after solidification to austenite (massive A). Under very high cooling rates there is no time for diffusion to occur.

B. Dendrite Tip Undercooling   Vitek et al. (27) attributed the change solidification mode, from primary ferrite to primary austenite, at high cooling rates to dendrite tip undercooling. Brooks and Thompson (37) explained this undercooling effect based on Figure 9.18. Alloy C_0 solidifies in the primary ferrite mode at low cooling rates. Under rapid cooling in laser or electron beam welding, however, the melt can undercool below the extended austenite liquidus ($C_{L\gamma}$), and it becomes thermodynamically possible for the melt to solidify as primary austenite. The closer C_0 is to the apex of the three-phase triangle, the easier sufficient undercooling can occur to switch the solidification mode from primary ferrite to primary austenite.

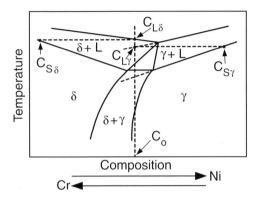

Figure 9.18   Vertical section of Fe–Cr–Ni phase diagram showing change in solidification from ferrite to austenite due to dendrite tip undercooling. Reprinted from Brooks and Thompson (37).

Figure 9.19   Weld centerline austenite in an autogenous gas–tungsten arc weld of 309 stainless steel solidified as primary ferrite. From Kou and Le (9).

Kou and Le (9) made autogenous gas–tungsten arc welds in 309 stainless steel, which has a composition close to the apex of the three-phase triangle, as shown in Figure 9.4*b*. At 2 mm/s (5 ipm) welding speed, primary ferrite was observed across the entire weld (similar to that shown in Figure 9.3*b*). At a higher welding speed of 5 mm/s (12 ipm), however, primary austenite was observed along the centerline, as shown in Figure 9.19. Electron probe microanalysis (EPMA) revealed no apparent segregation of either Cr or Ni near the weld centerline to cause the change in the primary solidification phase. From Equation (8.3) the growth rate $R = V \cos \alpha$, where α is the angle between the welding direction and the normal to the pool boundary. Because of the teardrop shape of the weld pool during welding (Figure 2.22), α drops to zero and R increases abruptly at the weld centerline. As such, the cooling rate (GR) increases abruptly at the weld centerline, as pointed out subsequently by Lippold (38).

Elmer et al. (34) calculated the dendrite tip undercooling for the alloys in Figure 9.16 under various electron beam travel speeds. An undercooling of 45.8°C was calculated at the travel speed of 175 mm/s, which is sufficient to depress the dendrite tip temperature below the solidus temperature (Figure 9.15). This helps explain why alloy 4 can change from primary ferrite solidification at low travel speeds to primary austenite solidification at much higher travel speeds.

9.1.5   Ferrite Dissolution upon Reheating

Lundin and Chou (39) observed ferrite dissolution in multiple-pass or repair austenitic stainless steel welds. This region exists in the weld metal of a previous deposited weld bead, adjacent to but not contiguous with the fusion zone of the deposited bead under consideration. Both the ferrite number and ductility are lowered in this region, making it susceptible to fissuring under strain. This is because of the dissolution of δ-ferrite in the region of the weld metal that is reheated to below the γ-solvus temperature. Chen and Chou (40) reported, in Figure 9.20, a significant ferrite loss in a 316 stainless steel weld

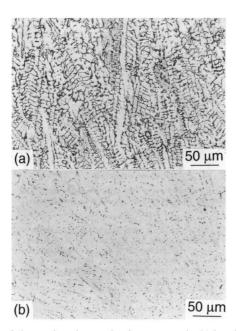

Figure 9.20   Effect of thermal cycles on ferrite content in 316 stainless steel weld: (*a*) as welded; (*b*) subjected to thermal cycle of 1250°C peak temperature three times after welding. Reprinted from Chen and Chou (40).

subjected to three postweld thermal cycles with a 1250°C peak temperature, which is just below the $\gamma + \delta$ two-phase region of about 1280–1425°C.

9.2 AUSTENITE-TO-FERRITE TRANSFORMATION IN LOW-CARBON, LOW-ALLOY STEEL WELDS

9.2.1 Microstructure Development

The dendrites or cells in the weld metal are not always discernible. First, significant solute partitioning does not occur during solidification if the partition ratio k is too close to 1. The miscrosegregation, especially solute segregation to the interdendritic or intercellular regions, in the resultant weld metal can be too little to bring out the dendritic or cellular structure in the grain interior even though the grain structure itself can still be very clear. Second, if solid-state diffusion occurs rapidly, microsegregation either is small or is homogenized quickly, and the dendrites or cells in the resultant weld metal can be unclear. Third, post-solidification phase transformations, if they occur, can produce new microstructures in the grain interior and/or along grain boundaries and the subgrain structure in the resultant weld metal can be overshadowed.

Several *continuous-cooling transformation (CCT) diagrams* have been sketched schematically to explain the development of the weld metal microstructure of low-carbon, low-alloy steels (41–45). The one shown in Figure 9.21 is based on that of Onsoien et al. (45). The hexagons represent the transverse cross sections of columnar austenite grains in the weld metal. As austenite (γ) is cooled down from high temperature, ferrite (α) nucleates at the grain boundary and grows inward. The grain boundary ferrite is also called "allotriomorphic" ferrite, meaning that it is a ferrite without a regular faceted

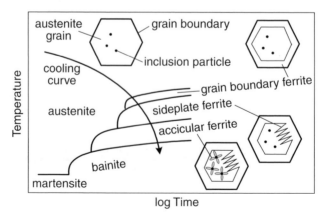

Figure 9.21 Continuous-cooling transformation diagram for weld metal of low-carbon steel.

shape reflecting its internal crystalline structure. At lower temperatures the mobility of the planar growth front of the grain boundary ferrite decreases and Widmanstatten ferrite, also called side-plate ferrite, forms instead. These side plates can grow faster because carbon, instead of piling up at the planar growth front, is pushed to the sides of the growing tips. Substitutional atoms do not diffuse during the growth of Widmanstatten ferrite. At even lower temperatures it is too slow for Widmanstatten ferrite to grow to the grain interior and it is faster if new ferrite nucleates ahead of the growing ferrite. This new ferrite, that is, acicular ferrite, nucleates at inclusion particles and has randomly oriented short ferrite needles with a basket weave feature.

Figure 9.22 shows the microstructure of the weld metal of a low-carbon, low-alloy steel (46). It includes in Figure 9.22a grain boundary ferrite (A),

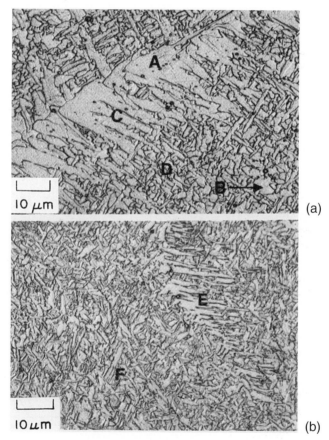

Figure 9.22 Micrographs showing typical weld metal microstructures in low-carbon steels: A, grain boundary ferrite; B, polygonal ferrite; C, Widmanstatten ferrite; D, acicular ferrite; E, upper bainite; F, lower bainite. Reprinted from Grong and Matlock (46).

Widmanstatten ferrite (C), and acicular ferrite (D) and in Figure 9.22b upper bainite (E) and lower bainite (F). A polygonal ferrite (B) is also found. Examination with transmission electron microscopy (TEM) is usually needed to identify the upper and lower bainite. The microstructure of a low-carbon steel weld containing predominately acicular ferrite is shown in Figure 9.23 and at a higher magnification in Figure 9.24 (47). The dark particles are inclusions.

9.2.2 Factors Affecting Microstructure

Bhadeshia and Suensson (48) showed in Figure 9.25 the effect of several factors on the development of microstructure of the weld metal: the weld metal composition, the cooling time from 800 to 500°C (Δt_{8-5}), the weld metal oxygen content, and the austenite grain size. The vertical arrows indicate the directions in which these factors increase in strength. This will be explained with the help of CCT curves.

A. Cooling Time   Consider the left CCT curves (broken lines) in Figure 9.26. As cooling slows down (Δt_{8-5} increases) from curve 1 to curve 2 and curve 3, and the transformation product can change from predominately bainite (Figure 9.25c), to predominately acicular ferrite (Figure 9.25b) to predominately grain boundary and Widmanstatten ferrite (Figure 9.25a).

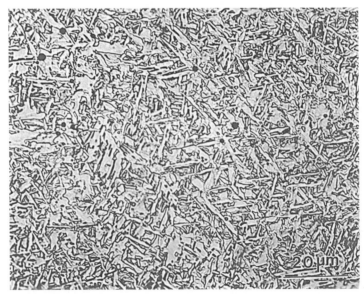

Figure 9.23   Predominately acicular ferrite microstructure of a low-carbon, low-alloy steel weld. Reprinted from Babu et al. (47).

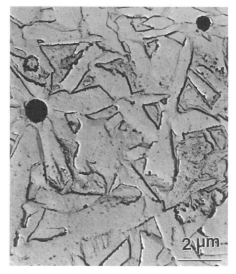

Figure 9.24 Acicular ferrite and inclusion particles in a low-carbon, low-alloy steel weld. Reprinted from Babu et al. (47).

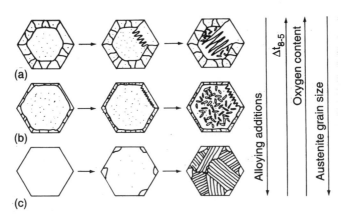

Figure 9.25 Schematic showing effect of alloy additions, cooling time from 800 to 500°C, weld oxygen content, and austenite grain size. Reprinted from Bhadeshia and Svensson (48).

B. Alloying Additions   An increase in alloying additions (higher hardenability) will shift the CCT curves toward longer times and lower temperatures. Consider now cooling curve 3 in Figure 9.26. The transformation product can change from predominately grain boundary and Widmanstatten ferrite (left CCT curves) to predominately acicular ferrite (middle CCT curves) to predominately bainite (right CCT curves). This is like what Figure 9.25 shows.

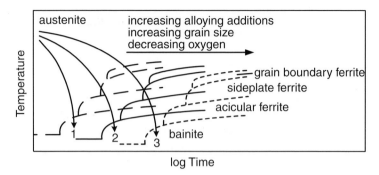

Figure 9.26 Effect of alloying elements, grain size, and oxygen on CCT diagrams for weld metal of low-carbon steel.

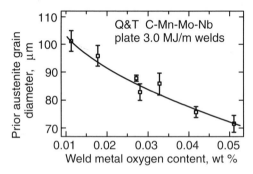

Figure 9.27 Prior austenite grain diameter as a function of weld metal oxygen content in submerged arc welds. Reprinted from Fleck et al. (49). Courtesy of American Welding Society.

C. Grain Size   Similar to the effect of alloying additions, an increase in the austenite grain size (less grain boundary area for ferrite nucleation) will also shift the CCT curves toward longer times and lower temperatures.

D. Weld Metal Oxygen Content   The effect of the weld metal oxygen content on the weld metal microstructure is explained as follows. First, as shown in Figure 9.27, Fleck et al. (49) observed in submerged arc welds that the austenite grain size before transformation decreases with increasing weld metal oxygen content. Liu and Olson (50) observed that increasing the weld metal oxygen content increased the inclusion volume fraction and decreased the average inclusion size. In fact, a large number of smaller size inclusions of diameters less than 0.1 μm was found. Since fine second-phase particles are known to increasingly inhibit grain growth by pinning the grain boundaries as the particles get smaller and more abundant (51), increasing the weld metal oxygen content should decrease the prior austenite grain size.

Therefore, the effect of decreasing the weld metal oxygen content is similar to that of increasing the prior austenite grain size. This is just like what Figure 9.25 shows.

Second, larger inclusions, which are favored by lower weld metal oxygen contents, can act as favorable nucleation sites for acicular ferrite (50). Appropriate inclusions appear to be in the size range 0.2–2.0 μm, and the mean size of about 0.4 μm has been suggested to be the optimum value (49, 51–53). Fox et al. (54) suggested in submerged arc welds of HY-100 steel that insufficient inclusion numbers are generated for the nucleation of acicular ferrite if the oxygen content is too low (<200 ppm). On the other hand, many small oxide inclusions (<0.2 μm) can be generated if the oxygen content is too high (>300 ppm). These inclusions, though too small to be effective nuclei for acicular ferrite, reduce the grain size and thus provide much grain boundary area for nucleation of grain boundary ferrite. As such, an optimum oxygen content can be expected for acicular ferrite to form. This is just like what Figure 9.25b shows.

The existence of an optimum oxygen content for acicular ferrite to form has also been reported by Onsoien et al. (45) in GMAW with oxygen or carbon dioxide added to argon, as shown clearly in Figure 9.28. With Ar–O_2 as the shielding gas, the shielding gas oxygen equivalent is the volume percentage of O_2 in the shielding gas. With Ar–CO_2 as the shielding gas, it becomes the volume percentage of CO_2 in the shielding gas that will produce the same oxygen content in the weld metal. As expected, the experimental results show that the higher the shielding gas oxygen equivalent, the more hardenability elements such as Mn and Si from the filler wire are oxidized. Consider again cooling curve 3 in Figure 9.26. As the shielding gas oxygen equivalent is reduced, the CCT curves can shift from left (broken lines) to middle (solid

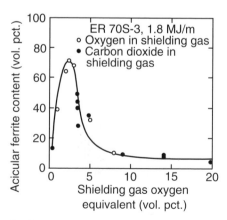

Figure 9.28   Acicular ferrite content as a function of shielding gas oxygen equivalent for gas–metal arc welds. Reprinted from Onsoien et al. (45). Courtesy of American Welding Society.

lines) and a predominately acicular microstructure is produced. However, as the shielding gas oxygen equivalent is reduced further, the CCT curves can shift from middle (solid lines) to right (dotted lines) and acicular ferrite no longer predominates.

Other factors have also been reported to affect amount of acicular ferrite in the weld metal. For example, it has been reported that acicular ferrite increases with increasing basicity index of the flux for submerged arc welding (54), Ti (55, 56), and Mn and Ni (57).

9.2.3   Weld Metal Toughness

Acicular ferrite is desirable because it improves the toughness of the weld metal (55, 56). As shown in Figure 9.29, Dallam et al. (57) observed that the

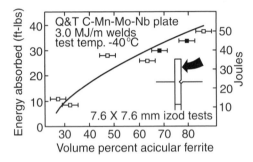

Figure 9.29 Subsize Charpy V-notch toughness values as a function of volume fraction of acicular ferrite in submerged arc welds. Reprinted from Fleck et al. (49). Courtesy of American Welding Society.

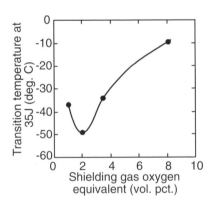

Figure 9.30 Weld metal Charpy V-notch toughness expressed as transition temperature as a function of shielding gas oxygen equivalent. Reprinted from Onsoien et al. (45). Courtesy of American Welding Society.

weld metal Charpy V-notch toughness in submerged arc welds increases with increasing volume fraction of acicular ferrite in the weld metal. The interlocking nature of acicular ferrite, together with its fine grain size, provides the maximum resistance to crack propagation by cleavage. The formation of grain boundary ferrite, ferrite side plates, or upper bainite is detrimental to weld metal toughness, since these microstructures provide easy crack propagation paths.

Onsoien et al. (45) tested the Charpy V-notch toughness of GMA weld metal using an energy absorption of 35 J as the criterion for measuring the transition temperature for ductile-to-brittle fracture. Figure 9.30 showed that the maximum toughness (minimum transition temperature)occurs at a shielding gas oxygen equivalent of about 2 vol %. This, as can be seen from Figure 9.28, essentially corresponds to the maximum amount of acicular ferrite in the weld metal, thus clearly demonstrating the beneficial effect of acicular ferrite on weld metal toughness. Ahlblom (58) has shown earlier a clear minimum in the plots of Charpy V-notch transition temperature versus weld metal oxygen content.

REFERENCES

1. David, S. A., Goodwin, G. M., and Braski, D. N., *Weld. J.*, **58:** 330s, 1979.
2. David, S. A., *Weld. J.*, **60:** 63s, 1981.
3. Lippold, J. C., and Savage, W. F., *Weld. J.*, **58:** 362s, 1974.
4. Lippold, J. C., and Savage, W. F., *Weld. J.*, **59:** 48s, 1980.
5. Cieslak, M. J., and Savage, W. F., *Weld. J.*, **59:** 136s, 1980.
6. Cieslak, M. J., Ritter, A. M., and Savage, W. F., *Weld. J.*, **62:** 1s, 1982.
7. Okagawa, R. K., Dixon, R. D., and Olson, D. L., *Weld. J.*, **62:** 204s, 1983.
8. *Metals Handbook*, Vol. 8, 8th ed., American Society for Metals, Metals Park, OH, 1973.
9. Kou, S., and Le, Y., *Metall. Trans.*, **13A:** 1141, 1982.
10. Brooks, J. A., and Garrison, Jr., W. M., *Weld. J.*, **78:** 280s, 1999.
11. Inoue, H., Koseki, T., Ohkita, S., and Fuji, M., *Sci. Technol. Weld. Join.*, **5:** 385, 2000.
12. Schaeffler, A. L., *Metal Prog.*, **56:** 680, 1949.
13. Delong, W. T., *Weld. J.*, **53:** 273s, 1974.
14. Sato, Y. S., Kokawa, H., and Kuwana, T., *Sci. Technol. Weld. Join.*, **4:** 41, 1999.
15. Lundin, C. D., Chou, C. P. D., and Sullivan, C. J., *Weld. J.*, **59:** 226s, 1980.
16. Kotecki, D. J., and Siewert, T. A., *Weld. J.*, **71:** 171s, 1992.
17. McCowan, C. N., Siewert, T. A., and Olson, D. L., *WRC Bull.*, **342:** 1–36, 1989.
18. Lake, F. B., Expansion of the WRC-1988 Ferrite Diagram and Nitrogen Prediction, Abstracts of Papers, 1988 AWS Convention, Detroit, MI, pp. 214–215.
19. Kotecki, D. J., *Weld. J.*, **78:** 180s, 1999.
20. Kotecki, D. J., *Weld. J.*, **79:** 346s, 2000.

21. Kotecki, D. J., *Weld Dilution and Martensite Appearance in Dissimilar Metal Welding*, IIW Document II-C-195-00, 2000.

22. Kotecki, D. J., private communications, Lincoln Electric Company, Cleveland, OH, 2001.

23. Balmforth, M. C., and Lippold, J. C., *Weld. J.*, **79:** 339s, 2000.

24. Vitek, J. M., Iskander, Y. S., and Oblow, E. M., *Weld. J.*, **79:** 33s, 2000.

25. Vitek, J. M., Iskander, Y. S., and Oblow, E. M., *Weld. J.*, **79:** 41s, 2000.

26. Vitek, J. M., and David, S. A., in *Trends in Welding Research in the United States*, Ed. S. A. David, American Society for Metals, Metals Park, OH, 1982.

27. Vitek, J. M., DasGupta, A., and David, S. A., *Metall. Trans.*, **14A:** 1833, 1983.

28. Katayama, S., and Matsunawa, A., *Proc. International Congress on Applications of Laser and Electro-Optics 84*, **44**: 60–67, 1984.

29. Katayama, S., and Matsunawa, A., in *Proc. International Congress on Applications of Laser and Electro-Optics 85*, San Francisco, 1985, IFS Ltd., Kempston, Bedford, UK, 1985, p. 19.

30. Olson, D. L., *Weld. J.*, **64:** 281s, 1985.

31. David, S. A., Vitek, J. M., and Hebble, T. L., *Weld. J.*, **66:** 289s, 1987.

32. Bobadilla, M., Lacaze, J., and Lesoult, G., *J. Crystal Growth*, **89:** 531, 1988.

33. Elmer, J. W., Allen, S. M., and Eagar, T. W., *Metall. Trans.*, **20A:** 2117, 1989.

34. Elmer, J. W., Eagar, T. W., and Allen, S. M., in *Weldability of Materials*, Eds. R. A. Patterson and K. W. Mahin, ASM International, Materials Park, OH, 1990, pp. 143–150.

35. Lippold, J. C., *Weld. J.*, **73:** 129s, 1994.

36. Koseki, T., and Flemings, M. C., *Metall. Mater. Trans.*, **28A:** 2385, 1997.

37. Brooks, J. A., and Thompson, A. W., *Int. Mater. Rev.*, **36:** 16, 1991.

38. Lippold, J. C., *Weld. J.*, **64:** 127s, 1985.

39. Lundin, C. D., and Chou, C. P. D., *Weld. J.*, **64:** 113s, 1985.

40. Chen, M. H., and Chou, C. P., *Sci. Technol. Weld. Join.*, **4:** 58, 1999.

41. Abson, D. J., and Dolby, R. E., *Weld. Inst. Res. Bull.*, **202:** July 1978.

42. Dolby, R. E., *Metals Technol.* **10:** 349, 1983.

43. Classification of Microstructure in Low Carbon Low Alloy Weld Metal, IIW Doc. IX-1282-83, 1983, International Institute of Welding, London, UK.

44. Vishnu, P. R., in *ASM Handbook*, Vol. 6: *Welding, Brazing and Soldering*, ASM International, Materials Park, OH, 1993, pp. 70–87.

45. Onsoien, M. I., Liu, S., and Olson, D. L., *Weld. J.*, **75:** 216s, 1996.

46. Grong, O., and Matlock, D. K., *Int. Metals Rev.*, **31:** 27, 1986.

47. Babu, S. S., Reidenbach, F., David, S. A., Bollinghaus, Th., and Hoffmeister, H., *Sci. Technol. Weld. Join.*, **4:** 63, 1999.

48. Bhadeshia, H. K. D. H., and Svensson, L. E., in *Mathematical Modelling of Weld Phenomena*, Eds. H. Cerjak and K. Easterling, Institute of Materials, 1993.

49. Fleck, N. A., Grong, O., Edwards, G. R., and Matlock, D. K., *Weld. J.*, **65:** 113s, 1986.

50. Liu, S., and Olson, D. L., *Weld. J.*, **65:** 139s, 1986.

51. Ashby, M. F., and Easterling, K. E., *Acta Metall.*, **30:** 1969, 1982.

52. Bhadesia, H. K. D. H., in *Bainite in Steels*, Institute of Materials, London, 1992, Chapter 10.

53. Edwards, G. R., and Liu, S., in *Proceedings of the first US-Japan Symposium on Advanced Welding Metallurgy*, AWA/JWS/JWES, San Francisco, CA, and Yokohama, Japan, 1990, pp. 213–292.

54. Fox, A. G., Eakes, M. W., and Franke, G. L., *Weld. J.*, **75:** 330s, 1996.

55. Dolby, R. E., Research Report No. 14/1976/M, Welding Institute, Cambridge 1976.

56. Glover, A. G., McGrath, J. T., and Eaton, N. F., in *S Toughness Characterization and Specifications for HSLA and Structural Steels.* ed. P. L. Manganon, Metallurgical Society of AIME, NY, pp. 143–160.

57. Dallam, C. B., Liu, S., and Olson, D. L., *Weld. J.*, **64:** 140s, 1985.

58. Ahlblom, B., Document No. IX-1322-84, International Institute of Welding, London, 1984.

FURTHER READING

1. Grong, O., and Matlock, D. K., *Int. Meter. Rev.*, **31:** 27, 1986.

2. Brooks, J. A., and Thompson, A. W., *Int. Mater. Rev.,* **36:** 16, 1991.

3. Vishnu, P. R., in *ASM Handbook*, Vol. 6: *Welding, Brazing and Soldering*, ASM International, Materials Park, OH, 1993, pp. 70–87.

4. Brooks, J. A., and Lippold, J. C., in *ASM Handbook*, Vol. 6: *Welding, Brazing and Soldering*, ASM International, Materials Park, OH, 1993, pp. 456–470.

PROBLEMS

9.1 **(a)** Construct pseudobinary phase diagrams for 55% and 74% Fe. Mark on the diagrams the approximate compositions of 310 (essentially Fe–25 Cr–20 Ni) and 304 (essentially Fe–18Cr–8Ni) stainless steels.

 (b) From the diagrams and the approximate compositions, indicate the primary solidification phases.

9.2 A 308 stainless-steel filler (essentially Fe–20Cr–10Ni) is used to weld 310 stainless steel. What is the primary solidification phase if the dilution ratio is about 60%?

9.3 A 304 stainless-steel sheet with a composition given below is welded autogenously with the GTAW process. The shielding gas is Ar-2% N_2, and the nitrogen content of the weld metal is about 0.13%. The contents of other alloying elements are essentially the same as those in the base metal.

 (a) Calculate the ferrite numbers for the base metal and the weld metal.

 (b) The weld metal exhibits the primary solidification phase of austenite, and the ferrite content measurements indicate essentially zero ferrite number. Is the calculated ferrite number for the weld metal consistent with the observed one? (Composition: 18.10Cr, 8.49Ni, 0.060C, 0.66Si, 1.76Mn, 0.36Mo, 0.012S, 0.036P, and 0.066N.)

9.4 A significant amount of ferrite is lost in a 316 stainless steel weld after being subjected to three postweld thermal cycles with a 1250°C peak temperature, which is just below the $\gamma + \delta$ two-phase region of about 1280 to 1425°C. Sketch a curve of ferrite number vs. temperature from 900 to 1400°C and explain it.

9.5 Kou and Le (9) quenched 309 stainless steel during autogenous GTAW. The weld metal side of the quenched pool boundary showed dendrites of δ-ferrite but the weld pool side showed dendrites of primary austenite. Explain why.

9.6 It has been observed in welding austenitic stainless steel with a teardrop-shaped weld pool that the weld metal solidifies with primary ferrite except near the centerline, where it solidifies as primary austenite. Sketch a curve of the growth rate R versus the distance y away from the weld centerline. How does your result explain the ferrite content change near the centerline?

10 Weld Metal Chemical Inhomogeneities

In this chapter we shall discuss chemical inhomogeneities in the weld metal, including solute segregation, banding, inclusions, and gas porosity. Solute segregation can be either microsegregation or macrosegregation. *Microsegregation* refers to composition variations across structures of microscopic sizes, for instance, dendrite arms or cells (Chapter 6). *Macrosegregation*, on the other hand, refers to variations in the local average composition (composition averaged over many dendrites) across structures of macroscopic sizes, for instance, the weld. Macrosegregation has been determined by removing samples across the weld metal with a small drill for wet chemical analysis (1). Microsegregation, on the other hand, has been determined by *electron probe microanalysis* (EPMA) (2–8) or *scanning transmission electron microscopy* (STEM) (9–14). The spacial resolution is lower in the former (e.g., about $1\,\mu$m) and higher in the latter (e.g., about $0.1\,\mu$m).

10.1 MICROSEGREGATION

Alloying elements with an equilibrium segregation coefficient $k < 1$ tend to segregate toward the boundary between cells or dendrite arms, and those with $k > 1$ tend to segregate toward the core of cells or dendrite arms (Chapter 6). Microsegregation can have a significant effect on the solidification cracking susceptibility of the weld metal (Chapter 11).

10.1.1 Effect of Solid-State Diffusion

Microsegregation can be reduced significantly by solid-state diffusion during and after solidification. Consequently, microsegregation measured after welding may not represent the true microsegregation during welding, which is more relevant to solidification cracking (Chapter 11).

Kou and Le (15) quenched stainless steels with liquid tin during autogenous GTAW in order to preserve the high-temperature microstructure around the pool boundary. In 430 ferritic stainless steel the dendrites were clear near the quenched pool boundary but became increasingly blurred away from it, suggesting homogenization of microsegregation by the solid-state diffusion.

Ferrite has a *body-centered-cubic* (bcc) structure, which is relatively open and thus easy for diffusion. In 310 austenitic stainless steel, however, the dendrites were clear even away from the quenched pool boundary, suggesting much less homogenization due to solid-state diffusion. Austenite has a *face-centered-cubic* (fcc) structure, which is more close packed than bcc and thus more difficult for diffusion.

Brooks and Garrison (8) quenched a precipitation-strengthened martensitic stainless steel with liquid tin during GTAW. They measured microsegregation across the columnar dendrites, as shown in Figure 10.1*a*, where the quenched weld pool is in the upper right corner of the micrograph. The weld metal solidified as a single-phase ferrite. As shown in Figure 10.1*b*, near the

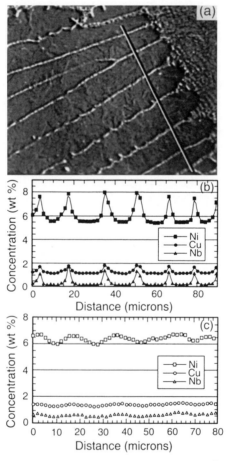

Figure 10.1 Microsegregation across columnar dendrites near quenched weld pool in a martensitic stainless steel: (*a, b*) near growth front; (*c*) 400 μm behind. Reprinted from Brooks and Garrison (8). Courtesy of American Welding Society.

solidification front segregation of Ni, Cu, and Nb toward the boundaries between dendrites is clear, the measured equilibrium segregation coefficients being 0.85, 0.9, and 0.36, respectively. The concentration peaks correspond to the dendrite boundaries. The microsegregation profiles at 400 μm behind the solidification front are shown in Figure 10.1c. As shown, microsegregation is reduced significantly by solid-state diffusion at elevated temperatures.

Instead of using the Scheil equation (Chapter 6), Brooks and Baskes (16–19) calculated weld metal microsegregation considering solid-state diffusion and Lee et al. (20) further incorporated the kinetics of dendrite coarsening. Figure 10.2a shows the calculated microsegregation in a Fe–23Cr–12Ni stainless steel that solidifies as ferrite (17, 18), where r is the radius of the growing cell and R is the final cell radius. Since $k > 1$ for Cr, the cell core ($r = 0$) is rich in Cr. The cell grows to about 60% of its full radius in 0.015s after solidification starts and 100% in 0.15s, with Cr diffusing away from the cell

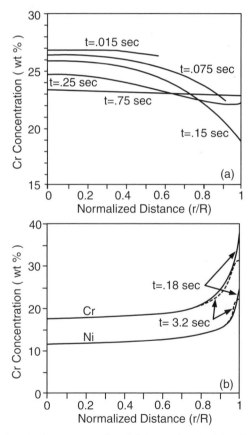

Figure 10.2   Calculated microsegregation: (a) Fe–23Cr–12Ni (solid-state diffusion significant); (b) Fe–21Cr–14Ni. Reprinted from Brooks (18).

core as solidification proceeds. Diffusion is fast in ferrite because of its relatively loosely packed bcc structure. Chromium diffusion continues after solidification is over (0.25 s) and the resultant cell is nearly completely homogenized in 0.75 s after solidification starts. Figure 10.2*b*, shows the calculated microsegregation in a Fe–21Cr–14Ni stainless steel that solidifies as austenite at the time of final solidification (0.18 s) and 3 s later. Both Cr and Ni are highly segregated to the cell boundary. Little solid-state diffusion occurs during cooling except near the cell boundary, where some diffusion occurs because of the steep concentration gradients. Diffusion is slow in austenite because of its more densely packed fcc structure.

Figure 10.3 shows the STEM microsegregation profiles across a dendrite arm in the weld metal of 308 austenitic stainless steel (about Fe–20Cr–10Ni) measured by David et al. (10). The primary solidification phase is δ-ferrite

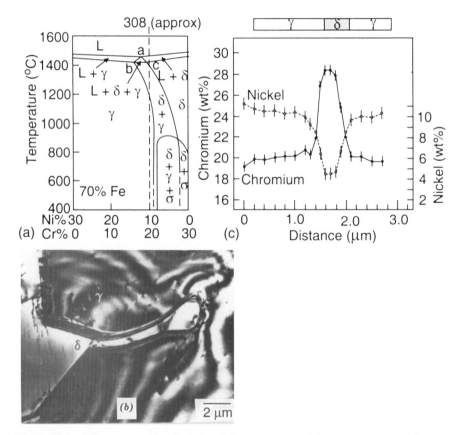

Figure 10.3 Microsegregation in 308 stainless steel weld: (*a*) phase diagram; (*b*) TEM micrograph of a dendrite arm; (*c*) microsegregation across the arm. Reprinted from David et al. (10). Courtesy of American Welding Society.

(Figure 10.3a), but the dendrite arm has transformed to γ except for the remaining δ-ferrite core (Figure 10.3b). Nickel ($k < 1$) segregates toward the dendrite boundary, while Cr ($k > 1$) segregates toward the dendrite core (Figure 10.3c). Similar microsegregation profiles have been measured in welds of 304L (11, 12) and 309 (13) stainless steels, and it has been confirmed that diffusion occurs during the $\delta \rightarrow \gamma$ transformation (6, 7, 10, 14).

10.1.2   Effect of Dendrite Tip Undercooling

In addition to solid-state diffusion, microsegregation can also be affected by the extent of dendrite tip undercooling. The difference between the equilibrium liquidus temperature T_L and the dendrite tip temperature T_t is the total undercooling ΔT, which can be divided into four parts:

$$\Delta T = \Delta T_C + \Delta T_R + \Delta T_T + \Delta T_K \qquad (10.1)$$

where ΔT_C = concentration-induced undercooling

ΔT_R = curvature-induced undercooling

ΔT_T = thermal undercooling

ΔT_K = kinetic undercooling

The solute rejected by the dendrite tip into the liquid can pile up and cause undercooling at the dendrite tip ΔT_C, similar to constitutional supercooling at a planar growth front (Chapter 6). The equilibrium liquidus temperature in a phase diagram is for a flat solid–liquid interface, and it is suppressed if the interface has a radius of curvature like a dendrite tip. Thermal undercooling is present where there is a significant nucleation barrier for the liquid to transform to solid. The kinetic undercooling, which is usually negligible, is associated with the driving force for the liquid atoms to become attached to the solid. It has been observed that the higher the velocity of the dendrite tip, V_t, the smaller the radius of the dendrite tip, R_t, and the larger the undercooling at the dendrite tip, ΔT (21). This is shown schematically in Figure 10.4. Models have been proposed for dendrite tip undercooling, including those by Burden and Hunt (22) and Kurz et al. (23).

Brooks et al. (19) calculated weld metal microsegregation in Fe–Nb alloys considering dendrite tip undercooling as well as solid-state diffusion. Composition analyses of tin-quenched gas–tungsten arc welds indicated a dendrite tip undercooling of 3.07°C in Fe–1.8Nb and 5.63°C in Fe–3.3Nb. Figure 10.5a show the calculated results for a Fe–3.3Nb alloy, based on the model of Kurz et al. (23) with $k = 0.28$ for Nb and $D_L = 6.8\,\text{cm}^2/\text{s}$. Undercooling results in a higher core Nb concentration during the initial stages of solidification (0.08 s). But as solidification proceeds to completion (1.2 s) and the weld cools (5.6 s), the concentration profiles with and without undercooling start to converge due to the significant effect of solid-state diffusion. It was, therefore, concluded that

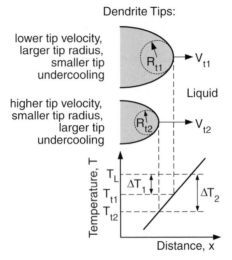

Figure 10.4 Dendrite tip undercooling.

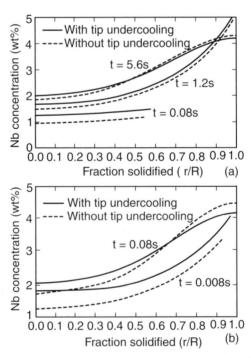

Figure 10.5 Calculated microsegregation in a Fe–3.3Nb weld: (*a*) slower cooling rate; (*b*) higher cooling rate. Reprinted from Brooks et al. (19).

although tip undercooling can result in a higher Nb concentration during initial solidification, its effect on the final microsegregation is small because of the overwhelming effect of solid-state diffusion. With a very high cooling rate of $10^{4}°C/s$ such as in high-energy beam welding (less time for solid-state diffusion), Figure 10.5*b* shows that the effect of undercooling can be important throughout solidification and subsequent cooling. Final microsegregation is significantly less with dendrite tip undercooling than without.

10.2   BANDING

10.2.1   Compositional and Microstructural Fluctuations

In addition to microsegregation across dendrites, microsegregation can also exist in the weld metal as a result of banding during weld pool solidification (Chapter 6). Banding in welds can cause perturbations in the solidification structure as well as the solute concentration (24). Figure 10.6 shows banding and rippling near the centerline of the as-welded top surface of a YAG laser weld (conduction mode) in a 304 stainless steel, with alternating bands of dendritic and planarlike structures. Figure 10.7 shows alternating bands of austenite (light) and martensite (dark) in an A36 steel welded with a E309LSi filler by GTAW (25). The composition of A36 is Fe–0.71Mn–0.29Cu–0.18Si–0.17C–0.13Ni–0.09Cr–0.05Mo and that of E309LSi is Fe–23.16Cr–13.77Ni–1.75Mn–0.79Si–0.20Cu–0.16Mo–0.11N–0.02C. The high hardness (smaller indentation marks) in the darker regions reflects the presence of martensite. The hydrogen crack in the martensite near but within

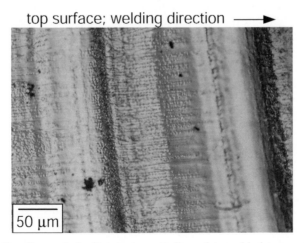

top surface; welding direction ⟶

50 μm

Figure 10.6   Banding and rippling near centerline of as-welded top surface of a 304 stainless steel YAG laser welded in conduction mode.

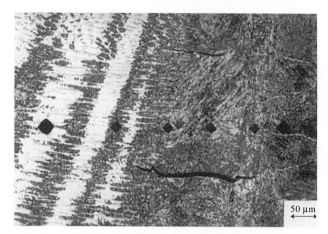

Figure 10.7 Banding near fusion boundary of a carbon steel welded with an austenitic stainless steel filler metal. Reprinted from Rowe et al. (25). Courtesy of American Welding Society.

the fusion boundary was promoted by the use of Ar–6% H_2 as the shielding gas.

10.2.2 Causes

Banding in the weld metal can occur due to a number of reasons. Fluctuations in the welding speed during manual welding or arc pulsing during pulsed arc welding can cause banding. However, even under steady-state welding conditions, banding can still occur, as evidenced by the surface rippling of the weld. Beside fluctuations in the welding speed and the power input, the following *mechanisms* also have been proposed: Solidification halts due to the rapid evolution of latent heat caused by high solidification rates during welding (26–28), oscillations of weld pool metal due to uncontrollable variations in arc stability and the downward stream of the shielding gas (29), and fluctuations in weld pool turbulence due to electromagnetic effects (30, 31).

10.3 INCLUSIONS AND GAS POROSITY

Inclusions and gas porosity tend to deteriorate the mechanical properties of the weld metal. Gas–metal and slag–metal reactions can produce gas porosity and inclusions in the weld metal and affect the weld metal properties (Chapter 3). Inclusions can also result from incomplete slag removal during multiple-pass welding (32), the large dark-etching particle near position D in Figure 10.8 being one example (33). Figure 10.9 shows trapped surface oxides as inclusions in the weld metal and their elimination by modifying the joint design (34).

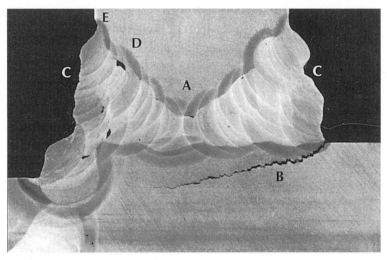

Figure 10.8 Multipass weld with slag inclusions (D) and other defects, including lack of fusion (A), lamellar tearing (B), poor profile (C), and undercut (E). Reprinted from Lochhead and Rodgers (33). Courtesy of American Welding Society.

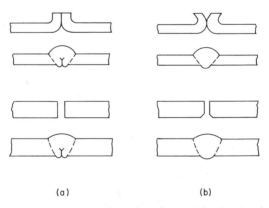

(a)   (b)

Figure 10.9 Joint designs and trapping of surface oxides in aluminum welding: (*a*) oxide trapped; (*b*) oxide not trapped (34).

Various ways of reducing gas porosity in the weld metal have already been described in Chapter 3. Gas pores can be round or interdendritic, as shown in Figure 10.10, which shows a gas–metal arc weld in 7075 aluminum made with a 4043 filler metal. Similar gas pores have been reported in other aluminum welds (35–37). Although round gas pores can be randomly distributed in the weld metal, they can also line up and form bands of porosity when banding is severe during weld metal solidification (35, 36). It is often difficult to tell whether interdendritic pores are due to gas formation or due to solidification

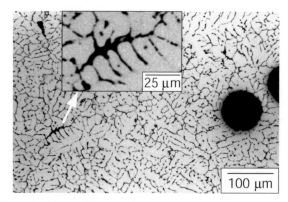

Figure 10.10 Porosity in aluminum weld showing both spherical and interdendritic gas pores. One interdendritic pore is enlarged for clarity.

shrinkage (21). However, if they are due to gas formation, they must have formed during the latter stages of solidification, where the dendritic structure has essentially been established.

10.4 INHOMOGENEITIES NEAR FUSION BOUNDARY

Dissimilar metal welding is often encountered in welding, where a filler metal different in composition from the base metal is used or where two base metals different in composition are welded together. In dissimilar metal welding, the region near the fusion boundary often differ significantly from the bulk weld metal in composition and sometimes even microstructure and properties. The region, first discovered by Savage et al. (38, 39), has been called the unmixed zone (38, 39), filler-metal-depleted area (40), partially mixed zone (41), intermediate mixed zone (42), and hard zone (43). It has been observed in various welds, including stainless steels, alloy steels, aluminum alloys, and superalloys (38–47).

10.4.1 Composition Profiles

Ornath et al. (46) determined the composition profiles across the fusion boundary of a low-alloy steel welded with a stainless steel filler of Fe–18Cr–8Ni–7Mn, as shown in Figure 10.11. Plotting the concentration against the distance from the fusion boundary according to Equation (6.17) (for solute segregation during the initial transient of solidification), they obtained a straight line. As such, they proposed that segregation rather than diffusion is responsible for the observed composition profiles. In other words, the composition profiles are caused by the rejection of Cr, Ni, and Mn into the melt by the solid weld metal during the initial stage of solidification.

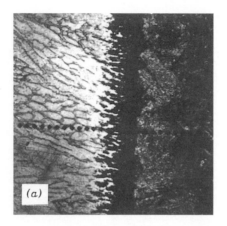

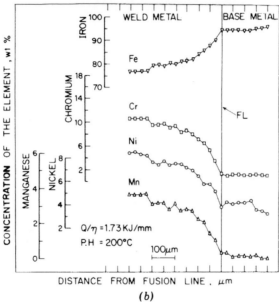

Figure 10.11   Fusion boundary of a low-alloy steel welded with an austenitic stainless steel electrode: (*a*) microstructure (magnification 55×); (*b*) segregation. Reprinted from Ornath et al. (46). Courtesy of American Welding Society.

Baeslack et al. (45) determined the composition profiles across the fusion boundary of a 304L stainless steel welded with a 310 stainless steel filler, as shown in Figure 10.12. Unlike the composition profiles shown in Figure 10.11, the weld metal next to the fusion line, labeled as the *unmixed zone* by the authors, has essentially the same composition as the base metal, suggesting stagnant melted base metal unmixed with the filler metal. Figure 10.13 is a schematic sketch for an unmixed zone. The arrows show the directions of local

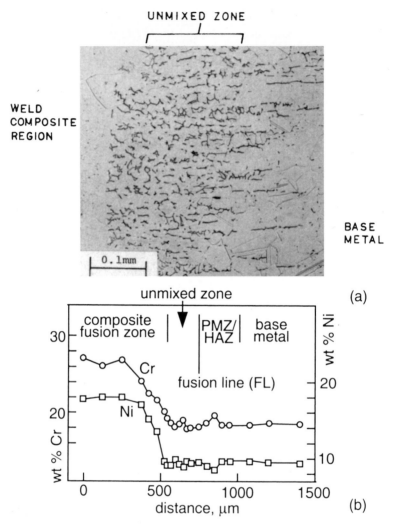

Figure 10.12   Unmixed zone in a 304L stainless steel welded with a 310 filler metal: (*a*) microstructure; (*b*) composition profiles. Reprinted from Baeslack et al. (45). Courtesy of American Welding Society.

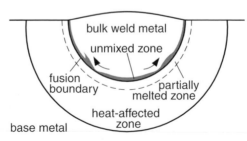

Figure 10.13   Zone of unmixed melted base metal along the fusion boundary.

fluid flow during welding, which is not strong enough for thorough mixing but strong enough to move parts of the melted base metal. Also shown in the figure are the partially melted zone (Chapters 12 and 13) and the heat-affected zone (Chapters 14–18).

10.4.2   Effect of Inhomogeneities

Inhomogeneities in the region along the fusion boundary have been reported to cause problems, including hydrogen cracking (48, 49), corrosion (50), and stress corrosion cracking (45). *Martensite* often exists in the region in carbon or alloy steels welded with austenitic stainless steel fillers. This is because the weld metal composition here can be within the martensite region of the constitutional diagrams (Chapter 9). This composition transition shown in Figure 10.11 covers compositions in the martensite range of the Schaeffler diagram, thus explaining the formation of martensite (46). Linnert (48) observed martensite and *hydrogen cracking* in weld metal along the fusion boundary of a Cr–Ni–Mo steel welded with a 20Cr–10Ni stainless steel filler. Savage et al. (49) reported similar hydrogen cracking in HY-80 welds. Rowe et al. (25) showed hydrogen cracking along the weld metal side of the fusion boundary in a A36 steel gas–tungsten arc welded with a ER308 stainless steel filler metal and an Ar-6% H_2 shielding gas.

Omar (43) welded carbon steels to austenitic stainless steels by SMAW. Figure 10.14 shows that the hard martensite layer in the weld metal along the carbon steel side of the fusion boundary can be eliminated by using a Ni-base filler metal plus preheating and controlling the interpass temperature. The austenitic stainless steel electrode E309, which has much less Ni and more Cr, did not work as well.

10.5   MACROSEGREGATION IN BULK WELD METAL

Weld pool convection (Chapter 4) can usually mix the weld pool well to minimize macrosegregation across the resultant weld metal. Houldcroft (1) found no appreciable macrosegregation in pure aluminum plates single-pass welded with Al–5% Cu filler. Similar results were observed in Al–1.0Si–1.0Mg plates single-pass welded using either Al–4.9% Si or Al–1.4% Si as the filler metal (1). However, in single-pass dissimilar-metal welding, macrosegregation can still occur if weld pool mixing is incomplete. In multiple-pass dissimilar-metal welding, macrosegregation can still occur even if weld pool mixing is complete in each pass.

10.5.1   Single-Pass Welds

Macrosegregation can occur in a dissimilar weld between two different base metals because of *insufficient mixing* in the weld pool. Matsuda et al. (51) showed Cu macrosegregation across an autogenous gas–tungsten arc weld between thin sheets of 1100 aluminum (essentially pure Al) and 2024

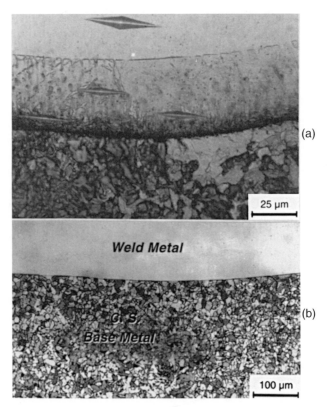

Figure 10.14   Carbon steel side of weld metal in a weld between a carbon steel and an austenitic stainless steel made with a Ni-based filler metal: (*a*) martensite along fusion boundary; (*b*) martensite avoided by preheating and controlling interpass temperature. Reprinted from Omar (43). Courtesy of American Welding Society.

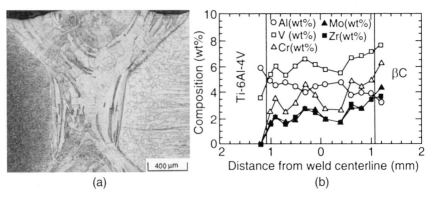

Figure 10.15   Macrosegregation in a laser beam weld between Ti–6Al–4V and Ti–3Al–8V–6Cr–4Mo–4Zr (βC): (*a*) transverse cross section; (*b*) composition profiles. Reprinted from Liu et al. (53). Courtesy of American Welding Society.

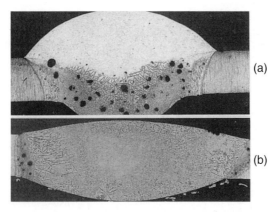

(a)

(b)

Figure 10.16 Powder metallurgy Al–10Fe–5Ce alloy gas–tungsten arc welded with Al–5Si filler metal: (*a*) AC; (*b*) DCEN. Reprinted from Metzger (54). Courtesy of American Welding Society.

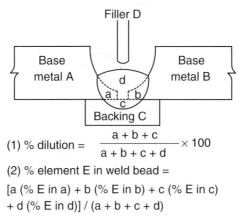

Figure 10.17 Filler metal dilution and composition in dissimilar-metal welding. Reprinted from Estes and Turner (55). Courtesy of American Welding Society.

aluminum (essentially Al–4.4Cu). Macrosegregation was reduced through enhanced mixing by magnetic weld pool stirring. In laser or electron beam welding, the welding speed can be too high to give the weld metal sufficient time to mix well before solidification (52, 53). Figures 10.15 shows macrosegregation in a laser weld between Ti–6Al–4V (left) and Ti–3Al–8V–6Cr–4Mo–4Zr (right) (53).

Macrosegregation due to lack of weld pool mixing has also been observed in GTAW of some *powder metallurgy alloys*. Such alloys are made by consolidation of rapidly solidified powder having some special properties, for instance, extended solubility of alloying elements. Figure 10.16 shows the lack

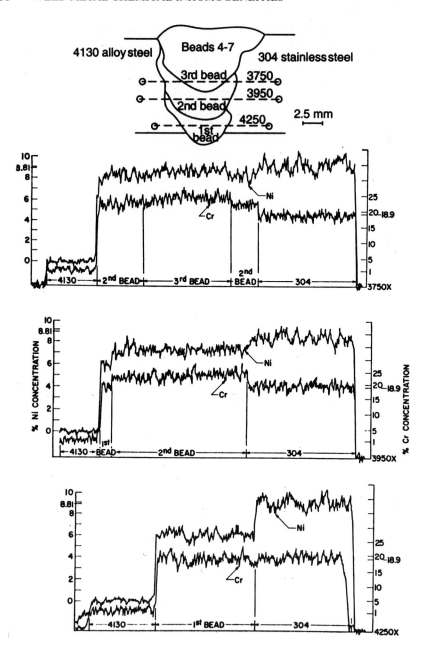

Figure 10.18   Macrosegregation in a multiple-pass weld between 4130 steel and 304 stainless steel. Reprinted from Estes and Turner (55). Courtesy of American Welding Society.

of mixing between an Al–10Fe–5Ce powder metallurgy base metal and a 4043 (Al–5Si) filler metal in AC GTAW and improved mixing and reduced gas porosity with DCEN GTAW (54). This may be because in GTAW weld penetration is higher with DCEN than with AC (Chapter 1).

10.5.2   Multiple-Pass Welds

Figure 10.17 shows the filler metal dilution and composition of the first bead (root pass) in a dissimilar-metal weld (55). The composition of the bead depends not only on the compositions of the base and filler metals but also on the extent of dilution. Apparently, the second bead to be deposited on top of the first bead will have a different composition from the first bead regardless of the extent of weld pool mixing.

Figure 10.18 shows the composition profiles across a multiple-pass weld between 4130 alloy steel and 304 stainless steel with a 312 stainless steel as the filler metal (55). Composition differences between beads are evident. For instance, the Cr content varies from 18% in the first bead to 25% in the third. As a result of such macrosegregation, the ferrite content (which affects the resistance to solidification cracking and corrosion) varies from one bead to another, as shown in Figure 10.19.

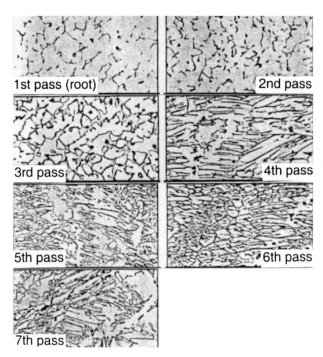

Figure 10.19   Variations in microstructure in a multiple-pass weld between 4130 steel and 304 stainless steel. Reprinted from Estes and Turner (55). Courtesy of American Welding Society.

REFERENCES

1. Houldcroft, R. T., *Br. Weld. J.,* **1:** 468, 1954.

2. Lippold, J. C., and Savage, W. F., *Weld. J.*, **58:** 362s, 1979.

3. Takalo, T., Suutala, N., and Moisio, T., *Metall. Trans.,* **10A:** 1173, 1979.

4. Suutala, N., Takalo, T., and Moisio, T., *Metall. Trans.,* **10A:** 1183, 1979.

5. Ciestak, M. J., and Savage, W. F., *Weld. J.*, **59:** 136s, 1980.

6. Suutala, N., Takalo, T., and Moisio, T., *Weld. J.*, **60:** 92s, 1981.

7. Leone, G. L., and Kerr, H. W., *Weld. J.*, **61:** 13s, 1982.

8. Brooks, J. A., and Garrison, W. M. Jr., *Weld. J.,* **78:** 280s, 1999.

9. Gould, J. C., Ph.D. Thesis, Carnegie-Mellon University, Pittsburgh, PA, 1983.

10. David, S. A., Goodwin, G. M., and Braski, D. N., *Weld. J.,* **58:** 330s, 1979.

11. Lyman, C. E., Manning, P. E., Duquette, D. J., and Hall, E., *Scan. Electron. Microsc.*, **1:** 213, 1978.

12. Lyman, C. E., *Weld. J.*, **58:** 189s, 1979.

13. Brooks, J. A., Ph.D. Thesis, Carnegie-Mellon University, Pittsburgh, PA, 1982.

14. Cieslak, M. J., Ritter, A. M., and Savage, W., *Weld. J.*, **61:** 1s, 1982.

15. Kou, S., and Le, Y., *Metall. Trans. A*, **13A:** 1141, 1982.

16. Brooks, J. A., and Baskes, M. I., in *Advances in Welding Science and Technology*, Ed. S. A. David, ASM International, Materials Park, OH, March 1986, p. 93.

17. Brooks, J. A., and Baskes, M. I., in *Recent Trends in Welding Science and Technology*, Eds. S. A. David and J. M. Vitek, ASM International, Materials Park, OH, March 1990, p. 153.

18. Brooks, J. A., in *Weldability of Materials*, Eds. R. A. Patterson and K. W. Mahin, ASM International, Materials Park, OH, March 1990, p. 41.

19. Brooks, J. A., Li, M., Baskes, M. I., and Yang, N. C. Y., *Sci. Technol. Weld. Join.*, **2:** 160, 1997.

20. Lee, J. Y., Park, J. M., Lee, C. H., and Yoon, E. P., in *Synthesis/Processing of Lightweight Metallic Materials II*, Eds. C. M. Ward-Close, F. H. Froes, S. S. Cho, and D. J. Chellman, The Minerals, Metals and Materials Society, Warrendale, PA 1996, p. 49.

21. Flemings, M. C., *Solidification Processing*, McGraw-Hill, New York, 1974.

22. Burden, M. H., and Hunt, J. D., *J. Crystal Growth*, **22:** 109, 1974.

23. Kurz, W., Giovanola, B., and Trivedi, R., *Acta Metall.*, **34:** 823, 1986.

24. Davies, G. J., and Garland, J. G., *Int. Metall. Rev.*, **20:** 83, 1975.

25. Rowe, M. D., Nelson, T. W., and Lippold, J. C., *Weld. J.*, **78:** 31s, 1999.

26. Gurev, H. S., and Stout, R. D., *Weld. J.*, **42:** 298s, 1963.

27. Cheever, D. L., and Howden, D. G., *Weld. J.*, **48:** 179s, 1969.

28. Morchan, B. A., and Abitdnar, A., *Automat. Weld.*, **21:** 4, 1968.

29. Ishizaki, K., *J. Jpn. Weld. Soc.*, **38:** 1963.

30. Jordan, M. F., and Coleman, M. C., *Br. Weld. J.*, **15:** 552, 1968.

31. Waring, J., *Aust. Weld. J.*, **11:** 15, 1967.

32. Gurney, T. R., *Fatigue of Welded Structures*, Cambridge University Press, Cambridge, 1968, p. 156.

33. Lochhead, J. C., and Rodgers, K. J., *Weld. J.*, **78:** 49, 1999.

34. *Inert Gas Welding of Aluminum Alloys*, Society of the Fusion Welding of Light Metals, Tokyo, Japan, 1971 (in Japanese).

35. D'Annessa, A. T., *Weld. J.*, **45:** 569s, 1966.

36. D'Annessa, A. T., *Weld. J.*, **49:** 41s, 1970.

37. D'Annessa, A. T., *Weld. J.*, **46:** 491s, 1967.

38. Savage, W. F., and Szekeres, E. S., *Weld. J.*, **46:** 94s, 1967.

39. Savage, W. F., Nippes, E. F., and Szekeres, E. S., *Weld. J.*, **55:** 260s, 1976.

40. Duvall, D. S., and Owczarski, W. A., *Weld. J.*, **47:** 115s, 1968.

41. Kadalainen, L. P., *Z. Metallkde.*, **70:** 686, 1979.

42. Doody, T., *Weld. J.*, **61:** 55, 1992.

43. Omar, A. A., *Weld. J.*, **67:** 86s, 1998.

44. Lippold, J. C., and Savage, W. F., *Weld. J.*, **59:** 48s, 1980.

45. Baeslack, W. A. III, Lippold, J. C., and Savage, W. F., *Weld. J.*, **58:** 168s, 1979.

46. Ornath, F., Soudry, J., Weiss, B. Z., and Minkoff, I., *Weld. J.*, **60:** 227s, 1991.

47. Albert, S. K., Gills, T. P. S., Tyagi, A. K., Mannan, S. L., Kulkarni, S. D., and Rodriguez, P., *Weld. J.*, **66:** 135s, 1997.

48. Linnert, G. E., *Welding Metallurgy*, Vol. 2. American Welding Society, Miami, FL, 1967.

49. Savage, W. F., Nippes, E. F., and Szekeres, E. S., *Weld. J.*, **55:** 276s, 1976.

50. Takalo, T., and Moisio, T., IIW Annual Assembly, Tel Aviv, 1975.

51. Matsuda, F., Ushio, M., Nakagawa, H., and Nakata, K., in *Proceedings of the Conference on Arc Physics and Weld Pool Behavior*, Vol. 1, Welding Institute, Arbington Hall, Cambridge, 1980, p. 337.

52. Seretsky, J., and Ryba, E. R., *Weld. J.*, **55:** 208s, 1976.

53. Liu, P. S., Baeslack III, W. A., and Hurley, J., *Weld. J.*, **73:** 175s, 1994.

54. Metzger, G. E., *Weld. J.*, **71:** 297s, 1992.

55. Estes, C. L., and Turner, P. W., *Weld. J.*, **43:** 541s, 1964.

FURTHER READING

1. Davies, G. J., *Solidification and Casting, Wiley*, New York, 1973.

2. Davies, G. J., and Garland, J. G., *Int. Metall. Rev.*, **20:** 83, 1975.

3. Flemings, M. C., *Solidification Processing*, McGraw-Hill, New York, 1974.

4. Savage, W. F., *Weld. World*, **18:** 89, 1980.

PROBLEMS

10.1   With the help of Schaeffler's diagram, show that martensite can form in the fusion zone at $70\,\mu m$ from the fusion boundary of the weld shown in Figure 10.11.

10.2 Butt welding of 5052 aluminum (Al–2.5Mg) with a single-V joint is carried out with 5556 filler (Al–5.1Mg). The dilution ratio of the first pass is 80%. In the second pass 40% of the material comes from the filler wire, 40% from the base metal, and 20% from the first pass. Calculate the compositions of the two passes, assuming uniform mixing in both.

10.3 Suppose that in the previous problem the workpiece composition is Fe–25Cr–20Ni and the filler composition is Fe–20Cr–10Ni. What is the difference in the ferrite content between the two passes based on Schaeffler's diagram.

10.4 Consider the pseudo-binary-phase diagram shown in Figure 10.3*a*. Sketch the Ni and Cr concentration profiles across a dendrite arm for an alloy that has a composition just to the left of point *b*.

10.5 Consider welding Ni to Ti. Can macrosegregation occur in LBW? Why or why not? Is the chance of macrosegregation higher or lower in GTAW than in LBW?

10.6 Explain why gas porosity can be severe in the GTAW of powder metallurgy alloy Al–10Fe–5Ce (Figure 10.16*a*). Explain why gas porosity can be significantly less with DCEN than with AC.

10.7 Consider banding in the YAG laser weld of 304 stainless steel (Figure 10.6). What could have caused banding in this weld? Is the growth rate higher during dendritic or planarlike solidification and why?

11 Weld Metal Solidification Cracking

Solidification cracking in the weld metal will be described, the metallurgical and mechanical factors affecting the crack susceptibility will be discussed, and the methods for reducing cracking will be presented.

11.1 CHARACTERISTICS, CAUSE, AND TESTING

11.1.1 Intergranular Cracking

Solidification cracking, which is observed frequently in castings and ingots, can also occur in fusion welding, as shown in Figure 11.1. Such cracking, as shown in Figure 11.2, is intergranular, that is, along the grain boundaries of the weld metal (1). It occurs during the terminal stage of solidification, when the tensile stresses developed across the adjacent grains exceed the strength of the almost completely solidified weld metal (2–4). The solidifying weld metal tends to contract because of both solidification shrinkage and thermal contraction. The surrounding base metal also tends to contract, but not as much, because it is neither melted nor heated as much on the average. Therefore, the contraction of the solidifying metal can be hindered by the base metal, especially if the workpiece is constrained and cannot contract freely. Consequently, tensile stresses develop in the solidifying weld metal. The severity of such tensile stresses increases with both the degree of constraint and the thickness of the workpiece.

The various *theories of solidification cracking* (4–7) are effectively identical and embody the concept of the formation of a coherent interlocking solid network that is separated by essentially continuous thin liquid films and thus ruptured by the tensile stresses (2). The fracture surface often reveals the dendritic morphology of the solidifying weld metal, as shown in Figure 11.3 for 308 stainless steel fractured under augmented strain during GTAW (8). If a sufficient amount of liquid metal is present near the cracks, it can "backfill" and "heal" the incipient cracks.

The *terminal stage of solidification* mentioned above refers to a fraction of solid, f_S, close to 1, and not necessarily a temperature near the lower limit of the solidification temperature range. Depending on how it varies with temperature in an alloy, f_S can be close to 1 and an essentially coherent inter-

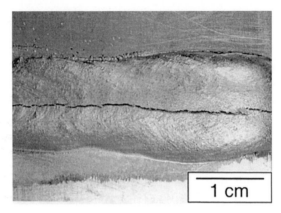

Figure 11.1   Solidification cracking in a gas–metal arc weld of 6061 aluminum.

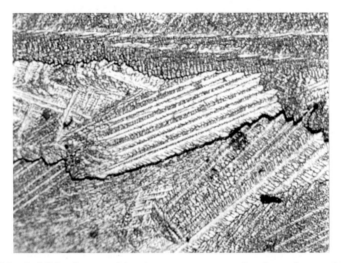

Figure 11.2   Solidification cracking in an autogenous bead-on-plate weld of 7075 aluminum (magnification 140×). From Kou and Kanevsky (1).

locking solid network can form even though the temperature is only slightly below the liquidus temperature. In fact, Matsuda et al. (9, 10) have reported that solidification cracking occurs in some carbon and stainless steels at temperatures slightly below their liquidus temperatures.

11.1.2   Susceptibility Testing

A. Houldcroft Test   This test, also called the fishbone test, is shown in Figure 11.4 (11, 12). It is often used for evaluating the solidification cracking

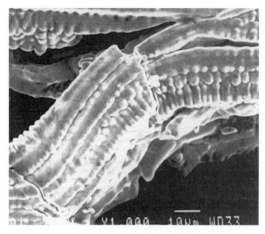

Figure 11.3 SEM fracture surface of a 308 stainless steel weld fractured under augmented strain during GTAW. Reprinted from Li and Messeler (8). Courtesy of American Welding Society.

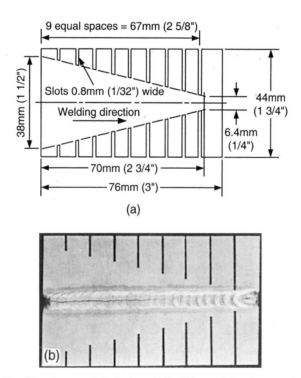

Figure 11.4 Houldcroft test: (*a*) design (11); (*b*) cracking in an aluminum alloy specimen. Reprinted from Liptax and Baysinger (12). Courtesy of American Welding Society.

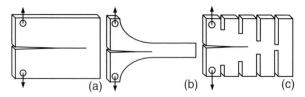

Figure 11.5   Three types of specimens cracking under tension: (*a*) uniform width along crack; (*b*) decreasing width along crack; (*c*) uniform width along crack but slotted.

susceptibility of sheet gage materials. The crack length from the starting edge of the test specimen is used to indicate the susceptibility to cracking (11–14).

The specimen is free from constraints and a progression of slots of varying depth allows the dissipation of stresses within it. In such a test, solidification cracking initiates from the starting edge of the test specimen and propagates along its centerline. As the heat source moves inward from the starting edge of the test specimen, solidification begins and the solidifying structure is torn apart because the starting edge continues to expand as a result of continued heat input to the specimen. The reason for using the slots is explained with the help of specimens cracking under tension, as shown in Figure 11.5. The crack may run all the way through the entire length of the specimen (Figure 11.5a). Reducing the width of the specimen can reduce the amount of stress along the length and bring the crack to a stop (Figure 11.5b). In order not to dramatically change the heat flow condition along the length of the weld, however, the material next to the slots is not cut off (Figure 11.5c).

B. Varestraint Test   This test was developed by Savage and Lundin (15). As shown in Figure 11.6a, an augmented strain is applied to the test specimen (usually 12.7 mm thick) by bending it to a controlled radius at an appropriate moment during welding (15). Both the amount of the applied strain and the crack length (either the total length of all cracks or the maximum crack length) serve as an index of cracking sensitivity. The specimen can also be bent transverse to the welding direction, that is, the *transverse Varestraint test*, as shown in Figure 11.6b (16). This may promote cracking inside the weld metal more than that outside. Figure 11.6c shows the cracking pattern observed in a Varestraint test specimen of a 444 Nb-stabilized ferritic stainless (17). Figure 11.7 compares the solidification cracking susceptibility of several materials by Varestraint testing (18).

Other methods have also been used for testing the solidification cracking susceptibility of weld metals (19, 20), including the circular patch test (21–23) and Sigamajig test (24).

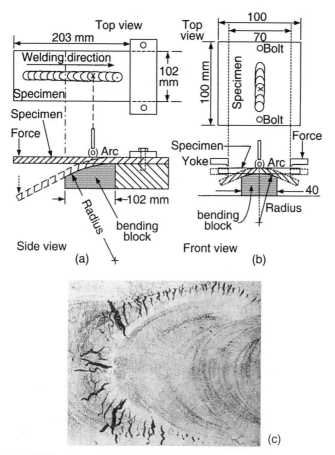

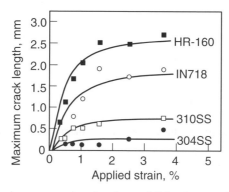

(c)

Figure 11.6 Solidification cracking tests: (*a*) Varestraint (15); (*b*) transverse Varestraint (16); (*c*) cracking in a ferritic stainless steel Varestraint specimen (17). (*c*) Reprinted from Krysiak et al. (17), courtesy of ASM International.

Figure 11.7 Varestraint test results showing solidification cracking susceptibility of several different materials. Reprinted from DuPont et al. (18). Courtesy of American Welding Society.

11.2 METALLURGICAL FACTORS

Metallurgical factors that have been known to affect the solidification crack-ing susceptibility of weld metals include (i) the solidification temperature range, (ii) the amount and distribution of liquid at the terminal stage of solid-ification, (iii) the primary solidification phase, (iv) the surface tension of the grain boundary liquid, and (v) the grain structure. All these factors are directly or indirectly affected by the weld metal composition. The first two factors are affected by microsegregation during solidification. Microsegregation in turn can be affected by the cooling rate during solidification. In fact, in austenitic stainless steels the primary solidification phase can also be affected by the cooling rate. These metallurgical factors will be discussed in what follows.

11.2.1 Solidification Temperature Range

Generally speaking, the wider the solidification (freezing) temperature range, the larger the $(S + L)$ region in the weld metal or the mushy zone and thus the larger the area that is weak and susceptible to solidification cracking. The solidification temperature range of an alloy increases as a result of either the presence of undesirable impurities such as sulfur (S) and phosphorus (P) in steels and nickel-base alloys or intentionally added alloying elements.

A. Effect of S and P   Impurities such as S and P are known to cause severe solidification cracking in carbon and low-alloy steels even at relatively low concentrations. They have a strong tendency to segregate at grain boundaries (25) and form low-melting-point compounds (FeS in the case of S), thus widen-ing the solidification temperature range. In addition, S and P can cause severe solidification cracking in nickel-base alloys (26–28) and ferritic stainless steels (29). In the case of austenitic stainless steels, their detrimental effect on solidi-fication cracking can be significantly affected by the primary solidification phase, as will be discussed subsequently.

Figure 11.8 shows the effect of various elements, including S and P, on the solidification temperature range of carbon and low-alloy steels (30). As shown, S and P tend to widen the solidification temperature range of steels tremen-dously. The wider the solidification temperature range, the more grain bound-ary area in the weld metal remains liquid during welding and hence susceptible to solidification cracking, as depicted in Figure 11.9.

B. Effect of Reactions during Terminal Solidification   Reactions, especially eutectic reactions, can often occur during the terminal stage of solidification and extend the solidification temperature range, for instance, in aluminum alloys and superalloys.

Dupont et al. (31–35) studied the effect of eutectic reactions on the solidification temperature range and solidification cracking. The materials studied include some superalloys and stainless steels containing niobium (Nb).

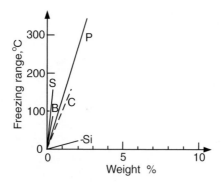

Figure 11.8 Effect of alloying elements on the solidification temperature range of carbon and low-alloy steels. Modified from *Principles and Technology of the Fusion Welding of Metals* (30).

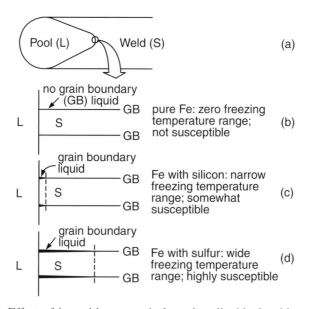

Figure 11.9 Effect of impurities on grain boundary liquid of weld metal: (*a*) weld metal near pool; (*b*) no liquid in pure Fe; (*c*) some liquid with a small amount of Si; (*d*) much more liquid with a small amount of sulfur.

Figure 11.10 shows the weld metal microstructure of a Nb-bearing superalloy (34). Two eutectic-type constituents are present in the microstructure, γ-NbC and γ-Laves. The liquid solidifies as the primary γ and two eutectic reactions occur during the terminal stage of solidification, first L $\rightarrow$ γ + NbC and then L $\rightarrow$ γ + Laves. The second eutectic reaction, when it occurs, takes place at a considerably lower temperature than the first, thus extending the solidification

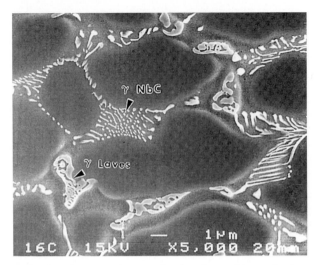

Figure 11.10   Scanning electron micrographs showing weld metal microstructure of a Nb-bearing superalloy. Reprinted from DuPont et al. (34).

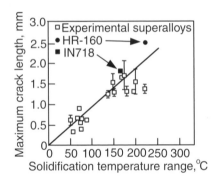

Figure 11.11   Maximum crack length as a function of solidification temperature range for Nb-bearing superalloys and two other materials. Reprinted from DuPont et al. (18). Courtesy of American Welding Society.

temperature range significantly. For the alloy shown in Figure 11.10, for instance, solidification starts at 1385.6°C (liquidus temperature), $L \rightarrow \gamma + NbC$ takes place at 1355.2°C and $L \rightarrow \gamma + Laves$ at a considerably lower temperature, 1248.2°C (31). The solidification temperature range is 137.4°C. As shown in Figure 11.11, the maximum crack length in Varestraint testing increases with increasing solidification temperature range (17).

Before leaving the subject of eutectic reactions, two stainless steel welds will be used here to illustrate the two eutectic reactions along the solidification path (35). The first weld, containing 0.48 wt % Nb and 0.010 wt % C, is an

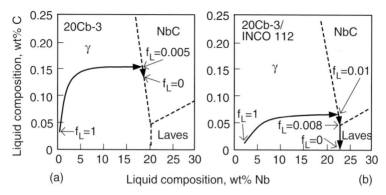

Figure 11.12 Solidification paths (solid lines) of Nb-stabilized austenitic stainless steels: (*a*) 20Cb-3; (*b*) 20Cb-3–INCO 112 composite. Reprinted from DuPont (35). Courtesy of American Welding Society.

autogenous gas–tungsten arc weld of a Nb-stabilized austenitic stainless steel 20Cb-3. To increase the Nb content, a relatively large composite weld was made in alloy 20Cb-3 with an INCO 112 filler wire containing 3.81 wt % Nb and then machined flat at the top surface for further welding. The second weld, containing 2.20 wt % Nb and 0.014 wt % C, is an autogenous gas–tungsten arc weld within the 20Cb-3–INCO 112 composite weld.

The solidification paths of the two welds are shown in Figures 11.12. As mentioned previously in Chapter 6, the arrows in the solidification path indicate the directions in which the liquid composition changes as temperature decreases. Here, solidification of the 20Cb-3 weld initiates by a primary $L \rightarrow \gamma$ reaction, as shown in Figure 11.12a. Because of the relatively high C–Nb ratio of the alloy, the interdendritic liquid becomes enriched in C until the γ/NbC twofold saturation line is reached. Here, the $L \rightarrow \gamma + NbC$ reaction takes over. However, since the fraction of liquid is already very small ($f_L = 0.005$), solidification is soon over ($f_L = 0$) at about 1300°C before the $L \rightarrow \gamma + Laves$ reaction takes place. In contrast, the lower C–Nb ratio of the 20Cb-3–INCO 112 composite weld caused the interdendritic liquid to become more highly enriched in Nb as shown in Figure 11.12b. The solidification path barely intersects the γ/NbC twofold saturation line ($f_L = 0.01$) before it reaches the γ/Laves twofold saturation line ($f_L = 0.008$). The $L \rightarrow \gamma + NbC$ reaction takes only briefly, and the $L \rightarrow \gamma + Laves$ reaction takes over until solidification is complete ($f_L = 0$) at about 1223°C, which is significantly lower than 1300°C.

11.2.2   Amount and Distribution of Liquid during Terminal Solidification

A. Amount of Liquid   Figures 11.13a–d show the effect of composition on the solidification cracking sensitivity of several aluminum alloys (36–41). Figure 11.13e shows the crack sensitivity in pulsed laser welding of Al–Cu

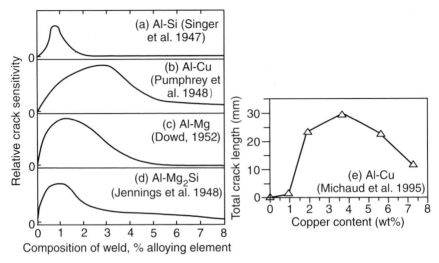

Figure 11.13   Effect of composition on crack sensitivity of some aluminum alloys. (*a–d*) From Dudas and Collins (40). (*e*) Reprinted from Michaud et al. (41).

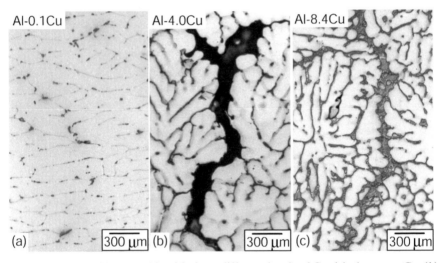

Figure 11.14   Aluminum welds with three different levels of Cu: (*a*) almost no Cu; (*b*) 4% Cu; (*c*) 8% Cu.

alloys (41). Figure 11.14*a* shows an aluminum weld with little Cu (alloy 1100 gas–metal arc welded with filler 1100), and there is no evidence of cracking. Figure 11.14*b* shows a crack in an aluminum weld with about 4% Cu (alloy 2219 gas–metal arc welded with filler 1100). Figure 11.14*c* shows a crack healed by the eutectic liquid in an aluminum weld with about 8% Cu (alloy 2219 gas–metal arc welded with filler 2319 plus extra Cu).

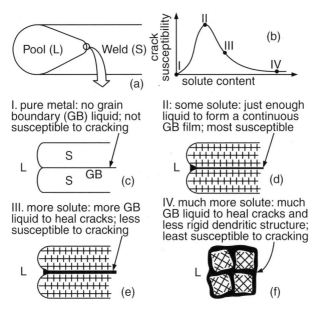

Figure 11.15 Effect of composition on crack susceptibility: (*a*) weld; (*b*) crack susceptibility curve; (*c*) pure metal; (*d*) low solute; (*e*) more solute; (*f*) much more solute.

As shown in Figure 11.13, the maximum crack sensitivity occurs somewhere between pure aluminum and highly alloyed aluminum (say no less than 6 wt % solute). The presence of a maximum in the crack susceptibility composition curve (Figure 11.13) is explained qualitatively with the help of Figure 11.15. Pure aluminum is not susceptible to solidification cracking because there is no low-melting-point eutectic present at the grain boundary to cause solidification cracking. In highly alloyed aluminum, on the other hand, the eutectic liquid between grains can be abundant enough to "heal" incipient cracks (3). Somewhere in between these two composition levels, however, the amount of liquid between grains can be just large enough to form a thin, continuous grain boundary film to make the materials rather susceptible to solidification cracking but without extra liquid for healing cracks. A fine equiaxed dendritic structure with abundant liquid between grains (Figure 11.15f) can deform more easily under stresses than a coarse columnar dendritic structure (Figure 11.15e) and thus has a lower susceptibility to cracking.

B. Calculation of Fraction Liquid Dupont et al. (31–35) calculated the fraction of the liquid (f_L) as a function of distance (x) within the mushy zone in Nb-bearing superalloys and austenitic stainless steels. The f_L–x curve provides the quantitative information for the amount and distribution of interdendritic liquid in the mushy zone. The two stainless steel welds discussed previously will be considered here, namely, the weld in alloy 20Cb-3 and the weld in the 20Cb-3–INCO 112 composite weld. According to DuPont et al. (35), Nb, Si, and C are treated as solutes in these alloys and, as an approximation, the remaining elements in the solid solution with γ are treated as the "γ-solvent."

Assume that the slopes of the liquidus surface with respect to Nb, Si, and C, that is, $m_{L,Nb}$, $m_{L,Si}$, and $m_{L,C}$, respectively, are constant. Equation (6.2) can be extended to find the temperature of a liquid on the liquidus surface T as follows:

$$T = T_m + m_{L,Nb}C_{L,Nb} + m_{L,Si}C_{L,Si} + m_{L,C}C_{L,C} \tag{11.1}$$

Assuming negligible solid diffusion but complete liquid diffusion for Nb and Si, the following equations can be written based on the Scheil equation [Equation (6.9)]:

$$C_{L,Nb} = C_{0,Nb}f_L^{(k_{Nb}-1)} \tag{11.2}$$

$$C_{L,Si} = C_{0,Si}f_L^{(k_{Si}-1)} \tag{11.3}$$

Since diffusion in solid and liquid is fast for a small interstitial solute such as C, the following equation can be written based on the equilibrium lever rule [Equation (6.6)]:

$$C_{L,C} = \frac{C_{0,C}}{f_L + k_C(1-f_L)} \tag{11.4}$$

Note that the equilibrium partition ratios k_{Nb}, k_{Si}, ans k_C have been assumed constant in Equations (11.2)–(11.4) for simplicity. Inserting these equations into Equation (11.1) yields the following relationship between temperature and fraction liquid:

$$T = T_m + m_{L,Nb}C_{0,Nb}f_L^{(k_{Nb}-1)} + m_{L,Si}C_{0,Si}f_L^{(k_{Si}-1)} + m_{L,C}\left(\frac{C_{0,C}}{f_L + k_C(1-f_L)}\right) \tag{11.5}$$

The cooling rate (ε or GR) can be determined from the secondary dendrite spacing (d). According to Equation (6.20),

$$d = at_f^n = b(\varepsilon)^{-n} = b(GR)^{-n} \tag{11.6}$$

For instance, from the dendrite arm spacing of the weld metal the cooling rate GR is about 250°C/s. The growth rate at the weld centerline R is the welding speed 3 mm/s. As such, the temperature gradient G is about 83°C/mm.

Assuming that G is constant in the mushy zone and taking $x = 0$ at the liquidus temperature of the alloy T_L,

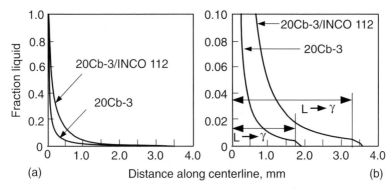

Figure 11.16 Fraction liquid as a function of distance within the mushy zone of 20Cb-3 and 20Cb-3–INCO 112 composite (a) and enlarged (b). Reprinted from DuPont (35). Courtesy of American Welding Society.

$$x = \frac{T_L - T}{G} \tag{11.7}$$

which can be used to find the temperature T at any distance x.

Equations (11.5) and (11.7) can be combined to determine how the liquid fraction f_L varies with distance x within the mushy zone, and the results are shown in Figure 11.16. The liquid fraction drops rapidly with distance near the pool boundary but slowly further into the mushy zone. Also, the mushy zone is significantly wider in the 20Cb-3–INCO 112 composite weld than in the 20Cb-3 weld due to the significantly larger solidification temperature range of the former. This explains why the former is more susceptible to solidification cracking.

C. Liquid Distribution   As shown previously in Figure 11.11 for Nb-bearing superalloys, the alloys with a narrower solidification temperature range are less susceptible to solidification cracking. In fact, some of these alloys have a wide solidification temperature range just like the more susceptible alloys. As in the more crack-susceptible alloys, the L → γ+ Laves reaction follows the L → γ+ NbC reaction during the terminal stage of solidification. However, as shown in Figure 11.17 for one of these less susceptible alloys, the amount of terminal liquid undergoing the L → γ+ Laves reaction in these alloys is small and remains isolated. This type of morphology, unlike the continuous grain boundary liquid undergoing the L → γ+ Laves reaction in the more crack-susceptible alloys, should be more resistant to crack propagation throughout the mushy zone. Since an isolated L → γ+ Laves reaction does not really contribute to solidification cracking, it should not have to be included in the solidification temperature range, and the lower bound of the effective solidification temperature range should more accurately be represented by the L → γ+ NbC reaction (31, 34).

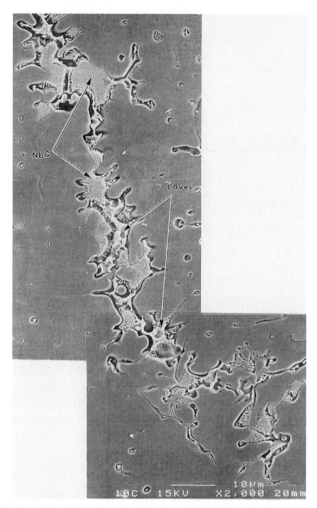

Figure 11.17 Scanning electron micrographs showing morphology of γ-NbC and γ-Laves constituents in solidification cracks of a Nb-bearing superalloy. Reprinted from DuPont et al. (34).

11.2.3 Ductility of Solidifying Weld Metal

The less ductile a solidifying weld metal is, the more likely it will crack during solidification. Nakata and Matsuda (16) used the transverse Varestraint test (Figure 11.6b) to determine the so-called *ductility curve*, as illustrated in Figure 11.18. At any given strain the ductility curve ranges from the liquidus temperature T_L to the temperature at the tip of the longest crack. To construct the curve, a strain of ε_1 is applied during welding, and the maximum crack length is examined after welding (Figure 11.18a). From the temperature distribution

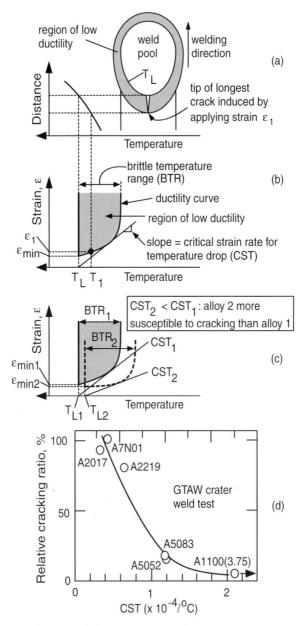

Figure 11.18 Ductility of solidifying weld metal: (*a*) temperature distribution (temperature increases from right to left); (*b*) ductility curve; (*c*) ductility curves with different crack susceptibility; (*d*) effect of CST on cracking. (*d*) From Nakata and Matsuda (16).

along the weld centerline measured with a thermocouple during welding, the temperature T_1 at the tip of the longest crack and hence the point (T_1, ε_1) can be determined (Figure 11.18b). By repeating the experiment for various applied strains and finding the corresponding crack tip temperatures, the ductility curve can be determined. The maximum crack length and hence the temperature range of the ductility curve first increase with increasing applied strain but then level off as the applied strain increases further. The widest temperature range covered by the ductility curve is called the *brittle temperature range* (BTR). The weld metal is "brittle" in the sense that it is much less ductile in this temperature range than either the weld pool or the completely solidified weld metal. The minimum strain required to cause cracking is called ε_{min}. The slope of the tangent to the ductility curve is called the *critical strain rate for temperature drop* (CST), that is, the critical rate at which the strain varies with temperature drop.

In general, the lower ε_{min}, the greater BTR or the smaller CST is, the greater the susceptibility to solidification cracking (Figure 11.18c). According to Nakata and Matsuda (16), CST best correlates with the cracking susceptibility of the weld metal (Figure 11.18d). It should be pointed out, however, that the values of CST in Figure 11.18d were not determined exactly as described above. Because the ε_{min} values were too small to be determined, a slow-bending transverse Varestraint test had to be used to modify the ductility curve.

For a given material cracking can be avoided if the strain rate for temperature drop can be kept below the critical value, that is, if $-d\varepsilon/dT < $ CTS. Mathematically, $-d\varepsilon/dT = (\partial\varepsilon/\partial t)/(-\partial T/\partial t)$, where t is time. The strain rate ($\partial\varepsilon/\partial t$) consists of both the self-induced tensile strain in the solidifying weld metal whose contraction is hindered by the adjacent base metal and the augmented strain if it is applied. $-\partial T/\partial t$ is the cooling rate, for instance, 150°C/s. In Varestraint testing an augmented strain is applied essentially instantaneously, and the high $\partial\varepsilon/\partial t$ and hence $-d\varepsilon/dT$ easily cause cracking. If the augmented strain were applied very slowly on a solidifying weld metal whose self-induced strain rate during solidification is low, $\partial\varepsilon/\partial t$ can be small enough to keep $-d\varepsilon/dT$ below the CST, and the weld metal can solidify without cracking. Cracking can also be avoided if the cooling rate $-\partial T/\partial t$ is increased dramatically to reduce $-d\varepsilon/dT$ to below the CST. Yang et al. (42) avoided solidification cracking in GTAW of 2024 aluminum sheets by directing liquid nitrogen behind the weld pool to increase the cooling rate. They also showed, with finite-element modeling of heat flow and thermal stresses, that the condition $-d\varepsilon/dT < $ CTS can be achieved by cooling the area right behind the weld pool.

Before leaving the subject of the ductility curve, the relationship between the BTR and the solidification temperature range needs to be discussed. Nakata and Matsuda (16) observed in aluminum alloys such as alloys 2017, 2024, and 2219 that the cooling curve showed a clear solidification temperature range. A distinct point of arrest (a short flat region) in the curve corresponding to the formation of eutectics was observed as well as a discontinuity at the liquidus temperature corresponding to the formation of the aluminum-rich solid. For these alloys the BTR was found to be the same as the solidification temperature range, that is, T_L to T_E. In other words, the region of low ductility behind the weld pool corresponds to the mushy zone discussed in previous chapters. On the other hand, for aluminum alloys such as alloys 5052, 5083, and 6061, the cooling curve did not show a distinct point of arrest, and the inflection point in the curve had to be taken as the "nominal" solidus temperature T_S. For these alloys the BTR was found significantly (about 20%) larger than the nominal solidification temperature range of T_L to T_S and was thus a more realistic representation of the nonequilibrium solidification temperature range of T_L to T_E. Presumably, eutectics still formed below T_S but not in quantities enough to show a distinct point of arrest in the cooling curve. It is not clear if *differential thermal analysis* was tried, which is known to detect eutectic temperatures well.

In the case of steels and stainless steels, the solidification temperature range is relatively narrow, and impurities such as P and S are found to enlarge the BTR, presumably by forming eutectics with a low melting point (16).

11.2.4 Primary Solidification Phase

For austenitic stainless steels the susceptibility to solidification cracking is much lower when the primary solidification phase is δ-ferrite rather than austenite (43–45). As the ratio of the Cr equivalent to the Ni equivalent increases, the primary solidification phase changes from austenite to δ-ferrite, and cracking is reduced. As shown by Takalo et al. (44) in Figure 11.19a for arc welding, this change occurs at $Cr_{eq}/Ni_{eq} = 1.5$. Similarly, as shown by Lienert (45) in Figure 11.19b for pulsed laser welding, this change occurs in the Cr_{eq}/Ni_{eq} range of 1.6–1.7. In both cases $Cr_{eq} = Cr + 1.37Mo + 1.5Si + 2Nb + 3Ti$ and $Ni_{eq} = Ni + 0.3Mn + 22C + 14.2N + Cu$. As discussed previously (Chapter 9), under high cooling rates in laser or electron beam welding, a weld metal that normally solidifies as primary ferrite can solidify as primary austenite because of undercooling, and this is consistent with the change occurring at a higher $Cr_{eq}–Ni_{eq}$ ratio in pulsed laser welding.

In general, austenitic stainless steels containing 5–10% δ-ferrite are significantly more resistant to solidification cracking than fully austenitic stainless steels (46). As mentioned previously in Chapter 9, ferrite contents significantly greater than 10% are usually not recommended, for the corrosion resistance will be too low (47–50). Furthermore, upon exposure to elevated temperatures (600–850°C), δ-ferrite can transform to brittle σ-ferrite and impair the mechanical properties of austenitic stainless steels, unless the ferrite content is kept low (51).

It is generally believed that, since harmful impurities such as sulfur and phosphorus are more soluble in δ-ferrite than in austenite (see Table 11.1), the concentration of these impurities at the austenite grain boundaries, and thus their damaging effect on solidification cracking, can be reduced if δ-ferrite is present in significant amounts (43, 52, 53). In addition, it is also believed that

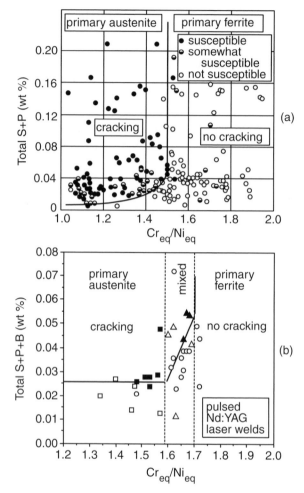

Figure 11.19 Solidification crack susceptibility of austenitic stainless steels: (*a*) arc welds; (*b*) laser welds. (*a*) From Takalo et al. (44). (*b*) Reprinted from Lienert (45).

TABLE 11.1   Solubility of Sulfur and Phosphorus in Ferrite and Austenite (wt%)

	In δ-Ferrite	In Austenite
Sulfure	0.18	0.05
Phosphorus	2.8	0.25

Source: Borland and Younger (52).

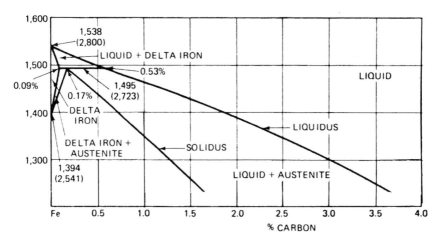

Figure 11.20   Iron–carbon phase diagram.

when δ-ferrite is the primary solidification phase, the substantial boundary area between δ-ferrite and austenite acts as a sink for sulfur and phosphorus. This decreases the concentration of such impurities at the austenite grain boundaries and, therefore, reduces solidification cracking (43, 54–57). Hull (46) suggests that the propagation of solidification cracks in cast austenitic stainless steels is halted by δ-ferrite due to the fact that the ferrite–austenite interface energy is lower than the austenite–austenite interface energy (58, 59). Solidification cracks stopped by δ-ferrite have also been observed in 309 stainless steel welds by Brooks et al. (60).

In addition to austenitic stainless steels, the primary solidification phase can also affect the solidification cracking susceptibility of carbon steels. According to the iron–carbon phase diagram shown in Figure 11.20, when the carbon content is greater than 0.53, austenite becomes the primary solidification phase and solidification cracking becomes more likely. In fact, the wider solidification temperature range at a higher carbon content further increases the potential for solidification cracking.

11.2.5   Surface Tension of Grain Boundary Liquid

The effect of the amount and distribution of the grain boundary liquid on the solidification cracking of weld metals has been described earlier in this section. The higher the surface tension of the grain boundary liquid, the larger its dihedral angle is. Figures 11.21a and b show the dihedral angle and distribution of the grain boundary liquid (61). Figure 11.21c shows the effect of the dihedral angle on the susceptibility to solidification cracking in several aluminum alloys evaluated in autogenous spot GTAW (16). As shown, except for alloy 1100, which is essentially pure aluminum and thus not susceptible to cracking, the

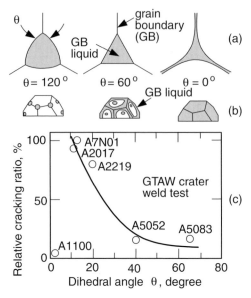

Figure 11.21 Grain boundary liquid: (*a*) dihedral angle; (*b*) distribution of liquid at grain boundary; (*c*) effect of dihedral angle on solidification cracking. (*c*) From Nakata and Matsuda (16).

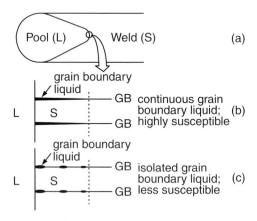

Figure 11.22 Effect of grain boundary liquid morphology on crack susceptibility: (*a*) weld; (*b*) continuous; (*c*) isolated.

susceptibility decreases with increasing dihedral angle of the grain boundary liquid.

The effect of the surface tension of the grain boundary liquid on solidification cracking is further depicted in Figure 11.22. If the surface tension between the solid grains and the grain boundary liquid is very low, a liquid film will

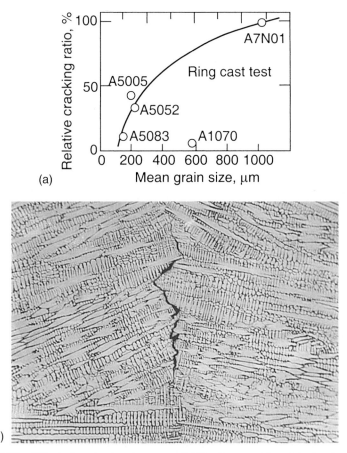

(a)

(b)

Figure 11.23 Effect of grain structure on solidification cracking: (*a*) aluminum alloys; (*b*) centerline cracking in a coarse-grain 310 stainless steel weld. (*a*) From Nakata and Matsuda (16). (*b*) From Kou and Le (65).

form between the grains and the solidification cracking susceptibility is high. On the other hand, if the surface tension is high, the liquid phase will be globular and will not wet the grain boundaries. Such discontinuous liquid globules do not significantly reduce the strength of the solid network and, therefore, are not as harmful. For example, FeS forms films at the grain boundaries of steels while MnS forms globules. Due to its globular morphology and higher melting point, MnS has been known to be far less harmful than FeS.

11.2.6 Grain Structure of Weld Metal

Fine equiaxed grains are often less susceptible to solidification cracking than coarse columnar grains (16, 62–64), as shown in Figure 11.23*a* (16) for several aluminum alloys. Alloy A1070 is not susceptible to solidification cracking because it is essentially pure aluminum. Fine equiaxed grains can deform to accommodate contraction strains more easily, that is, it is more ductile, than columnar grains. Liquid feeding and healing of incipient cracks can also be more effective in fine-grained material. In addition, the grain boundary area is much greater in fine-grained material and, therefore, harmful low-melting-point segregates are less concentrated at the grain boundary.

It is interesting to note that, due to the steep angle of abutment between columnar grains growing from opposite sides of the weld pool, welds made with a teardrop-shaped weld pool tend to be more susceptible to centerline solidification cracking than welds made with an elliptical-shaped weld pool. A steep angle seems to favor the head-on impingement of columnar grains growing from opposite sides of the weld pool and the formation of the continuous liquid film of low-melting-point segregates at the weld centerline. As a result, centerline solidification cracking occurs under the influence of transverse contraction stresses. Centerline cracking is often observed in welding. Figure 11.23b is an example in an autogenous gas–tungsten arc weld of a 310 stainless steel made with a teardrop-shaped weld pool (65).

11.3 MECHANICAL FACTORS

11.3.1 Contraction Stresses

So far, the metallurgical factors of weld solidification cracking have been described. But without the presence of stresses acting on adjacent grains during solidification, no cracking can occur. Such stresses, as already mentioned, can be due to thermal contraction or solidification shrinkage or both. Austenitic stainless steels have relatively high thermal expansion coefficients (as compared with mild steels) and, therefore, are often prone to solidification cracking.

Aluminum alloys have both high thermal expansion coefficients and high solidification shrinkage (5). As a result, solidification cracking can be rather serious in some aluminum alloys, especially those with wide solidification temperature ranges.

11.3.2 Degree of Restraint

The degree of restraint of the workpiece is another mechanical factor of solidification cracking. For the same joint design and material, the greater the restraint of the workpiece, the more likely solidification cracking will occur.

Figure 11.24 illustrates the effect of workpiece restraint on solidification cracking (66). As shown, solidification cracking occurred in the second (left-hand-side) weld of the inverse "T" joint due to the fact that the degree

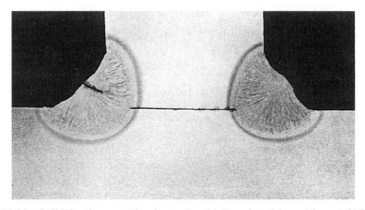

Figure 11.24 Solidification cracking in steel weld. Reprinted from Linnert (66). Courtesy of American Welding Society.

of restraint increased significantly after the first (right-hand-side) weld was made.

11.4 REDUCING SOLIDIFICATION CRACKING

11.4.1 Control of Weld Metal Composition

Weld metals of crack-susceptible compositions should be avoided. In autogenous welding no filler metal is used, and the weld metal composition is determined by the base-metal composition. To avoid or reduce solidification cracking, base metals of susceptible compositions should be avoided. When a base metal of a crack-susceptible composition has to be welded, however, a *filler metal* of a proper composition can be selected to adjust the weld metal composition to a less susceptible level.

A. Aluminum Alloys   According to Figure 11.13, an Al–3% Cu alloy can be rather crack susceptible during welding. If the Cu content of the base metal is raised to above 6%, solidification cracking can be significantly reduced. In fact, 2219 aluminum, one of the most weldable Al–Cu alloys, contains 6.3% Cu. When a filler metal is used, the weld metal composition is determined by the composition of the base metal, the composition of the filler metal, and the dilution ratio. The dilution ratio, as mentioned previously, is the ratio of the amount of the base metal melted to the total amount of the weld metal. Again using Al–3% Cu as an example, the weld metal Cu content can be increased by using 2319 filler metal, which is essentially an Al–6.3% Cu alloy. If the joint design and heat input are such that the dilution ratio is low, the weld metal copper content can be kept sufficiently high to avoid solidification cracking. Figure 11.25 shows the approximate dilution in three typical joint designs (40).

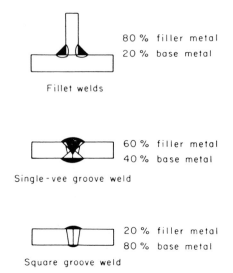

Figure 11.25 Approximate dilution of three weld joints by base metal in aluminum welding. Reprinted from Dudas and Collins (40). Courtesy of American Welding Society.

TABLE 11.2   Guid to Choice of Filler Metals for Minimizing Solidification Cracking in Welds of High-Strength Aluminum Alloys

Base Metals	7000 (Al–Zn–Mg–Cu)	7000 (Al–Zn–Mg)	6000 (Al–Mg–Si)	5000 (Al–Mg)	2000 (Al–Cu)
2000 (Al–Cu)	NR[a]	NR	NR	NR	4043 4145 2319
5000 (Al–Mg)	5356	5356 5556 5183	5356 5556 5183	5356 5556 5183	—[b]
6000 (Al–Mg–Si)	5356	5356 5556 5183	4043 4643 5356	—	—
7000 (Al–Zn–Mg)	5356	5356 5556	—	—	—
7000 (Al–Zn–Mg–Cu)	5356 5556	—	—	—	—

[a] NR, not recommend.
[b] Charts that recommend filler choice for many applications are available from filler metal suppliers.
Source: Dudas and Collins (40).

Table 11.2 is a guide to choice of filler metals for minimizing solidification cracking in welds of high-strength aluminum alloys (40). Experimental data such as those in Figure 11.26 from solidification cracking testing by ring casting can also be useful (67, 68).

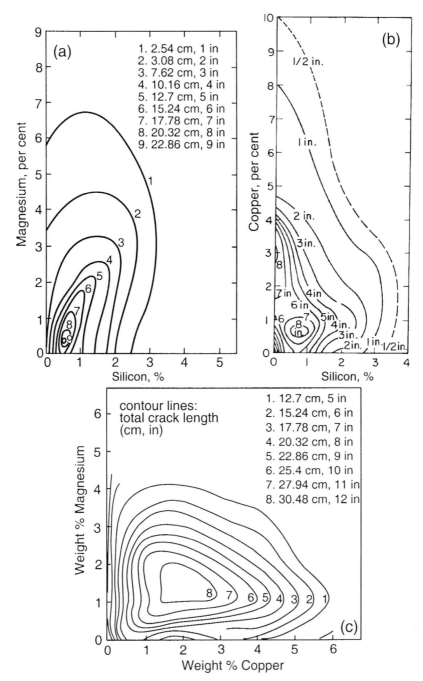

Figure 11.26 Solidification cracking susceptibility of aluminum alloys: (*a*) Al–Mg–Si; (*b*) Al–Cu–Si (*c*) Al–Mg–Cu. (*a*, *b*) Modified from Jennings et al. (67). (*c*) Modified from Pumphrey and Moore (68).

Minor alloying elements have also been found to affect the solidification cracking susceptibility of aluminum alloys. For example, the Fe–Si ratio has been found to significantly affect the solidification cracking susceptibility of 3004 and other Al–Mg alloys (69, 70); therefore, proper control of the content of minor alloying elements can be important in some materials.

B. Carbon and Low-Alloy Steels   The weld metal manganese content can significantly affect solidification cracking. It is often kept high enough to ensure the formation of MnS rather than FeS. This, as described previously, is because the high melting point and the globular morphology of MnS tend to render sulfur less detrimental. Figure 11.27 shows the effect of the *Mn–S ratio* on the solidification cracking tendency of carbon steels (71). At relatively low carbon levels the solidification cracking tendency can be reduced by increasing the Mn–S ratio. However, at higher carbon levels (i.e., 0.2–0.3% C), increasing the Mn–S ratio is no longer effective (72). In such cases lowering the weld metal carbon content, if permissible, is more effective.

One way of lowering the weld metal carbon content is to use *low-carbon electrodes*. In fact, in welding high-carbon steels one is often required to make the first bead (i.e., the root bead) with a low-carbon electrode. This is because, as shown in Figure 11.28, the first bead tends to have a higher dilution ratio and a higher carbon content than subsequent beads (73). A high carbon content is undesirable because it promotes not only the solidification cracking of the weld metal but also the formation of brittle martensite and, hence, the postsolidification cracking of the weld metal. Therefore, in welding steels of very high carbon contents (e.g., greater than 1.0% C), extra steps should be taken to avoid introducing excessive amounts of carbon from the base metal into the weld metal. As shown in Figure 11.29, one way to achieve this is to

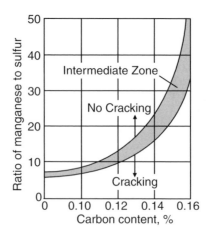

Figure 11.27   Effect of Mn–S ratio and carbon content on solidification cracking susceptibility of carbon steel weld metal. Reprinted from Smith (71).

Figure 11.28 Schematic sketch of multipass welding. Note that the root pass has the highest dilution ratio. From Jefferson and Woods (73).

310 Stainless steel

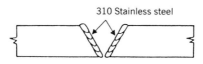

Figure 11.29 Buttering the groove faces of very high carbon steel with a 310 stainless wire steel before welding. From Jefferson and Woods (73).

"butter" the groove faces of the base metal with austenitic stainless steel (such as 25–20 stainless) electrodes before welding (73). In welding cast irons, pure nickel electrodes have also been used for *buttering*. In any case, the surface layers remain in the ductile austenitic state, and the weld can then be completed either with stainless electrodes or with other cheaper electrodes.

C. Nb-Bearing Stainless Steels and Superalloys The *C–Nb ratio* of the weld metal can affect the susceptibility to solidification cracking significantly (17, 31–35). A high C–Nb ratio can reduce solidification cracking by avoiding the low-temperature $L \rightarrow \gamma +$ Laves reaction, which can widen the solidification temperature range.

D. Austenitic Stainless Steels As mentioned previously, it is desirable to maintain the weld *ferrite content* at a level of 5–10% in order to obtain sound welds. The quantitative relationship between the weld ferrite content and the weld metal composition in austenitic stainless steels has been determined by Schaeffler (74), DeLong (75), Kotecki (76, 77), Balmforth and Lippold (78), and Vitek et al. (79, 80). These constitution diagrams have been shown previously in Chapter 9. Alloying elements are grouped into ferrite formers (Cr, Mo, Si, and Cb) and austenite formers (Ni, C, and Mn) in order to determine the corresponding chromium and nickel equivalents for a given alloy.

Example: Consider the welding of a 1010 steel to a 304 stainless steel. For convenience, let us assume the dilution ratio is 30%, half from 304 stainless steel and half from 1010 steel, as shown in Figure 11.30. Estimate the ferrite content and the solidification cracking susceptibility of a weld made with a type 310 electrode that contains 0.12% carbon and a weld made with a type 309 ELC (extra low carbon) electrode that contains 0.03% carbon.

For the weld made with the 310 stainless steel electrode, the weld metal composition can be calculated as follows:

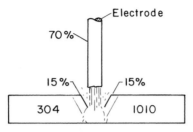

Figure 11.30 Welding 304 stainless steel to 1010 carbon steel.

Element	Electrode	×70%	Type 304	×15%	1010 Steel	×15%	Weld Metal
Cr	26.0	18.2	18.0	2.7	0	0	20.9
Ni	21.0	14.7	8.0	1.2	0	0	15.9
C	0.12	0.084	0.05	0.0075	0.10	0.015	0.1065
Mn	1.75	1.23	2.0	0.30	0.4	0.06	1.59
Si	0.4	0.28	—	—	0.2	0.03	0.31

According to the WRC-1992 diagram shown in Figure 9.11, the chromium and nickel equivalents of the weld metal are as follows:

$$\text{Chromium equivalent} = 20.9$$
$$\text{Nickel equivalent} = 15.9 + 35 \times 0.1065 = 19.6$$

Based on the diagram, the weld metal is fully austenitic and, therefore, is susceptible to solidification cracking. If the 309 ELC electrode is used, the weld metal composition can be calculated as follows:

Element	Electrode	×70%	Type 304	×15%	1010 Steel	×15%	Weld Metal
Cr	24	16.8	18.0	2.7	0	0	19.5
Ni	13	9.1	8.0	1.2	0	0	10.3
C	0.03	0.021	0.05	0.0075	0.10	0.015	0.0435
Mn	1.98	1.39	2.0	0.30	0.4	0.06	1.75
Si	0.4	0.28	0	0	0.2	0.03	0.31

Therefore, the chromium and nickel equivalents of the weld metal are

$$\text{Chromium equivalent} = 19.5$$
$$\text{Nickel equivalent} = 10.3 + 35 \times 0.0435 = 11.8$$

According to the WRC-1992 diagram, the weld metal now is austenitic with a ferrite number of about 8 and, therefore, should be much more resistant to solidification cracking.

It should be emphasized that neither the constitution diagrams nor the magnetic measurements of the weld metal ferrite content (such as those determined by Magne–Gage readings) reveal anything about the weld metal solidification. In fact, the primary solidification phase and the quantity of δ-ferrite at high temperatures (i.e., during solidification) are more important than the amount of ferrite retained in the room temperature microstructure in determining the sensitivity to solidification cracking (81). Also, the amount of harmful impurities such as sulfur and phosphorus should be considered in determining the weldability of a material; a material with a higher ferrite content can be more susceptible to solidification cracking than another material with a lower ferrite content if the impurity level of the former is also higher. The cooling rate during solidification is another factor that the constitution diagrams fail to recognize (Chapter 9).

11.4.2   Control of Solidification Structure

A. Grain Refining   As mentioned previously, welds with coarse columnar grains are often more susceptible to solidification cracking than those with fine equiaxed grains. It is, therefore, desirable to grain refine the weld metal. In fact, both 2219 aluminum and 2319 filler metal are designed in such a way that they not only have a non-crack-sensitive copper content but also have small amounts of grain refining agents such as Ti and Zr to minimize solidification cracking. Dudas and Collins (40) produced grain refining and eliminated solidification cracking in a weld made with an Al–Zn–Mg filler metal by adding small amounts of Zr to the filler metal. Garland (82) has grain refined welds of aluminum–magnesium alloys by vibrating the arc during welding, thereby reducing solidification cracking.

B. Magnetic Arc Oscillation   Magnetic arc oscillation has been reported to reduce solidification cracking in aluminum alloys, HY-80 steel, and iridium alloys (83–88). Kou and Le (86–88) studied the effect of magnetic arc oscillation on the grain structure and solidification cracking of aluminum alloy welds. As already shown in Figure 7.30, transverse arc oscillation at low frequencies can produce alternating columnar grains. Figure 11.31 demonstrates that this type of grain structure can be effective in reducing solidification cracking (86). As illustrated in Figure 11.32, columnar grains that reverse their orientation at regular intervals force the crack to change its direction periodically, thus making crack propagation difficult (87). Figure 11.33 shows the effect of the oscillation frequency on the crack susceptibility of 2014 aluminum welds (86). As shown, a minimum crack susceptibility exists at a rather low frequency, where alternating grain orientation is most pronounced. This frequency can vary with the welding speed.

As shown in Figure 11.34, arc oscillation at much higher frequencies than 1 Hz, though ineffective in 2014 aluminum, is effective in 5052 aluminum (87).

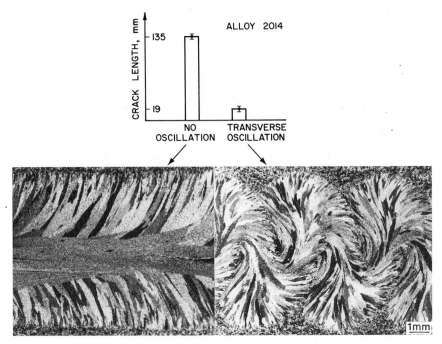

Figure 11.31   Effect of transverse arc oscillation (1 Hz) on solidification cracking in gas–tungsten arc welds of 2014 aluminum. Reprinted from Kou and Le (86).

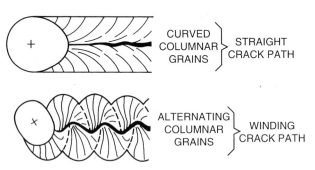

Figure 11.32   Schematic sketches showing effect of arc oscillation on solidification cracking. From Kou and Le (87).

As shown in Figure 11.35, this is because grain refining occurs in alloy 5052 welds at high oscillation frequencies (88). Heterogeneous nucleation is believed to be mainly responsible for the grain refining, since a 0.043 wt % Ti was found in the 5052 aluminum used.

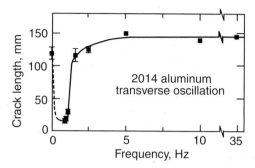

Figure 11.33   Effect of oscillation frequency on solidification cracking in gas–tungsten arc welds of 2014 aluminum. From Kou and Le (86).

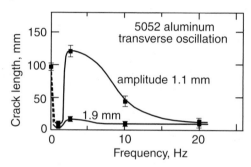

Figure 11.34   Effect of arc oscillation frequency on the solidification cracking in gas–tungsten arc welds of 5052 aluminum. From Kou and Le (87).

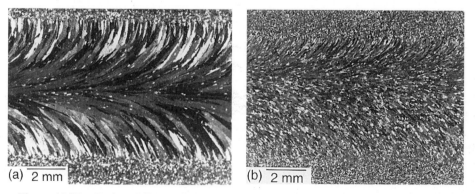

Figure 11.35   Grain structure of gas–tungsten arc welds of 5052 aluminum: (*a*) no arc oscillation; (*b*) 20 Hz transverse arc oscillation. Reprinted from Kou and Le (87).

11.4.3   Use of Favorable Welding Conditions

A. Reducing Strains   As mentioned previously in Chapter 1, the use of high-intensity heat sources (electron or laser beams) significantly reduces the distortion of the workpiece and hence the thermally induced strains. Less restraint and proper preheating of the workpiece can also help reduce strains.

Dyatlov and Sidoruk (89) and Nikov (90) found that preheating the workpiece decreased the magnitude of strains induced by welding. Sekiguchi and Miyake (91) reduced solidification cracking in steel plates by preheating. Hernandez and North (92) positioned additional torches behind and along the side of the welding head and inhibited solidification cracking in aluminum alloy sheets. It was suggested that the local heating decreased the amount of plastic straining resulting from the welding operation and produced a less stressful situation behind the weld pool.

B. Improving Weld Geometry   The weld bead shape can also affect solidification cracking (93). When a *concave* single-pass fillet weld cools and shrinks, the outer surface is stressed in tension, as show in Figure 11.36*a*. The outer surface can be considered as being pulled toward the toes and the root. However, by making the outer surface *convex*, as shown in Figure 11.36*b*, pulling toward the root actually compresses the outer surface and offsets the tension caused by pulling toward the toes. Consequently, the tensile stresses along the outer surface are reduced, and the tendency for solidification cracking to initiate from the outer surface is lowered. It should be pointed out, however, that excessive convexity can produce stress concentrations and induce fatigue cracking (Section 5.3) or hydrogen cracking (Section 17.4) at the toes. In multiple-pass welding, as illustrated in Figure 11.37, solidification cracking can also initiate from the weld surface if the weld passes are too wide and concave (93).

The weld width-to-depth ratio can also affect solidification cracking. As depicted in Figure 11.38, deep narrow welds with a low width-to-depth ratio can be susceptible to weld centerline cracking (93) This is because of the steep angle of the abutment between columnar grains growing from opposite sides

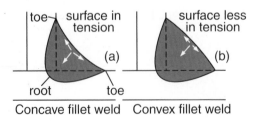

Figure 11.36   Effect of weld bead shape on state of stress at center of outer surface: (*a*) concave fillet weld; (*b*) convex fillet weld. Modified from Blodgett (93).

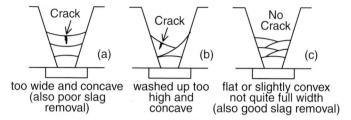

too wide and concave   washed up too   flat or slightly convex
(also poor slag   high and   not quite full width
removal)   concave   (also good slag removal)

Figure 11.37 Effect of weld bead shape on solidification in multipass weld: (*a*) concave; (*b*) concave; (*c*) convex. Modified from Blodgett (93).

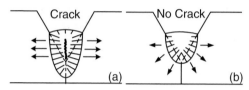

Figure 11.38 Effect of weld depth–width ratio on centerline cracking: (*a*) ratio too high; (*b*) ratio correct. Modified from Blodgett (93).

of the weld pool. This type of cracking is often observed in deep and narrow welds produced by EBW or SAW.

11.5   CASE STUDY: FAILURE OF A LARGE EXHAUST FAN

Figure 11.39*a* shows a schematic sketch of a large six-bladed exhaust fan for drawing chemical mist from a chemical processing chamber (94). The hub of the fan had a diameter of about 178 mm (7 in.) and a washer-shaped vertical fin. Two flat-bar spokes were welded to the fin at each 60° interval and provided the support for the fan blades. The fan acted only as the drawing force to pull the mist from the process area; it did not come into direct contact with the mist. A "scrubber" was positioned between the fan and the process unit to remove the chemical from the exhausted air mass. The spokes and blades were fabricated from type 316 austenitic stainless steel bars and plates, and the electrodes used for welding were type E316 covered electrodes. The fan measured about 1.5 m (5 ft) between blade tips, weighed about 320 kg (700 lb), and rotated at 1200 rpm in service. It failed after 12 weeks service time.

Figure 11.39*b* shows the fractured fillet weld joining the underside of the blade to the supporting spoke. The weld metal was nonmagnetic, that is, containing little δ-ferrite, and was therefore susceptible to solidification cracking. Note the jagged nature of the microfissures following the columnar grain structure above the arrowhead, as compared with the smooth crack typical of fatigue failure below the arrowhead. In fact, the fracture surface of

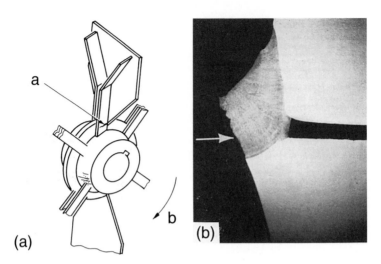

Figure 11.39 Failure of a large welded exhaust fan (94): (*a*) location of failure; (*b*) failed component.

the failed spoke member exhibited the "clam shell" markings typical of fatigue failure.

In summary, the fully austenitic weld metal produced by type E316 electrodes suffered from solidification cracking, which in turn initiated fatigue cracking and led to final failure. Type 309 or 308 electrodes, which contain more δ-ferrite to resist solidification cracking, could have been used to avoid solidification cracking in the weld metal (94).

REFERENCES

1. Kou, S., and Kanevsky, Y., unpublished research, Carnegie-Mellon University, Pittsburgh, PA, 1980.
2. Davies, G. J., and Garland, J. G., *Int. Metal Rev.*, **20:** 83, 1975.
3. Lees, D. C. G., *J. Inst. Metals*, **72:** 343, 1946.
4. Singer, A. R. E., and Jennings, P. H., *J. Inst. Metals*, **73:** 273, 1947.
5. Flemings, M. C., *Solidification Processing*, McGraw-Hill, New York, 1974.
6. Bishop, H. F., Ackerlind, C. G., and Pellini, W. S., *Trans. AFS*, **65:** 247, 1957.
7. Borland, J. C., *Br. Weld. J.*, **7:** 508, 1960.
8. Li, L., and Messler, Jr., R. W., *Weld. J.*, **78:** 387s, 1999.
9. Matsuda, F., Nakagawa, H., and Sorada, K., *Trans. JWSI*, **11:** 67, 1982.
10. Matsuda, F., Nakagawa, H., Kohomoto, H., Honda, Y., and Matsubara, Y., *Trans. JWSI*, **12:** 73, 1983.
11. Houldcroft, P. T., *Br. Weld. J.*, **2:** 471, 1955.
12. Liptax, J. A., and Baysinger, F. R., *Weld. J.*, **47:** 173s, 1968.

13. Garland, J. G., and Davies, G. J., *Metals Const. Br. Weld. J.*, vol. 1, p. 565, December 1969.

14. Rogerson, J. H., Cotterell, B., and Borland, J. C., *Weld. J.*, **42:** 264s, 1963.

15. Savage, W. F., and Lundin, C. D., *Weld. J.*, **44:** 433s, 1965.

16. Nakata, K., and Matsuda, F., *Trans. JWRI*, **24:** 83, 1995.

17. Krysiak, K. F., Grubb, J. F., Pollard, B., and Campbell, R. D., in *Welding, Brazing, and Soldering*, Vol. 6, ASM International, Materials Park, OH, December 1993, p. 443.

18. DuPont, J. N., Michael, J. R., and Newbury, B. D., *Weld. J.*, **78:** 408s, 1999.

19. Stout, R. D., in *Weldability of Steels*, 3rd ed., Eds. R. D. Stout and W. D. Doty, Welding Research Council, New York, 1978, p. 252.

20. Messler, R. W., Jr., *Principles of Welding, Processes, Physics, Chemistry and Metallurgy*, Wiley, New York, 1999, pp. 557–589.

21. Cieslak, M. J., *Weld. J.*, **66:** 57s, 1987.

22. David, S. A., and Woodhouse, J. J., *Weld. J.*, **66:** 129s, 1987.

23. Nelson, T. W., Lippold, J. C., Lin, W., and Baeslack III, *Weld. J.*, **76:** 110s, 1997.

24. Goodwin, G. M., *Weld. J.*, **66:** 33s, 1987.

25. Hondros, E. D., and Seah, M. P., *Int. Metals Rev.*, **222:** 12, 1977.

26. Pease, G. R., *Weld. J.*, **36:** 330s, 1957.

27. Canonico, D. A., Savage, W. F., Werner, W. J., and Goodwin, G. M., *paper presented at Effects of Minor Elements* on the Weldability of High-Nickel Alloys, Welding Research Council, NY, 1969, p. 68.

28. Savage, W. F., Nippes, E. F., and Goodwin, G. M., *Weld. J.*, **56:** 245s, 1977.

29. Kah, D. H., and Dickinson, D. W., *Weld. J.*, **60:** 135s, 1981.

30. *Principles and Technology of the Fusion Welding of Metals*, Vol. 1, Mechanical Engineering Publishing Co., Peking, China, 1979 (in Chinese).

31. DuPont, J. N., Robino, C. V., and Marder, A. R., *Weld. J.*, **77:** 417s, 1998.

32. DuPont, J. N., Robino, C. V., Marder, A. R., Notis, M. R., and Michael, J. R., *Metall. Mater. Trans. A.*, **29A:** 2785, 1998.

33. DuPont, J. N., Robino, C. V., and Marder, A. R., *Acta Mater.*, **46:** 4781, 1998.

34. DuPont, J. N., Robino, C. V., and Marder, A. R., *Sci. Technol. Weld. Join.*, **4:** 1, 1999.

35. DuPont, J. N., *Weld. J.*, **78:** 253s, 1999.

36. Singer, A. R. E., and Jennings, P. H., *J. Inst. Metals*, **73:** 197, 1947.

37. Pumphrey, W. I., and Lyons, J. V., *J. Inst. Metals*, **74:** 439, 1948.

38. Dowd, J. D., *Weld. J.*, **31:** 448s, 1952.

39. Jennings, P. H., Singer, A. R. E., and Pumphrey, W. L., *J. Inst. Metals*, **74:** 227, 1948.

40. Dudas, J. H., and Collins, F. R., *Weld. J.*, **45:** 241s, 1966.

41. Michaud, E. J., Kerr, H. W., and Weckman, D. C., in *Trends in Welding Research*, Eds. H. B. Smartt, J. A. Johnson, and S. A. David, ASM International, Materials Park OH, June 1995, p. 154.

42. Yang, Y. P., Dong, P., Zhang, J., and Tian, X., *Weld. J.*, **79:** 9s, 2000.

43. Gueussier, A., and Castro, R., *Rev. Metall.*, **57:** 117, 1960.

44. Takalo, T., Suutala, N., and Moisio, T., *Metall. Trans.*, **10A:** 1173, 1979.

45. Lienert, T. J., in *Trends in Welding Research*, Eds. J. M. Vitek, S. A. David, J. A. Johnson, H. B. Smartt, and T. DebRoy, ASM International, Materials Park, OH, June 1998, p. 726.

46. Hull, F. C., *Weld. J.*, **46:** 399s, 1967.

47. Baeslack, W. A., Duquette, D. J., and Savage, W. F., *Corrosion*, **35:** 45, 1979.

48. Baeslack, W. A., Duquette, D. J., and Savage, W. F., *Weld. J.*, **57:** 175s, 1978.

49. Baeslack, W. A., Duquette, D. J., and Savage, W. F., *Weld. J.*, **58:** 83s, 1979.

50. Manning, P. G., Duquette, D. J., and Savage, W. F., *Weld. J.*, **59:** 260s, 1980.

51. Thomas, R. G., *Weld. J.*, **57:** 81s, 1978.

52. Borland, J. C., and Younger, R. N., *Br. Weld. J.*, **7:** 22, 1960.

53. Thomas, R. D. Jr., *Metals Prog.*, **70:** 73, 1956.

54. Medovar, B. I., *Avtomaticheskaya Svarka*, **6:** 3, 1953.

55. Curran, R. M., and Rankin, A. W., *Weld. J.*, **34:** 205, 1955.

56. Borland, J. C., *Br. Weld. J.*, **7:** 558, 1960.

57. Rollason, E. C., and Bystram, M. C. T., *J. Iron Steel Inst.*, **169:** 347, 1951.

58. Smith, C. S., *Interphase Imperfections in Nearly Perfect Crystals*, Wiley, New York, 1952, pp. 377–401.

59. Taylor, J. W., *J. Inst. Metals*, **86:** 456, 1958.

60. Brooks, J. A., Thompson, A. W., and Williams, J. C., in *Physical Metallurgy of Metal Joining*, Eds. R. Kossowsky and M. E. Glickstein, Metallurgical Society of AIME, Warrendale, PA, 1980, p. 117.

61. Smith, C. R., *Trans. AIME*, **175:** 15, 1948.

62. Matsuda, F., Nakata, K., Tsukamoto, K., and Arai, K., *Trans. JWRI*, **12:** 93, 1983.

63. Matsuda, F., Nakata, K., Tsukamoto, K., and Uchiyama, T., *Trans. JWRI*, **13:** 57, 1984.

64. Dvornak, M. J., Frost, R. H., and Olson, D. L., *Weld. J.*, **68:** 327s, 1989.

65. Kou, S., and Le, Y., unpublished research, Carnegie-Mellon University, Pittsburgh, PA, 1981.

66. Linnert, G. E., *Welding Metallurgy*, Vol. 2, 3rd ed., American Welding Society, New York, 1967.

67. Jennings, P. H., Singer, A. R. E., and Pumphrey, W. I., *J. Inst. Metals*, **74:** 227, 1948.

68. Pumphrey, W. I., and Moore, D. C., *J. Inst. Metals*, **73:** 425, 1948.

69. Savage, W. F., Nippes, E. F., and Varsik, J. D., *Weld. J.*, **58:** 45s, 1979.

70. Evancho, J. W., and Baker, C. L., *Weld Crack Susceptibility of Al-Mg Alloys*, ALCOA Report, ALCOA Technology Center, ALCOA Center, PA, March 1980.

71. Smith, R. B., in *Welding, Brazing, and Soldering*, Vol. 6, ASM International, Materials Park, OH, December 1993, p. 642.

72. Borland, J. C., *Br. Weld. J.*, **8:** 526, 1961.

73. Jefferson, T. B., and Woods, G., *Metals and How to Weld Them*, James Lincoln Arc Welding Foundation, Cleveland, OH, 1961.

74. Schaeffler, A. L., *Metal. Prog.*, **56:** 680, 1949.

75. DeLong, W. T., *Weld. J.*, **53:** 273s, 1974.

76. Kotecki, D. J., *Weld. J.*, **78:** 180s, 1999.

77. Kotecki, D. J., *Weld. J.*, **79:** 346s, 2000.

78. Balmforth, M. C., and Lippold, J. C., *Weld. J.*, **79:** 339s, 2000.

79. Vitek, J. M., Iskander, Y. S., and Oblow, E. M., *Weld. J.*, **79:** 33s, 2000.

80. Vitek, J. M., Iskander, Y. S., and Oblow, E. M., *Weld. J.*, **79:** 41s, 2000.

81. Cieslak, M. J., and Savage, W. F., *Weld. J.*, **60:** 131s, 1981.

82. Garland, J. G., *Metal Const. Br. Weld. J.*, **21:** 121, 1974.

83. Tseng, C., and Savage, W. F., *Weld. J.*, **50:** 777, 1971.

84. David, S. A., and Liu, C. T., in *Grain Refinement in Castings and Welds*, Eds. G. J. Abbaschian and S. A. David, Metals Society of AIME, Warrendale, PA, 1983, p. 249.

85. Scarbrough, J. D., and Burgan, C. E., *Weld. J.*, **63:** 54, 1984.

86. Kou, S., and Le, Y., *Metall. Trans.*, **16A:** 1887, 1985.

87. Kou, S., and Le, Y., *Metall. Trans.*, **16A:** 1345, 1985.

88. Kou, S., and Le, Y., *Weld. J.*, **64:** 51, 1985.

89. Dyatlov, V. I., and Sidoruk, V. S., *Autom. Weld.*, **3:** 21, 1966.

90. Nikov, N. Y., *Weld. Production*, **4:** 25, 1975.

91. Sekiguchi, H., and Miyake, H., *J. Jpn. Weld. Soc.*, **6**(1): 59, 1975.

92. Hernandez, I. E., and North, T. H., *Weld. J.*, **63:** 84s, 1984.

93. Blodgett, O. W., *Weld. Innovation Q.*, **2**(3): 4, 1985.

94. *Fatigue Fractures in Welded Constructions*, Vol. 11, International Institute of Welding, London, 1979.

FURTHER READING

1. Davies, G. J., and Garland, J. G., *Int. Metall. Rev.*, **20:** 83, 1975.

2. Flemings, M. C., *Solidification Processing*, McGraw-Hill, New York, 1974.

PROBLEMS

11.1   Compare the solidification temperature range and fraction eutectic of Al–3.0Cu with those of Al–6.0Cu. For simplicity, assume Scheil's equation is a valid approximation and both the solidus and liquidus lines are straight in the Al–Cu system. $C_{SM} = 5.65$, $C_E = 33$, $T_E = 548°C$, and $T_m = 660°C$ (pure Al).

11.2   Centerline cracking is often observed in deep-penetration electron or laser beam welds. Explain why.

11.3   Fillet welds of 5052 Al (essentially Al–2.5Mg) are made with 5556 filler (essentially Al–5.1Mg). What are the approximate dilution ratio and weld metal composition? Is the weld metal susceptible to solidification cracking?

11.4   Solidification cracking in 2014 aluminum sheets can be reduced significantly by using low-frequency transverse arc oscillation. Low-frequency circular and longitudinal arc oscillations, however, are less effective. Explain why.

11.5 Low-frequency arc pulsation during autogenous GTAW of aluminum alloys, such as 6061 and 2014, is often found detrimental rather than beneficial in controlling solidification cracking. Explain why.

11.6 A structural steel has a nominal composition of 0.16 C, 1.4 Mn, 0.4 Si, 0.022 S, and 0.016 P. Because of macrosegregation of carbon during ingot casting, some of the steel plates produced contained as much as 0.245% C. Severe solidification cracking was reported in welds of these steel plates. Explain why. The problem was solved by using a different filler wire. Comment on the carbon content of the new filler wire.

11.7 Consider welding 1018 steel to 304 stainless steel by GMAW (dilution normally between 30 and 40%). Assume approximately equal contribution to weld metal dilution from each side of the joint. Will the weld metal be susceptible to solidification cracking if ER308L Si is used as the electrode? Will the weld metal be dangerously close to the martensite boundary? 1018 steel: 0.18C–0.02Cr–0.03Ni–0.01Mo–0.07Cu–0.01N; 304 stainless steel: 0.05C–18.30Cr–8.80Ni–0.05Mo–0.08Cu–0.04N; ER308L Si: 0.03C–19.90Cr–10.20Ni–0.21Mo–0.19Cu–0.06N.

11.8 Repeat the previous problem but with ER309L Si as the electrode. ER309L Si: 0.02C–24.10Cr–12.70Ni–0.13Mo–0.16Cu–0.05N.

11.9 Stainless steels normally considered resistant to solidification cracking in arc welding based on the constitution diagram (such as Schaeffler or WRC 1992), for instance, 304L, 316L, and 321Mo, can become rather susceptible in laser or electron beam welding. Explain why.

11.10 It has been observed in 1100 aluminum alloy that the Ti or Zr content of the alloy, up to about 0.5 wt %, can significantly affect its susceptibility to solidification cracking. Explain how Ti or Zr affects weld metal solidification cracking of the alloy and why.

11.11 It has been reported that autogenous gas–tungsten arc welds of Invar (Fe–36 wt % Ni) are rather susceptible to solidification cracking. Explain why based on a constitution diagram. The addition of Ti (e.g., 0.5–1.0%) has been found to change the Mn sulfide films along the grain boundaries to tiny Ti sulfide particles entrapped between dendrite arms. What is the effect of the Ti addition on solidification cracking?

PART III
The Partially Melted Zone

12 Formation of the Partially Melted Zone

Severe liquation can occur in the partially melted zone during welding. Several fundamental liquation-related phenomena are discussed in this chapter, including liquation mechanisms, solidification of the grain boundary (GB) liquid, and the resultant GB segregation.

12.1 EVIDENCE OF LIQUATION

The partially melted zone (PMZ) is the area immediately outside the weld metal where liquation can occur during welding. Figure 12.1a shows a portion of the PMZ in a gas–metal arc weld of a 6061 aluminum made with a 4145 filler metal. The presence of dark-etching GBs along the fusion boundary in this micrograph is an indication of GB liquation. The microstructure inside the white rectangle is enlarged in Figure 12.1b. The liquated and resolidified material along the GB consists of a dark-etching eutectic GB and a lighter etching α (Al-rich) band along the GB.

Figure 12.2 shows the PMZ microstructure in alloy 2219, which is essentially Al–6.3Cu (1). The liquated and resolidified material along the GB consists of a dark-etching eutectic GB and a light-etching α band along the GB. The α bands here appear lighter than those in Figure 12.1b because of the use of a different etching solution. As shown, the large dark-etching eutectic particles within grains are surrounded by a light-etching α phase, thus indicating that liquation also occurs within grains.

The formation of the PMZ in 2219 aluminum is explained in Figure 12.3. As shown in the phase diagram (Figure 12.3a), the composition of alloy 2219 is $C_0 = 6.3\%$ Cu. As shown by the thermal cycles (Figure 12.3b), the material at position b is heated up to between the eutectic temperature T_E and the liquidus temperature T_L during welding. Therefore, the material becomes a solid-plus-liquid mixture $(\alpha + L)$, that is, it is partially melted. The material at position a is completely melted while that at position c is not melted at all. A similar explanation for the formation of the PMZ, in fact, has been given previously in Figure 7.12.

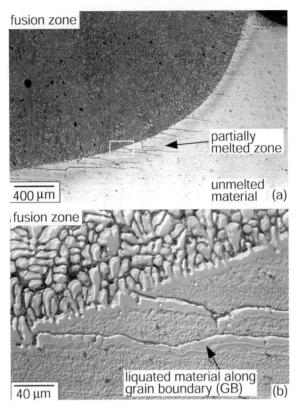

Figure 12.1   PMZ microstructure of gas–metal arc weld in 6061 aluminum made with 4145 aluminum filler wire. Rectangular area in (*a*) enlarged in (*b*).

12.2   LIQUATION MECHANISMS

Huang et al. (1–5) have conducted a series of studies on liquation and solidification in the PMZ of aluminum welds. Figure 12.4 shows five different PMZ liquation mechanisms. The phase diagram shown in Figure 12.4*a* is similar to the Al-rich side of the Al–Cu phase diagram. Here, A_xB_y is an intermetallic compound, such as Al_2Cu in the case of Al–Cu alloys. Alloy C_1 is within the solubility limit of the α phase, C_{SM}, and alloy C_2 is beyond it. In the as-cast condition both alloys C_1 and C_2 usually consist of an α matrix and the eutectic $\alpha + A_xB_y$ along GBs and in between dendrite arms.

12.2.1   Mechanism 1: A_xB_y Reacting with Matrix

This mechanism is shown in Figure 12.4*b*. Here alloy C_2, for instance, alloy 2219, consists of an α matrix and A_xB_y particles at any temperature up to the

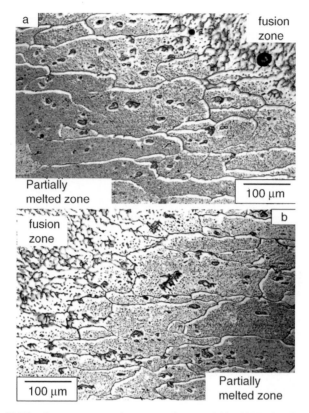

Figure 12.2 PMZ microstructure of gas–metal arc weld in 2219 aluminum made with 2319 aluminum filler wire: (*a*) left of weld; (*b*) right of weld. Reprinted from Huang and Kou (1). Courtesy of American Welding Society.

eutectic temperature T_E, regardless of the heating rate to T_E during welding. At T_E liquation is initiated by the eutectic reaction $A_x B_y + \alpha \rightarrow L$.

The liquation mechanism for alloy 2219 can be explained with the help of Figure 12.5. The base metal contains large and small θ (Al_2Cu) particles both within grains and along GBs (Figure 12.5*a*), as shown by the SEM image at the top. At the border line between the PMZ and the base metal (Figure 12.5*b*), the material is heated to the eutectic temperature T_E. Regardless of the heating rate θ (Al_2Cu) particles are always present when T_E is reached. Consequently, liquation occurs by the eutectic reaction $\alpha + \theta \rightarrow L_E$, where L_E is the liquid of the eutectic composition C_E. Upon cooling, the eutectic liquid solidifies into the eutectic solid without composition changes and results in eutectic particles and some GB eutectic (Figure 12.5*e*).

Above T_E, that is, inside the PMZ, liquation intensifies. The α matrix surrounding the eutectic liquid dissolves and the liquid increases in volume

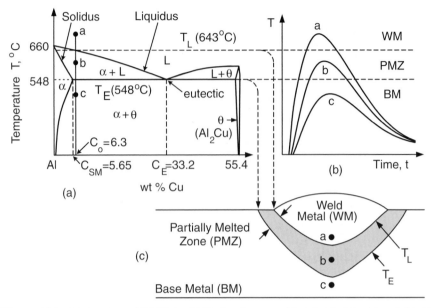

Figure 12.3 Formation of PMZ in 2219 aluminum weld: (*a*) Al-rich side of Al–Cu phase diagram; (*b*) thermal cycles; (*c*) transverse cross section.

(Figure 12.5*c*). This causes the liquid composition to change from eutectic to hypoeutectic ($<C_E$), as dictated by the phase diagram (Figure 12.4*a*). Upon cooling, the hypoeutectic liquid solidifies first as Cu-depleted α and last as Cu-rich eutectic when its composition increases to C_E, as also dictated by the phase diagram. This results in a band of Cu-depleted α along the GB eutectic and a ring of Cu-depleted α surrounding each large eutectic particle (Figure 12.5*f*), as shown in the optical micrograph at the bottom.

12.2.2   Mechanism 2: Melting of Eutectic

This mechanism is also shown in Figure 12.4*b*. Here alloy C_2, for instance, alloy 2219 in the as-cast condition, consists of an α matrix and the eutectic $\alpha + A_xB_y$ along GBs and in between dendrite arms. Regardless of the heating rate to T_E during welding, the eutectic is always present before T_E is reached. At T_E liquation is initiated by the melting of the eutectic.

12.2.3   Mechanism 3: Residual A_xB_y Reacting with Matrix

This mechanism is shown in Figure 12.4*c*. This is the well-known *constitutional liquation* mechanism proposed by Pepe and Savage (6, 7). Here alloy C_1 still contains residual A_xB_y particles just before reaching the eutectic temperature

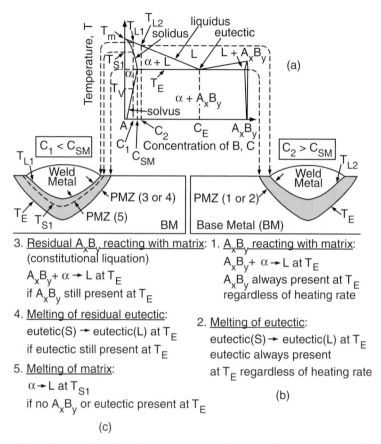

Figure 12.4 Five mechanisms for liquation in PMZ of aluminum alloys: (*a*) phase diagram; (*b*) two mechanisms for an alloy beyond the solid solubility limit (C_{SM}); (*c*) three mechanisms for an alloy within the solid solubility limit.

T_E. Examples of such A_xB_y include the titanium sulfide inclusion in 18% Ni maraging steel, Ni_2Nb Laves phase in Inconel 718, carbides in Ni-base superalloys, and several intermetallic compounds in aluminum alloys.

When alloy C_1 is heated very slowly to above the solvus temperature T_V, A_xB_y dissolves completely in the α matrix by solid-state diffusion, and the alloy becomes a homogeneous α solid solution of composition C_1. However, when alloy C_1 is heated rapidly to above T_V, as often the case in welding, A_xB_y does not have enough time to dissolve completely in the α matrix because solid-state diffusion is slow. That is, A_xB_y remains in the α matrix above T_V. Consequently, upon heating to the eutectic temperature T_E, the residual A_xB_y reacts with the surrounding α matrix and forms the liquid eutectic C_E at the

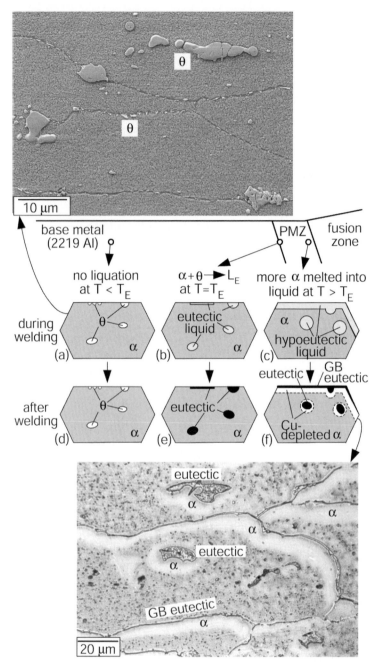

Figure 12.5 Microstructure evolution in PMZ of 2219 aluminum with SEM image of the base metal on the top and optical micrograph of the PMZ at the bottom. Reprinted from Huang and Kou (1). Courtesy of American Welding Society.

interface. Further heating to above T_E would allow additional time for further dissolution of A_xB_y and further formation of the liquid phase. Therefore, it is clear that localized melting should be possible with rapid heating rates at temperatures significantly below the equilibrium solidus temperature T_{S1}.

Figure 12.6 shows the microstructure of the PMZ of a resistance spot weld of 18-Ni maraging steel (6). This micrograph reveals the four stages leading to GB liquation. The first stage, visible at point A, shows a rodlike titanium sulfide inclusion beginning to form a thin liquid film surrounding the inclusion. The second stage is visible at point B, which is closer to the fusion boundary and hence experiences a higher peak temperature than at point A. As shown, liquation is more extensive and an elliptical liquid pool surrounds the remaining small gray inclusion. Still closer to the fusion boundary, at point C, no more solid inclusion remains in the liquid pool, and penetration of GBs by the liquid phase is evident along the GBs intersecting with the liquid phase. Finally, at position D, GB penetration by the liquid phase is so extensive that the GBs are liquated.

Constitutional liquation has also been observed in several nickel-base alloys (8–11), such as Udimet 700, Waspaloy, Hastelloy X, and Inconel 718, and in 347 stainless steel (12). Constitutional liquation can be initiated by the interaction between the matrix and particles of carbide or other intermetallic compounds. Examples include M_6C in Hastelloy X, MC carbide in Udimet 700, Waspaloy and Inconel 718, and Ni_2Nb Laves phase in Inconel 718. Figure 12.7 shows the PMZ of an Inconel 718 weld (13). Constitutional liquation occurs by the eutectic reaction between the Ni_2Nb Laves phase and the nickel matrix.

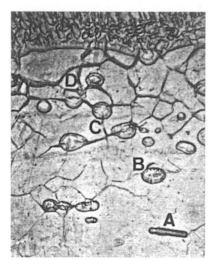

Figure 12.6 PMZ of an electric resistance spot weld of 18-Ni maraging steel showing constitutional liquation. The fusion zone is at the top. Magnification 385×. Reprinted from Pepe and Savage (6). Courtesy of American Welding Society.

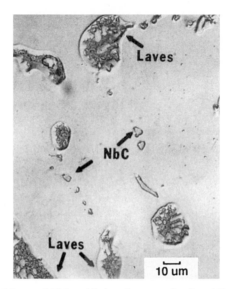

Figure 12.7   PMZ of Inconel 718 weld showing constitutional liquation due to Laves eutectic reaction. Reprinted from Kelly (13).

Constitutional liquation alone, however, is not enough to cause the liquid to penetrate most GBs in the PMZ. For this to occur in Ni-base alloys, GB migration is also needed. Figure 12.8 shows schematically the formation of GB films in the PMZ due to the simultaneous occurrence of constitutional liquation and GB migration (6). The microstructure at location d_0 is representative of the as-received plate. At location d_1, significant grain growth occurs above the effective grain coarsening temperature. Meanwhile, some of the moving GBs intersect with the solute-rich pools formed by constitutional liquation, thus allowing the solute-rich liquid to penetrate these GBs. At location d_2 more grain growth occurs and sufficient solute-rich liquid penetrates the GBs and forms GB films. These GBs are pinned due to the wetting action of the films. No further grain growth is expected until either the solute-rich liquid phase is dissipated by homogenization or the local temperature decreases to below the effective solidus of the solute-rich liquid to cause it to solidify. If insufficient time is available to dissipate the solute-rich liquid GB films before the local temperature decreases to below the effective solidus of the liquid, the liquid GB films will solidify as a solute-rich GB network. A "ghost" GB network will thus remain fixed when grain growth resumes, as shown by the dashed lines at location d_3. Grain growth will continue until either an equilibrium GB network is formed or the temperature decreases to below the effective grain coarsening temperature. Figure 12.9 shows the ghost GB network near the fusion boundaries of gas–tungsten arc welds of 18-Ni maraging steel (6) and a Ni-base superalloy 690 (14).

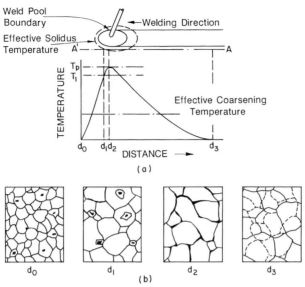

Figure 12.8 Schematic representation of constitutional liquation and formation of ghost GB network. Reprinted from Pepe and Savage (6). Courtesy of American Welding Society.

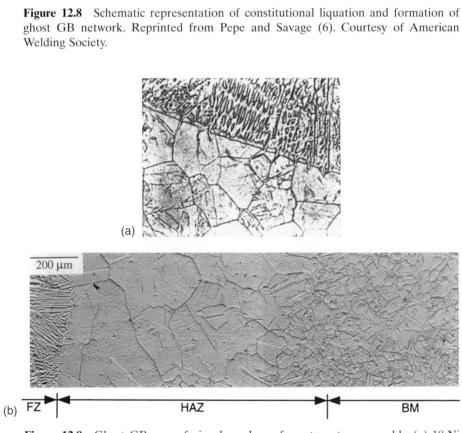

Figure 12.9 Ghost GBs near fusion boundary of gas–tungsten arc welds: (*a*) 18-Ni maraging steel; magnification 125×. Reprinted from Pepe and Savage (6). Courtesy of American Welding Society. (*b*) Ni-base superalloy 690. Reprinted from Lee and Kuo (14).

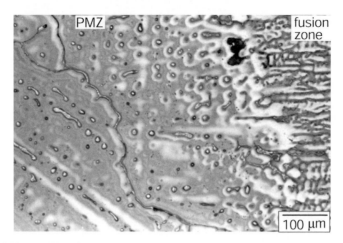

Figure 12.10 PMZ microstructure of gas–metal arc weld in as-cast Al–4.5Cu. Reprinted from Huang et al. (4).

12.2.4 Mechanism 4: Melting of Residual Eutectic

This mechanism is also shown in Figure 12.4c. Here alloy C_1, for instance, an as-cast Al–4.5Cu alloy, still contains the residual eutectic $\alpha + A_xB_y$ along GBs and in between dendrite arms when the eutectic temperature T_E is reached. If the alloy is heated very slowly to above the solvus temperature T_V, the eutectic can dissolve completely in the α matrix by solid-state diffusion. However, if it is heated rapidly to above T_V, as in welding, the eutectic does not have enough time to dissolve completely in the α matrix because solid-state diffusion takes time. Consequently, upon further heating to the eutectic temperature T_E, the residual eutectic melts and becomes liquid eutectic. Above T_E the surrounding α phase dissolves in the liquid and the liquid becomes hypoeutectic. Upon cooling, the hypoeutectic liquid solidifies first as solute-depleted α and last as solute-rich eutectic when its composition increases to C_E. Figure 12.10 shows the PMZ of a gas–metal arc weld of an as-cast Al–4.5% Cu alloy (4). Liquation is evident at the prior eutectic sites along the GB and in between dendrite arms. A light-etching α band is present along the eutectic GB. Likewise, light-etching α rings surround the eutectic particles near the fusion boundary.

12.2.5 Mechanism 5: Melting of Matrix

This mechanism is also shown in Figure 12.4c. Here alloy C_1 contains neither A_xB_y particles nor the $\alpha + A_xB_y$ eutectic when the eutectic temperature T_E is reached. An alloy Al–4.5Cu solution heat treated before welding is an example. Slow heating to T_E can also cause complete dissolution of A_xB_y or eutectic in the α matrix before T_E, but this usually is not likely in welding. The

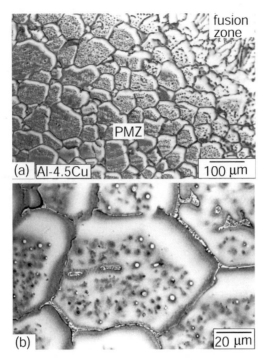

Figure 12.11 PMZ microstructure of gas–metal arc weld in cast Al–4.5Cu homogenized before welding (a) and magnified (b). Reprinted from Huang et al. (4).

PMZ ranges from the solidus temperature (T_{S1}) to the liquidus temperature (T_{L1}), instead of from the eutectic temperature to the liquidus temperature, as in all previous cases. Figure 12.11 shows the PMZ microstructure in an Al–4.5% Cu alloy solution heat treated and quenched before GMAW (4). Eutectic is present both along GBs and within grains in the PMZ even though the base metal is free of eutectic. Another example is a 6061-T6 aluminum containing submicrometer-size Mg_2Si precipitate, which is reverted in the α matrix before reaching T_E during GMAW (5).

12.2.6   Mechanism 6: Segregation-Induced Liquation

Lippold et al. (15) proposed a segregation-induced liquation mechanism for austenitic and duplex stainless steels. In such a mechanism the alloy and/or impurity elements that depress the melting point segregate to GBs, lower the melting point, and cause GB liquation. In other words, GB segregation takes place first and GB liquation next. This is opposite to all the other mechanisms described in this chapter, where liquation takes place first and segregation occurs during the solidification of the liquated material, as will be discussed subsequently. They suggested that such GB segregation can be

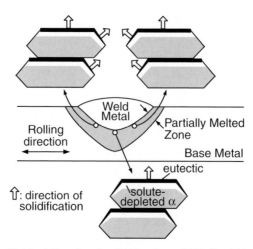

Figure 12.12   Directional solidification of GB liquid in PMZ.

caused by (i) equilibrium diffusion of atoms of the elements to GBs, (ii) GB sweeping of such atoms into migrating GBs during grain growth, and (iii) "pipeline" diffusion of such atoms along GBs in the fusion zone that are continuous across the fusion boundary into the PMZ. More details are available elsewhere (15, 16).

12.3   DIRECTIONAL SOLIDIFICATION OF LIQUATED MATERIAL

Huang et al. (1, 4) observed that the GB liquid has a tendency to solidify essentially *upward and toward the weld* regardless of its location with respect to the weld, as shown schematically in Figure 12.12. This directional solidification is caused by the high-temperature gradients toward the weld during welding. It has been generally accepted that the GB liquid between two neighboring grains solidifies from both grains to the middle between them. However, the micrographs in Figure 12.2 show that it solidifies from one grain to the other — in the direction upward and toward the weld. This, in fact, suggests GB migration in the same direction.

However, if the grains in the PMZ are very thin or very long, there may not be much GB area facing the weld. Consequently, solidification of the GB liquid is still directional but just upward (5).

12.4   GRAIN BOUNDARY SEGREGATION

As the GB liquid solidifies, solute atoms are rejected by the solid into the liquid if the equilibrium partition coefficient $k < 1$, such as in the phase diagram shown in Figure 12.3*a*. The GB liquid solidifies first as a solute-depleted α but finally as eutectic when the liquid composition reaches C_E.

Figure 12.13 depicts the GB segregation that develops during solidification of the GB liquid in the PMZ (5). Take alloy 2219 as an example. Let C_0 be the concentration of Cu in the base metal (6.3%). Theoretically, the measured concentration of the GB eutectic, C_e, is the eutectic composition C_E (33%) if the GB eutectic is normal and θ or Al$_2$Cu (55%) if it is divorced. A normal eutectic refers to a eutectic with a composite-like structure of $\alpha + \theta$. A divorced eutectic, on the other hand, refers to a eutectic that nucleates upon an existing α matrix and thus looks like θ alone. In practice, with EPMA (electron probe microanalysis) the value of C_e can be less than C_E if the GB is thinner than the volume of material excited by the electrons, that is, the surrounding α of low-Cu is included in the composition analysis. This volume depends on the voltage used in EPMA and the material being analyzed.

In the absence of back diffusion, the concentration of the element at the starting edge of the α strip should be kC_0, where k is the equilibrium partition ratio of the element. The dashed line shows the resultant GB segregation of the element. However, if back diffusion of the solute from the growth front into the solute-depleted α strip is significant, the concentration of the element at the starting edge of the α strip will be greater than kC_0, as the solid line indicates.

Severe GB segregation of alloying elements has been observed in the PMZ of gas–metal arc welds of alloys 2219, 2024, 6061, and 7075 (2, 5). Figure 12.14 shows that in alloy 2219 the composition varies from about 2% Cu at the starting edge of the α strip to 30% Cu at the GB eutectic (2). The 2–3% Cu content of the α strip is significantly lower than that of the base metal ($C_0 = 6.3\%$ Cu),

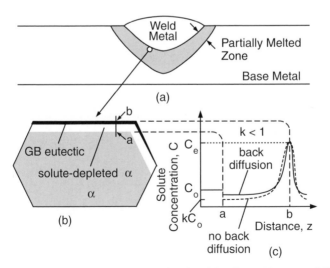

Figure 12.13   Grain boundary segregation in PMZ. From Huang and Kou (5).

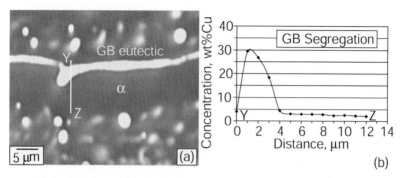

Figure 12.14 Grain boundary segregation in PMZ of 2219 aluminum weld: (*a*) electron micrograph; (*b*) composition profile. Reprinted from Huang and Kou (2). Courtesy of American Welding Society.

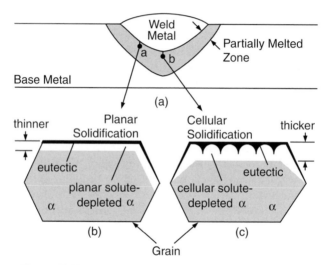

Figure 12.15 Solidification modes of GB liquid in PMZ.

thus confirming that the α strip is Cu depleted. The 2% Cu content at the starting edge of the α strip is higher than kC_0 ($1.07\% = 0.17 \times 6.3\%$), thus implying back diffusion of Cu into the α strip during GB solidification. The 30% Cu concentration at the GB is close to the 33% Cu composition of a normal eutectic.

12.5 GRAIN BOUNDARY SOLIDIFICATION MODES

As shown in Figures 12.1, 12.2, and 12.11, the α band along the eutectic GB is planar, namely, without cells or dendrites. This suggests that the solidification mode of the GB liquid is *planar*. The vertical temperature gradient G and the vertical growth rate R were determined in the PMZ of alloy 2219 and the upward solidification of the GB liquid was analyzed (3). The ratio G/R for planar GB solidification was found to be on the order of $10^5 \,^\circ\text{Cs/cm}^2$, which is close to that required for planar solidification of Al–6.3% Cu.

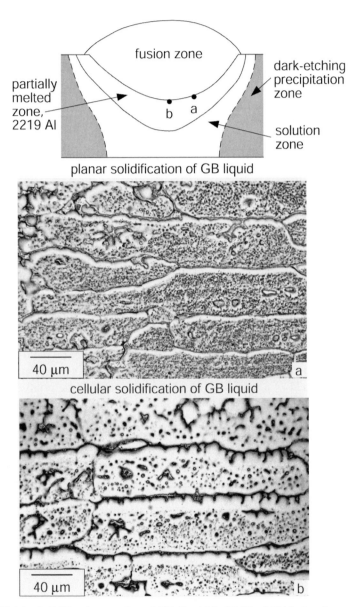

Figure 12.16 Solidification modes of GB liquid in PMZ of 2219 aluminum weld: (*a*) planar; (*b*) cellular. Reprinted from Huang and Kou (3). Courtesy of American Welding Society.

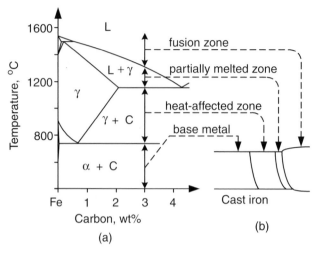

Figure 12.17 PMZ in a cast iron.

Although planar solidification of the GB liquid predominates in the PMZ, *cellular* solidification can also occur (3, 5). These cellular α bands share two common characteristics. First, they are often located near the weld bottom. Second, on average, they appear significantly thicker than the planar α bands nearby. These characteristics, depicted in Figure 12.15, can be because of the lower vertical temperature gradient G in the area or backfilling of liquid from the weld pool (3, 5). Since a thicker GB liquid has to solidify faster, the vertically upward solidification rate R is higher. The lower G/R in the area suggests a greater chance for constitutional supercooling and hence cellular instead of planar solidification. Figure 12.16 shows the PMZ microstructure in a gas–metal arc weld of alloy 2219 (3). However, it should be pointed out that planar solidification changes to cellular solidification gradually, and a thinner GB liquid may not have enough room for the transition to take place. Therefore, planar GB solidification may not necessarily mean that G/R is high enough to avoid cellular solidification.

12.6 PARTIALLY MELTED ZONE IN CAST IRONS

Figure 12.17 shows the PMZ in a cast iron weld, where γ, α, and C represent austenite, ferrite, and graphite, respectively. This area tends to freeze as white iron due to the high cooling rates and becomes very hard (17).

REFERENCES

1. Huang, C., and Kou, S., *Weld. J.*, **79:** 113s, 2000.
2. Huang, C., and Kou, S., *Weld. J.*, **80:** 9s, 2001.

3. Huang, C., and Kou, S., *Weld. J.*, **80:** 46s, 2001.
4. Huang, C., Kou, S., and Purins, J. R., in *Proceedings of Merton C. Flemings Symposium on Solidification and Materials Processing*, Eds. R. Abbaschian, H. Brody, and A. Mortensen, Minerals, Metals and Materials Society, Warrendale, PA, 2001, p. 229.
5. Huang, C., and Kou, S., *Weld. J.*, in press.
6. Pepe, J. J., and Savage, W. F., *Weld. J.*, **46:** 411s, 1967.
7. Pepe, J. J., and Savage, W. F., *Weld. J.*, **49:** 545s, 1970.
8. Owczarski, W. A., Duvall, D. S., and Sullivan, C. P., *Weld. J.*, **45:** 145s, 1966.
9. Duvall, D. S., and Owczarski, W. A., *Weld. J.*, **46:** 423s, 1967.
10. Savage, W. F., and Krantz, B. M., *Weld. J.*, **45:** 13s, 1966.
11. Thompson, R. G., and Genculu, S., *Weld. J.*, **62:** 337s, 1983.
12. Dudley, R., Ph.D. Thesis, Rensselaer Polytechnic Institute, Troy, NY, 1962.
13. Kelly, T. J., in *Weldability of Materials*, Eds. R. A. Patterson and K. W. Mahin, ASM International, Materials Park, OH, 1990, p. 151.
14. Lee, H. T., and Kuo, T. Y., *Sci. Technol. Weld. Join.*, **4:** 94, 1999.
15. Lippold, J. C., Baselack III, W. A., and Varol, I., *Weld. J.*, **71:** 1s, 1992.
16. Lippold, J. C., in *Technology and Advancements and New Industrial Applications in Welding*, Proceedings of the Taiwan International Welding Conference '98, Eds. C. Tsai and H. Tsai, Tjing Ling Industrial Research Institute, National University, Taipei, Taiwan, 1998.
17. Bushey, R. A., in *ASM Handbook*, Vol. 6: *Welding, Brazing and Soldering*, ASM International, Materials Park, OH, 1993, p. 708.

PROBLEMS

12.1 Sheets of alloy 2219 (Al–6.3Cu) 1.6mm thick are welded with the GTAW process using the following welding conditions: $I = 60$ A, $E = 10$ V, $U = 3$ mm/s. Suppose the arc efficiency is 70%. Estimate the PMZ width using the Adams two-dimensional equation (Chapter 2). Assume $T_L = 645°C$ and $T_E = 548°C$.

12.2 The heating rate in resistance spot welding can be much faster than that in GTAW. Is constitutional liquation expected to be more severe in the PMZ of 18-Ni-250 maraging steel in resistance spot welding or GTAW? Explain why.

12.3 It has been pointed out that Fe_3C tends to dissociate upon heating more easily than most other alloy carbides. Also, the diffusion rate of the interstitial solute, carbon, is much faster than substitutional solutes, for example, sulfur in the case of titanium sulfide inclusion. Do you expect a plain carbon eutectoid steel containing fine Fe_3C particles to be susceptible to constitutional liquation when heated to the eutectic temperature under normal heating rates during welding (say less than 500°C/s)? Why or why not?

12.4   It has been reported that by replacing Cb with Ta in Ni-base superalloy 718, the Laves eutectic reaction temperature increases from 1185 to 1225°C. Does the width of the PMZ induced by constitutional liquation (and liquation-induced cracking) increase or decrease in the alloy and why?

12.5   It has been suggested that in welding cast irons reducing the peak temperatures and the duration at the high temperatures is the most effective way to reduce PMZ problems. Is the use of a low-melting-point filler metal desirable in this respect? Is the use of high preheat temperature (to prevent the formation of martensite) desirable in this respect?

12.6   Based on the PMZ microstructure of the as-cast alloy Al–4.5Cu shown in Figure 12.10, what can be said about the solidification direction and mode of the grain boundary liquid and why?

12.7   Based on the PMZ microstructure of the homogenized alloy Al–4.5Cu shown in Figure 12.11, what can be said about the solidification direction and mode of the grain boundary liquid and why?

13 Difficulties Associated with the Partially Melted Zone

The partially melted zone (PMZ) can suffer from liquation cracking, loss of ductility, and hydrogen cracking. Liquation cracking, that is, cracking induced by grain boundary liquation in the PMZ during welding, is also called PMZ cracking or hot cracking. The causes of these problems and the remedies, especially for liquation cracking, will be discussed in this chapter. Liquation cracking and ductility loss are particularly severe in aluminum alloys. For convenience of discussion, the nominal compositions of several commercial aluminum alloys and filler wires are listed in Table 13.1 (1).

13.1 LIQUATION CRACKING

Figure 13.1 shows the longitudinal cross section at the bottom of a gas–metal arc weld of 2219 aluminum made with a filler wire of 1100 aluminum. The rolling direction of the workpiece is perpendicular to the plane of the micrograph. Liquation cracking in the PMZ is *intergranular* (2–6). Liquation cracking can also occur along the fusion boundary (3). The presence of a liquid phase at the intergranular fracture surface can be either evident (4) or unclear (5, 6).

13.1.1 Crack Susceptibility Tests

The susceptibility of the PMZ to liquation cracking can be evaluated using several different methods, such as Varestraint testing, circular-patch testing and hot-ductility testing, etc.

A. Varestraint Testing This is usually used for partially penetrating welds in plates (Chapter 11). In brief, the workpiece is subjected to augmented strains during welding, and the extent of cracking in the PMZ is used as the index for the susceptibility to liquation cracking (7–12).

B. Circular-Patch Testing This is usually used for fully penetrating welds in thin sheets. A relatively high restraint is imposed on the weld zone transverse to the weld (5). The fixture design shown in Figure 13.2 was used for liquation

TABLE 13.1   Nominal Compositions of Some Commercial Aluminum Alloys

	Si	Cu	Mn	Mg	Cr	Ni	Zn	Ti	Zr	Fe
1100	—	0.12								
2014	0.8	4.4	0.8	0.5						
2024	—	4.4	0.6	1.5						
2219	—	6.3	0.3	—	—	—	—	0.06	0.18	
2319	—	6.3	0.3	—	—	—	—	0.15	0.18	
4043	5.2									
5083	—	—	0.7	4.4	0.15					
5356	—	—	0.12	5.0	0.12	—	—	0.13		
6061	0.6	0.28	—	1.0	0.2					
6063	0.4	—	—	0.7						
6082	0.9	—	0.5	0.7	—	—	—	—	—	0.3
7002	—	0.75	—	2.5	—	—	3.5			
7075	—	1.6	—	2.5	0.23	—	5.6			

Source: Aluminum Association (1).

Figure 13.1   PMZ cracking in 2219 aluminum welded with filler metal 1100.

cracking test of aluminum welds (13). The specimen is sandwiched between two copper plates (the upper one having a big round opening for welding) and tightened by tightening the bolts against the stainless steel base plate. A similar design was used by Nelson et al. (14) for assessing solidification cracking in steel welds. Cracking is at the outer edge of the weld, not the inner. This is because contraction of the weld during cooling is hindered by the restraint, thus rendering the outer edge in tension and the inner edge in compression. Figure 13.3*a* shows a circular weld made in 6061 aluminum with a 1100 aluminum filler. The three cracks in the photo all initiate from the PMZ near the outer edge of the weld and propagate into the fusion zone along the welding

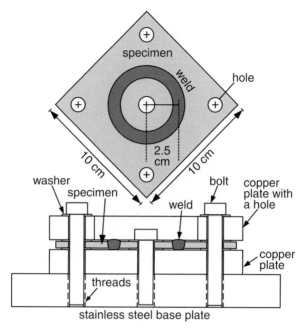

Figure 13.2   Schematic sketch of a circular-patch test.

direction (clockwise). Figure 13.3*b* shows a circular weld between 2219 aluminum and 1100 aluminum made with a 1100 aluminum filler (13). A long PMZ crack runs along the outer edge of the weld.

C. Hot Ductility Testing   This has been used extensively for evaluating the hot-cracking susceptibility of nickel-base alloys (15, 16). It is most often performed on a Gleeble weld simulator (Chapter 2), which is also a tensile testing instrument. The specimen is resistance heated according to a predetermined thermal cycle resembling that in the PMZ. It is tensile tested, for instance, at a stroke rate of 5 cm/s, at predetermined temperatures along the thermal cycle, either during heating to the peak temperature of the thermal cycle or during cooling from it. Several different criteria have been used for interpreting hot ductility curves (17). For instance, one of the criteria is based on the ability of the material to reestablish ductility, that is, how fast ductility recovers during cooling from the peak temperature. If ductility recovers right below the peak temperature, the alloy is considered crack resistant, such as that shown in Figure 13.4*a* for a low-B Cabot alloy 214 (6). On the other hand, if ductility recovers well below the peak temperature, the alloy is considered crack sensitive, such as that shown in Figure 13.4*b* for a high-B Cabot 214. The mechanism that boron affects liquation cracking in Ni-base superalloys is not clear (18).

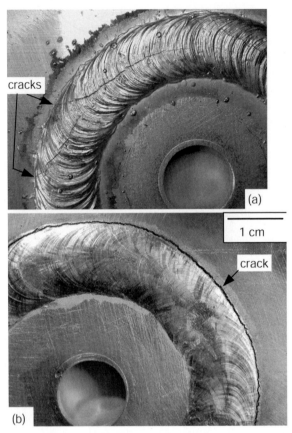

Figure 13.3 Cracking in circular-patch welds: (a) 6061 aluminum made with a 1100 filler wire; (b) 2219 aluminum (outside) welded to 1100 aluminum (inside) with a 1100 filler wire. From Huang and Kou (13).

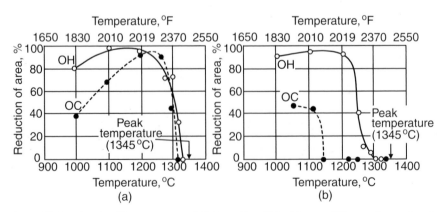

Figure 13.4 Hot ductility response of Cabot alloy 214 with two different boron levels: (a) 0.0002 wt% B; (b) 0.003 wt% B. OH, testing done on heating to 1345°C; OC, on cooling from 1345°C. From Cieslak (6). Reprinted from *ASM Handbook*, vol. 6, ASM International.

13.1.2 Mechanisms of Liquation Cracking

Figure 13.5a is a schematic showing the formation of liquation cracking in the PMZ of a full-penetration aluminum weld (13). Since the PMZ is weakened by grain boundary liquation, it cracks when the solidifying weld metal contracts and pulls it. Most aluminum alloys are susceptible to liquation cracking. This is because of their wide PMZ (due to wide freezing temperature range and high thermal conductivity), large solidification shrinkage (solid density significantly greater than liquid density), and large thermal contraction (large thermal expansion coefficient). The solidification shrinkage of aluminum is as high as 6.6%, and the thermal expansion coefficient of aluminum is roughly twice that of iron base alloys. Figure 13.5b shows liquation cracking in an alloy 6061 circular weld (Figure 13.3a). The light etching α band along the grain boundary is a clear evidence of the grain boundary liquid that weakened the PMZ during welding.

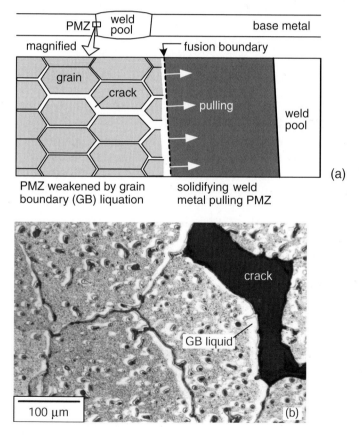

Figure 13.5 Formation of PMZ cracking in a full-penetration aluminum weld: (a) schematic; (b) PMZ cracking in 6061 aluminum. From Huang and Kou (13).

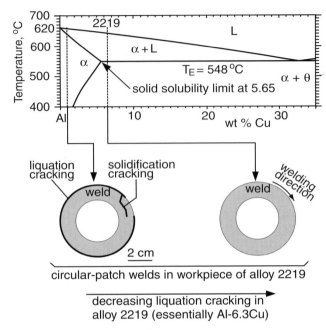

Figure 13.6 Effect of weld metal composition on PMZ cracking in 2219 aluminum. From Huang and Kou (13).

Figure 13.6 shows the effect of the weld metal composition on liquation cracking in 2219 aluminum, which is essentially Al-6.3Cu (13). The circular-patch weld on the right is identical to the PMZ (or the base metal) in composition, that is, Al-6.3Cu. No liquation cracking occurs. The circular-patch weld on the left (same as that in Figure 13.3b), however, has a significantly lower Cu content than the PMZ, and liquation cracking was severe. This effect of the weld metal composition will be explained as follows.

Since the cooling rate during welding is too high for equilibrium solidification, it is inappropriate to discuss liquation cracking based on the solidus temperature from an equilibrium phase diagram. From Equation (6.13) for nonequilibrium solidification, the fraction of liquid f_L at any given temperature T can be expressed as follows:

$$f_L = \left(\frac{(-m_L)C_o}{T_m - T} \right)^{1/1-k} \tag{13.1}$$

where m_L (<0) is the slope of the liquidus line in the phase diagram, C_o the solute content of the alloy, T_m the melting point of pure aluminum, and k the equilibrium partition ratio. Therefore, at any temperature T the lower C_o, the smaller f_L is, that is, the stronger the solid/liquid mixture becomes. Consider the circular-patch weld on the left in Figure 13.6. The weld metal has

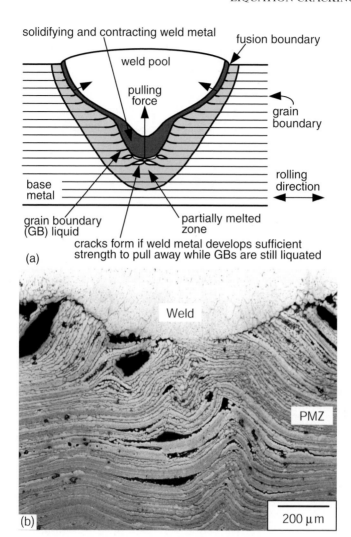

Figure 13.7   Weld metal pulling and tearing PMZ: (a) schematic sketch; (b) 7075 aluminum welded with filler 1100. From Huang and Kou (13).

a significantly lower C_o (Al-0.95Cu) than the PMZ (Al-6.3Cu). This suggests that at any temperature T the weld metal is significantly stronger than the PMZ, thus causing liquation cracking. As for the circular-patch on the right, the weld metal and the PMZ have the same C_o and hence the similar strength level. As such, no liquation cracking occurs.

Figure 13.7a is a schematic showing the formation of liquation cracking in the PMZ of a partial-penetration GMA weld of an aluminum alloy. The papillary (nipple) type penetration pattern shown in the figure is common in GMAW of aluminum alloys with Ar shielding, where spray transfer is the mode of filler metal transfer through the arc (13). The welding direction is perpendicular to the rolling direction. The weld metal in the papillary penetration, as indicated by its very fine cell spacing of the solidification microstructure, solidifies rapidly. This suggests that the rapidly solidifying and thus contracting weld metal in the papillary penetration pulls the PMZ that is weakened by grain boundary liquation. Figure 13.7b shows the transverse cross section near the bottom of a GMA weld of 7075 aluminum made with a filler wire of 1100 aluminum. As shown, the weld metal pulls and tears the PMZ near the tip of the papillary penetration (13).

13.2   LOSS OF STRENGTH AND DUCTILITY

As mentioned in the previous chapter, Huang and Kou (19–21) studied liquation in the PMZ of 2219 aluminum gas–metal arc welds and found both a Cu-depleted α band next to the Cu-rich GB eutectic and a Cu-depleted α ring surrounding each large Cu-rich eutectic particle in the grain interior. Results of microhardness testing showed that the Cu-depleted α was much softer than the Cu-rich eutectic. This suggests that the liquated material solidifies with severe segregation and results in a weak PMZ microstructure with a soft ductile α and a hard brittle eutectic right next to each other. Under tensile loading, the α yields without much resistance while the eutectic fractures badly.

Figure 13.8 shows the tensile testing results of a weld made perpendicular to the rolling direction (20). The maximum load and elongation before failure are both much lower in the weld specimen than in the base-metal specimen, as shown in Figure 13.8a. Fracture of eutectic is evident both along the GB and at large eutectic particles in the grain interior, as shown in Figures 13.8b and c. The fluctuations in the tensile load in Figure 13.8a are likely to be associated with the fracture of the eutectic.

13.3   HYDROGEN CRACKING

Savage et al. (22) studied hydrogen-induced cracking in HY-80 steel. They observed intergranular cracking in the PMZ and the adjacent region in the fusion zone where mixing between the filler and the weld metal is incomplete, as shown in Figure 13.9.

It was pointed out that the creation of liquated films on the GBs in the PMZ provides preferential paths along which hydrogen from the weld metal can diffuse across the fusion boundary. This, according to Savage et al. (22), is because liquid iron can dissolve approximately three to four times more nascent hydrogen than the solid, thus making the liquated GBs serve as "pipelines" along which hydrogen from the weld metal can readily diffuse across the fusion boundary. When these segregated films resolidify, they not only are left supersaturated with hydrogen but also exhibit a higher hardenability due to solute segregation. Consequently, they serve as preferred nucleation sites for hydrogen-induced cracking.

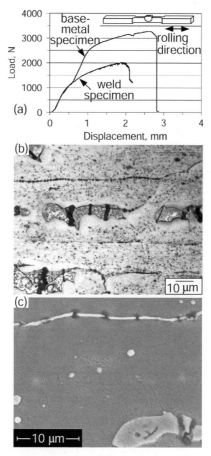

Figure 13.8   Results of tensile testing of a gas–metal arc weld of 2219 aluminum made perpendicular to the rolling direction. Reprinted from Huang and Kou (20). Courtesy of American Welding Society.

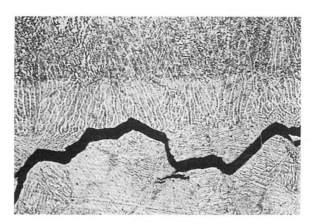

Figure 13.9   Hydrogen-induced cracking in the PMZ of HY-80 steel. Reprinted from Savage et al. (22). Courtesy of American Welding Society.

13.4   REMEDIES

Remedies for the problems associated with the PMZ can be grouped into four categories: filler metal, heat source, degree of restraint, and base metal. These will be discussed below.

13.4.1   Filler Metal

Liquation cracking can be reduced by selecting the proper filler metal. Metzger (23) reported the significant effect of the weld metal composition on liquation cracking in aluminum alloys. Liquation cracking occurred in 6061 aluminum welds produced with Al–Mg fillers at high dilution ratios but not in welds made with Al–Si fillers at any dilution ratios. Metzger's study has been confirmed by subsequent studies on alloys 6061, 6063, and 6082 (5, 10–12, 24–26).

Gittos and Scott (5) studied liquation cracking in alloy 6082 welded with 5356 and 4043 fillers using the circular-patch test. Like Metzger (23), Gittos and Scott (5) observed liquation cracking welds made with the 5356 filler at high dilution ratios (about 80%) but not in welds made with the 4043 filler at any dilution ratios. When it occurred, liquation cracking was along the outer edge of the weld and no cracking was observed along the inner edge.

Gittos and Scott (5) proposed the criterion of $T_{WS} > T_{BS}$ for liquation cracking to occur, where T_{WS} and T_{BS} are the solidus temperatures of the weld metal and the base metal, respectively. They assumed that if the weld metal composition is such that $T_{WS} > T_{BS}$, then the PMZ will solidify before the weld metal and thus resist tensile strains arising from weld metal solidification. The weld metal solidus temperature T_{WS} and the base-metal solidus temperature T_{BS} were taken from Figure 13.10, which shows the solidus temperatures in the Al-rich corner of the ternary Al–Mg–Si system (27). They found the variations in T_{WS} and T_{BS} with the dilution ratio shown in Figure 13.11a to be consistent with the results of their circular-patch testing.

Katoh and Kerr (10, 11) and Miyazaki et al. (12) studied liquation cracking in 6000 alloys, including 6061, using Varestraint testing. Longitudinal liquation cracking occurred when alloy 6061 was welded with a 5356 filler but not with a 4043 filler. They measured the solidus temperatures of the base metals and filler metals by differential thermal analysis. The solidus temperature T_{BS} of alloy 6061 was 597°C. Contrary to the $T_{WS} > T_{BS}$ cracking criterion proposed by Gittos and Scott (5), Miyazaki et al. (12) found $T_{WS} < T_{BS}$ whether the filler metal was 5356 or 4043 (12). This is shown in Figure 13.11b. It was proposed that the base metal of 6061 aluminum probably liquated at 559°C by constitutional liquation induced by the Al–Mg_2Si–Si ternary eutectic.

It should be noted that when attempting to avoid liquation cracking in the PMZ by choosing a proper filler metal, the solidification cracking susceptibility of the fusion zone still needs to be checked. Solidification cracking–composition diagrams (Figure 11.26) can be useful for this purpose (28).

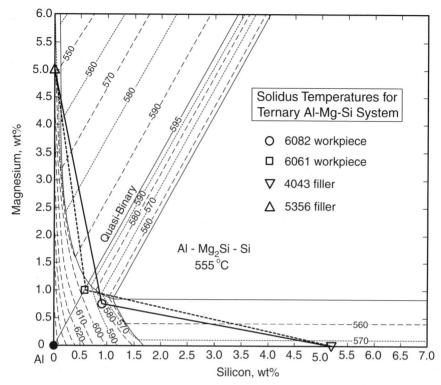

Figure 13.10 Ternary Al–Mg–Si phase diagram showing the solidus temperature. Modified from Phillips (27).

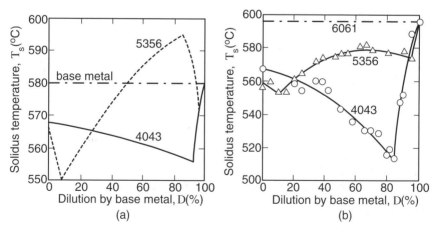

Figure 13.11 Variation of weld metal solidus temperature with dilution: (*a*) in 6082 aluminum (5); (*b*) in 6061 aluminum (12). (*a*) from Gittos et al. (5) and (*b*) from Miyazaki et al. (12). Reprinted from *Welding Journal*, Courtesy of American Welding Society.

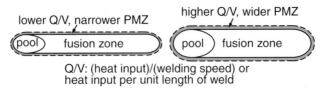

Figure 13.12   Effect of heat input on width of PMZ.

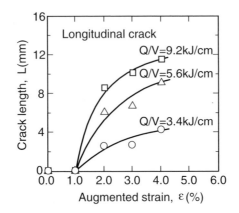

Figure 13.13   Effect of heat input on liquation cracking in Varestraint testing of 6061 aluminum welded with 5356 filler metal. Modified from Miyazaki et al. (12). Courtesy of American Welding Society.

13.4.2 Heat Source

The size of the PMZ and hence the extent of PMZ liquation can be reduced by reducing the heat input, as illustrated in Figure 13.12. Figure 13.13 shows the effect of the heat input on liquation cracking in Varestraint testing of gas–metal arc welds of alloy 6061 made with a 5356 filler metal (12). To minimize the difficulties associated with the PMZ, the heat input can be kept low by using multipass welding or low-heat-input welding processes (such as EBW and GTAW) when possible.

Kou and Le (29) reduced GB liquation and thus liquation cracking in the PMZ of 2014 aluminum alloy by using transverse arc oscillation (1 Hz) during GTAW. The extent of GB melting is significantly smaller with arc oscillation. With the same welding speed, the resultant speed of the heat source is increased by transverse arc oscillation (Chapter 8). This results in a smaller weld pool and a narrower PMZ.

13.4.3 Degree of Restraint

Liquation cracking and hydrogen-induced cracking in the PMZ are both caused by the combination of a susceptible microstructure and the presence of tensile stresses. The sensitivity of the PMZ to both types of cracking can be reduced by decreasing the degree of restraint and hence the level of tensile stresses.

13.4.4   Base Metal

Liquation cracking can be reduced by selecting the proper base metal for welding if it is feasible. The base-metal composition, grain structure, and microsegregation can affect the susceptibility of the PMZ to liquation cracking significantly.

A. Impurities   When impurities such as sulfur and phosphorus are present, the freezing temperature range can be widened rather significantly (Chapter 11). The widening of the freezing temperature range is due to the lowering of the incipient melting temperature, which is effectively the same as the liquation temperature in the sense of liquation cracking. The detrimental effect of sulfur and phosphorus on the liquation cracking of nickel-base alloys has been recognized (15, 16). The effect of minor alloying elements on the liquation temperature of 347 stainless steel is shown in Figure 13.14 (30).

B. Grain Size   The coarser the grains are, the less ductile the PMZ becomes. Furthermore, the coarser the grains are, the less the GB area is and hence the more concentrated the impurities or low-melting-point segregates are at the GB, as shown in Figure 13.15. Consequently, a base metal with coarser grains is expected to be more susceptible to liquation cracking in the PMZ, as shown in Figure 13.16 by Varestraint testing the gas–tungsten arc welds of 6061 aluminum (12). Thompson et al. (31) showed the effect of the grain size on liquation cracking in Inconel 718 caused by constitutional liquation. Guo et al. (32) showed in Figure 13.17 the effect of both the grain size and the boron content on the total crack length of electron beam welded Inconel 718 specimens. Figure 13.18 shows liquation cracking in gas–metal arc welds of two Al–4.5% Cu alloys of different grain sizes (33). Cracking is much more severe with coarse grains.

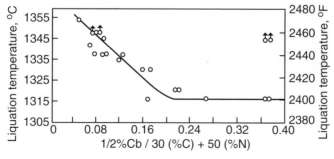

Figure 13.14   Effect of minor alloying elements on liquation temperature of 347 stainless steel. From Cullen and Freeman (30).

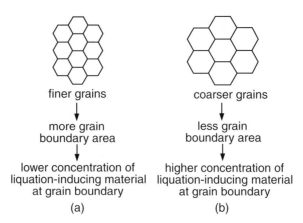

Figure 13.15 Effect of grain size on concentration of liquation-causing material at grain boundaries.

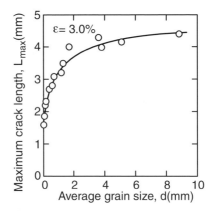

Figure 13.16 Effect of grain size on liquation cracking in Varestraint testing of 6061 aluminum gas–tungsten arc welds. Reprinted from Miyazaki et al. (12). Courtesy of American Welding Society.

C. Grain Orientation Lippold et al. (34) studied liquation cracking in the PMZ of 5083 aluminum alloy and found that PMZ cracking was more severe in welds made transverse to the rolling direction than those made parallel to the rolling direction. It was suggested that in the latter the elongated grains produced by the action of rolling were parallel to the weld and, therefore, it was more difficult for cracks to propagate into the base metal.

D. Microsegregation In the welding of as-cast materials, the PMZ is particularly susceptible to liquation cracking because of the presence of

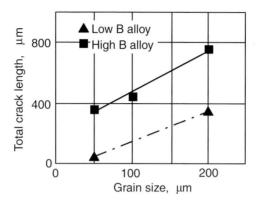

Figure 13.17   Effect of grain size and boron content on liquation cracking in PMZ of Inconel 718 electron beam welds. Reprinted from Guo et al. (32).

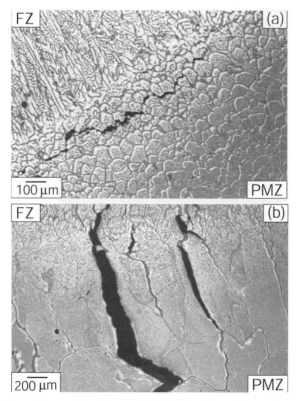

Figure 13.18   Liquation cracking in two Al–4.5% Cu alloys: (*a*) small grains; (*b*) coarse grains. Reprinted from Huang et al. (33).

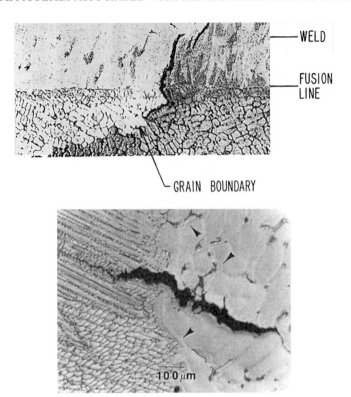

Figure 13.19   Liquation cracking originated from PMZ extending into the fusion zone. (*a*) Cast 304 stainless steel. Reprinted from Apblett (35). Courtesy of American Welding Society. (*b*) Cast corrosion-resistant austenitic stainless steel. Reprinted from Cieslak (36).

low-melting-point GB segregates. Upon heating during welding, excessive GB liquation occurs in the PMZ, making it highly susceptible to liquation cracking. Figure 13.19 shows liquation cracking in a cast 304 stainless steel (35) and a cast corrosion-resistant austenitic stainless steel (36). It initiates from the PMZ and propagates into the fusion zone.

REFERENCES

1. Aluminum Association, *Aluminum Standards and Data*, Aluminum Association, Washington, DC, 1982, p. 15.
2. Kreischer, C. H., *Weld. J.*, **42:** 49s, 1963.
3. Dudas, J. H., and Collins, F. R., *Weld. J.*, **45:** 241s, 1966.
4. Thompson, R. G., in *ASM Handbook*, Vol. 6, ASM International, Materials Park, OH, 1993, p. 566.

5. Gittos, N. F., and Scott, M. H., *Weld. J.*, **60:** 95s, 1981.

6. Cieslak, M. J., in *ASM Handbook*, Vol. 6: *Welding, Brazing and Soldering*, ASM International, Materials Park, OH, 1993, p. 88.

7. Savage, W. F., and Dickinson, D. W., *Weld. J.*, **51:** 555s, 1972.

8. Savage, W. F., and Lundin, C. D., *Weld. J.*, **44:** 433s, 1965.

9. Lippold, J. C., Nippes, E. F., and Savage, W. F., *Weld. J.*, **56:** 171s, 1977.

10. Katoh, M., and Kerr, H. W., *Weld. J.*, **66:** 360s, 1987.

11. Kerr, H. W., and Katoh, M., *Weld. J.*, **66:** 251s, 1987.

12. Miyazaki, M., Nishio, K., Katoh, M., Mukae, S., and Kerr, H. W., *Weld. J.*, **69:** 362s, 1990.

13. Huang, C., and Kou, S., *Weld. J.*, submitted for publication.

14. Nelson, T. W., Lippold, J. C., Lin, W., and Baselack III, W. A., *Weld. J.*, **76:** 110s, 1997.

15. *Effects of Minor Elements on the Weldability of High-Nickel Alloys*, Welding Research Council, 1969.

16. *Methods of High-Alloy Weldability Evaluation*, Welding Research Council, 1970.

17. Yeniscavich, W., in *Methods of High-Alloy Weldability Evaluation*, p. 1.

18. Kelly, T. J., in *Weldability of Materials*, Eds. R. A. Patterson and K. W. Mahin, ASM International, Materials Park, OH, 1990, p. 151.

19. Huang, C., and Kou, S., *Weld. J.*, **79:** 113s, 2000.

20. Huang, C., and Kou, S., *Weld. J.*, **80:** 9s, 2001.

21. Huang, C., and Kou, S., *Weld. J.*, **80:** 46s, 2001.

22. Savage, W. F., Nippes, E. F., and Szekeres, E. S., *Weld. J.*, **55:** 276s, 1976.

23. Metzger, G. E., *Weld. J.*, **46:** 457s, 1967.

24. Gitter, R., Maier, J., Muller, W., and Schwellinger, P., in *Proceedings of the Fifth International Conference on Aluminum Weldments*, Eds. D. Kosteas, R. Ondra, and F. Ostermann, Technische Universita Munchen, Munchen, 1992, pp. 4.1.1–4.1.13.

25. Powell, G. L. F., Baughn, K., Ahmed, N., Dalton J. W., and Robinson, P., in *Proceedings of International Conference on Materials in Welding and Joining*, Institute of Metals and Materials Australasia, Parkville, Victoria, Australia, 1995.

26. Ellis, M. B. D., Gittos, M. F., and Hadley, I., *Weld. Inst. J.*, **6:** 213, 1997.

27. Philips, H. W. L., *Annotated Equilibrium Diagrams of Some Aluminum Alloy Systems*, Institute of Metals, London, 1959, p. 67.

28. Jennings, P. H., Singer, A. R. E., and Pumphrey, W. I., *J. Inst. Metals*, **74:** 227, 1948.

29. Kou, S., and Le, Y., *Weld. J.*, **64:** 51, 1985.

30. Cullen, T. M., and Freeman, J. W., *J. Eng. Power*, **85:** 151, 1963.

31. Thompson, R. G., Cassimus, J. J., Mayo, D. E., and Dobbs, J. R., *Weld. J.*, **64:** 91s, 1985.

32. Guo, H., Chaturvedi, M. C., and Richards, N. L., *Sci. Technol. Weld. Join.*, **4:** 257, 1999.

33. Huang, C., Kou, S., and Purins, J. R., in *Proceedings of Merton C. Flemings Symposium on Solidification and Materials Processing*, Eds. R. Abbaschian, H. Brody, and A. Mortensen, Minerals, Metals and Materials Society, Warrendale, PA, 2001, p. 229.

34. Lippold, J. C., Nippes, E. F., and Savage, W. F., *Weld. J.*, **56:** 171s, 1977.

35. Apblett, W. R., and Pellini, W. S., *Weld. J.*, **33:** 83s, 1954.

36. Cieslak, M. J., in *ASM Handbook*, Vol. 6: *Welding, Brazing and Soldering*, ASM International, Materials Park, OH, 1993, p. 495.

PROBLEMS

13.1 Hot-ductility testing was performed on an 18% Ni maraging steel following a thermal cycle with a peak temperature of 1400°C. The on-heating part of the testing showed that the ductility dropped to zero at 1380°C (called the nil ductility temperature), and the on-cooling part showed that the ductility recovered from zero to about 7% at 1360°C. Is this maraging steel very susceptible to liquation cracking? Explain why or why not. Do you expect the specimen tensile tested on heating at 1380°C to exhibit brittle intergranular fracture of ductile transgranular dimple fracture? Why? What do you think caused PMZ liquation in this maraging steel?

13.2 (a) The effect of the carbon content and the Mn–S ratio on weld metal solidification cracking in steels has been described in Chapter 11. It has been reported that a similar effect also exists in the liquation cracking of the PMZ of steels. Explain why. (b) Because of the higher strength of HY-130 than HY-80, its chemical composition should be more strictly controlled if liquation cracking is to be avoided. Assume the following contents: HY-80: ≤0.18 C; 0.1–0.4 Mn; ≤0.025 S; ≤0.025 P; HY-130: ≤0.12 C; 0.6–0.9 Mn; ≤0.010 S; ≤0.010 P. Do these contents suggest a more strict composition control in HY-130?

13.3 Sulfur can form a liquid with nickel that has a eutectic temperature of 635°C. Do you expect high-strength alloy steels containing Ni (say more than 2.5%) to be rather susceptible to liquation cracking due to sulfur? Explain why or why not.

13.4 Low-transverse-frequency arc oscillation (Figure 8.17) has been reported to reduce PMZ liquation. Sketch both the weld and the PMZ behind the weld pool and show how this can be true.

13.5 Consider the circular-patch weld in Figure 13.3b. Will liquation cracking occur if the outer piece is alloy 1100 (essentially pure aluminum) and the inner piece (the circular patch) is alloy 2219 (Al-6.3 Cu)? Explain why or why not.

13.6 Like aluminum alloy 7075, alloy 2024 is very susceptible to liquation cracking. In GMAW of alloy 2024 do you expect liquation cracking to be much more severe with filler metal 4043 or 1100? Why?

13.7 In a circular-patch test alloy 2219 (Al-6.3Cu) is welded with alloy 2319 (Al-6.3Cu) plus extra Cu as the filler metal. The resultant composition of the weld metal is about Al-8.5Cu. Do you expect liquation cracking to occur? Explain why or why not.

13.8 Do you expect liquation cracking to occur in autogenous GTAW of 7075? Why or why not?

PART IV
The Heat-Affected Zone

14 Work-Hardened Materials

Metals can be strengthened in several ways, including solution hardening, work hardening, precipitation hardening, and transformation hardening. The effectiveness of the last three methods can be reduced significantly by heating during welding in the area called the *heat-affected zone* (HAZ), where the peak temperatures are too low to cause melting but high enough to cause the microstructure and properties of the materials to change significantly. Solution-hardening materials are usually less affected unless they have been work hardened and thus will not be discussed separately. This chapter shall focus on recrystallization and grain growth in the HAZ of work-hardened materials, which can make the HAZ much weaker than the base metal.

14.1 BACKGROUND

When a metal is cold worked and plastically deformed, for instance, cold rolled or extruded, numerous dislocations are generated. These dislocations can interact with each other and form dislocations tangles. Such dislocation tangles hinder the movement of newly generated dislocations and, hence, further plastic deformation of the metal. In this way, a metal is strengthened or hardened by cold working. This strengthening mechanism is called *work hardening*.

14.1.1 Recrystallization

Most of the energy expended in work hardening appears in the form of heat but, as shown in Figure 14.1, a finite fraction is stored in the material as strain energy (1). When a work-hardened material is annealed, the deformed grains in the material tend to recrystallize by forming fresh, strain-free grains that are soft, just like grains that have not been deformed. The *stored strain energy* is the driving force for recrystallization of a work-hardened material (2), and this energy is released as fresh, strain-free grains form. Figure 14.2 shows the various stages of recrystallization in a work-hardened brass (3). Slip bands, which have formed during severe work hardening, serve as the nucleation sites for new grains.

The extent of recrystallization increases with increasing annealing temperature and time (4). Therefore, it can be expected that the strength or hardness

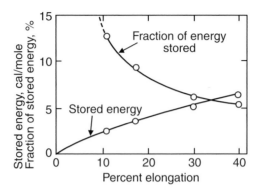

Figure 14.1 Stored energy and fraction of stored energy as a function of tensile elongation during cold working of high-purity copper. From Gordon (1).

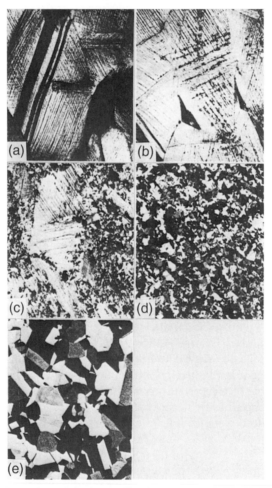

Figure 14.2 Deformation and recrystallization structures of α-brass: (*a*) 33% cold reduction; (*b*) short anneal at 580°C; (*c*) longer anneal; (*d*) completely recrystallized; (*e*) grain growth. From Burke (3).

of a work-hardened material tends to decrease with increasing annealing temperature and time. Figure 14.3 shows the hardness of a work-hardened cartridge brass as a function of annealing temperature (5).

Table 14.1 summarizes the recrystallization temperature for various metals. For most metals the recrystallization temperature is around 40–50% of their melting point in degrees Kelvin (6). It should be pointed out that the recrystallization temperature of a metal can be affected by the degree of work hardening and the purity level (2).

In fact, before recrystallization takes place, there exists a period of time during which certain properties of the work-hardened material, for instance, the electrical resistivity, tend to recover without causing any microstructural changes. This phenomenon is called recovery. However, since the mechanical properties of the material, such as strength or hardness, do not change significantly during recovery, recovery is not important in welding.

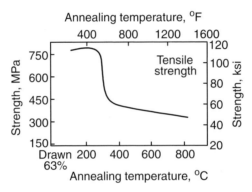

Figure 14.3   Strength of cartridge brass (Cu–35Zn) cold rolled to an 80% reduction in area and then annealed at various temperatures for 60 min. Reprinted from *Metals Handbook* (5).

TABLE 14.1   Recrystallization Temperatures for Various Metals

Metal	Minimum Recrystallization Temperature (°C)	Melting Temperature (°C)
Aluminum	150	660
Magnesium	200	659
Copper	200	1083
Iron	450	1530
Nickel	600	1452
Molybdenum	900	2617
Tantalum	1000	3000

Source: Brick et al. (6).

14.1.2   Grain Growth

Upon completion of recrystallization, grains begin to grow. The driving force for grain growth is the *surface energy*. The total grain boundary area and thus the total surface energy of the system can be reduced if fewer and coarser grains are present. This can be illustrated by the growth of soap cells in a flat container (7), as shown in Figure 14.4. It should be pointed out that since the driving force for grain growth is the surface energy rather than the stored strain energy, grain growth is not limited to work-hardened materials.

Like recrystallization, the extent of grain growth also increases with increasing annealing temperature and time. Figure 14.5 shows grain growth in cold-rolled brass as a function of temperature and time (5).

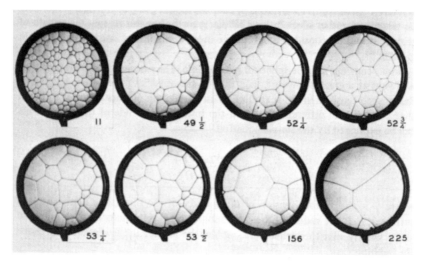

Figure 14.4   Growth of soap cells in a flat container. The numbers indicate growth time in minutes. From Smith (7).

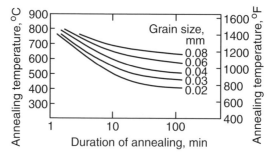

Figure 14.5   Grain growth of Cu–35Zn brass cold rolled to 63% reduction in area. Reprinted from *Metals Handbook* (5).

It is worth noting that carbide and nitride particles can inhibit grain growth in steels by hindering the movement of grain boundaries (2). These particles, if not dissolved during welding, tend to inhibit grain growth in the HAZ.

14.2   RECRYSTALLIZATION AND GRAIN GROWTH IN WELDING

The effect of work hardening is completely gone in the fusion zone because of melting and is partially lost in the HAZ because of recrystallization and grain growth. These strength losses should be taken into account in structural designs involving welding.

14.2.1   Microstructure

Figure 14.6 shows the weld microstructure of a work-hardened 304 stainless steel (8). The microstructure of the same material before work hardening is also included for comparison (Figure 14.6*a*). Recrystallization (Figure 14.6*d*) and grain growth (Figure 14.6*e*) are evident in the HAZ. Figure 14.7 shows grain growth in the HAZ of a molybdenum weld (9). Severe HAZ grain growth can result in coarse grains in the fusion zone because of epitaxial growth (Chapter 7). *Fracture toughness* is usually poor with coarse grains in the HAZ and the fusion zone.

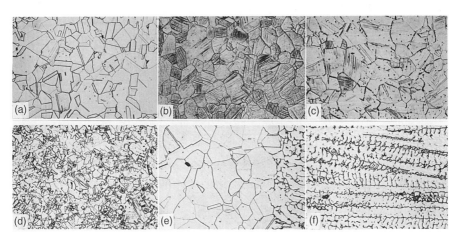

Figure 14.6   Microstructure across the weld of a work-hardened 304 stainless steel: (*a*) before work hardening; (*b*) base metal; (*c*) carbide precipitation at grain boundaries; (*d*) recrystallization; (*e*) grain growth next to fusion boundary; (*f*) fusion zone. Magnification 137×. Reprinted from *Metals Handbook* (8).

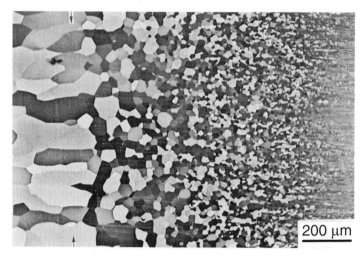

Figure 14.7 Grain growth in electron beam weld of molybdenum, arrows indicating fusion boundary. Reprinted from Wadsworth et al. (9). Copyright 1983 with permission from Elsevier Science.

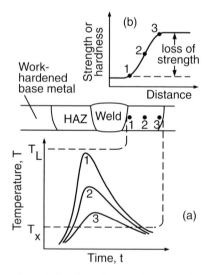

Figure 14.8 Softening of work-hardened material caused by welding: (*a*) thermal cycles; (*b*) strength or hardness profile.

14.2.2 Thermal Cycles

The loss of strength in the HAZ can be explained with the help of thermal cycles, as shown in Figure 14.8. The closer to the fusion boundary, the higher the peak temperature becomes and the longer the material stays above

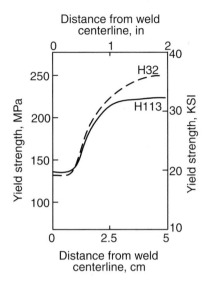

Figure 14.9 Yield strength profiles across welds of two work-hardened 5083 aluminum plates. Reprinted from Cook et al. (10). Courtesy of American Welding Society.

the effective recrystallization temperature, T_x. Under rapid heating during welding, the recrystallization temperature may increase because recrystallization requires diffusion and diffusion takes time. Since the strength of a work-hardened material decreases with increasing annealing temperature and time, the strength or hardness of the HAZ decreases as the fusion boundary is approached. Figure 14.9 shows the HAZ strength profiles of two work-hardened 5083 aluminum plates (10). It appears that the harder the base metal, the greater the strength loss is.

Grain growth in the HAZ can also be explained with the help of thermal cycles, as shown in Figure 14.10. The closer to the fusion boundary, the higher the peak temperature becomes and the longer the material stays at high temperatures. Since grain growth increases with increasing annealing temperature and time (Figure 14.5), the grain size in the HAZ increases as the fusion boundary is approached.

14.3 EFFECT OF WELDING PARAMETERS AND PROCESS

The effect of welding parameters on the HAZ strength is explained in Figure 14.11. Both the size of the HAZ and the retention time above the effective recrystallization temperature T_x increase with increasing heat input per unit length of the weld, that is, the ratio of heat input to welding speed. Consequently, the loss of strength in the HAZ becomes more severe as the heat input per unit length of the weld is increased. Figure 14.12 shows the effect of

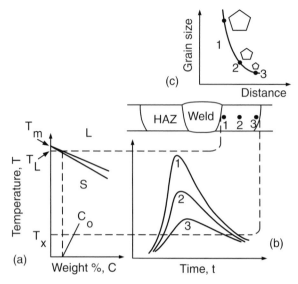

Figure 14.10   Grain growth in HAZ: (*a*) phase diagram; (*b*) thermal cycles; (*c*) grain size variations.

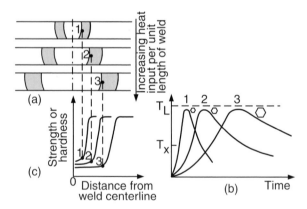

Figure 14.11   Effect of heat input per unit length of weld on: (*a*) width of HAZ (shaded), (*b*) thermal cycles near fusion boundary, and (*c*) strength or hardness profiles.

welding parameters on the HAZ strength of a work-hardened 5356-H321 aluminum alloy (11).

Finally, Figure 14.13 shows the effect of the welding process on the HAZ microstructure of a work-hardened 2219 aluminum (12). Because of the low heat input and the high cooling rate in EBW, very little recrystallization is observed in the HAZ of the work-hardened material. On the other hand,

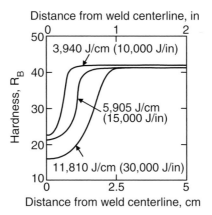

Figure 14.12   Effect of heat input per unit length of weld on HAZ hardness in a work-hardened 5356 aluminum. Reprinted from White et al. (11). Courtesy of American Welding Society.

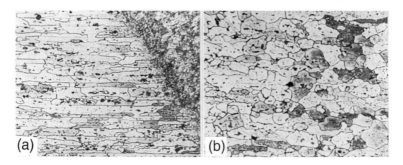

Figure 14.13   Microstructure near fusion boundary of a work-hardened 2219-T37 aluminum: (*a*) electron beam weld; (*b*) gas–tungsten arc weld. Magnification 80×. Reprinted from *Metals Handbook* (12).

because of the higher heat input and lower cooling rate in GTAW, recrystallization and even some grain growth are observed in the HAZ.

REFERENCES

1. Gordon, P., *Trans. AIME,* **203:** 1043, 1955.

2. Reed-Hill, R. E., *Physical Metallurgy Principles*, 2d ed., Van Nostrand, New York, 1973.

3. Burke, J. E., in *Grain Control in Industrial Metallurgy*, American Society for Metals, Cleveland, OH, 1949.

4. Decker, B. F., and Harker, D., *Trans. AIME,* **188:** 887, 1950.

5. *Metals Handbook*, 8th ed., Vol. 2, American Society for Metals, Metals Park, OH, 1972, p. 285.

6. Brick, R. M., Pense, A. W., and Gordon, R. B., *Structure and Properties of Engineering Materials*, 4th ed., McGraw-Hill, New York, 1977, p. 81.

7. Smith, C. S., ASM Seminar, Metal Interfaces, ASM, Metals Park, OH, 1952, p. 65.

8. *Metals Handbook*, 8th ed., Vol. 7, American Society for Metals, Metals Park, OH, 1972, p. 135.

9. Wadsworth, J., Morse, G. R., and Chewey, P. M., *Mater. Sci. Eng.*, **59:** 257 (1983).

10. Cook, L. A., Channon, S. L., and Hard, A. R., *Weld. J.*, **34:** 112, 1955.

11. White, S. S., Manchester. R. E., Moffatt, W. G., and Adams, C. M., *Weld. J.*, 39: 10s, 1960.

12. *Metals Handbook*, 8th ed., Vol. 7, American Society for Metals, Metals Park, OH, 1972, p. 268.

FURTHER READING

1. Reed-Hill, R. E., *Physical Metallurgy Principles*, 2nd ed., Van Nostrand, New York, 1973.

2. Brick, R. M., Pense, A. W., and Gordon, R. B., *Structure and Properties of Engineering Materials*, 4th ed., McGraw-Hill, New York, 1977.

PROBLEMS

14.1 A 301 stainless steel sheet work-hardened to about 480 Knoop hardness was welded, and in the HAZ the hardness droped to a minimum of about 240. Explain the loss of strength in the HAZ. The weld reinforcement was machined off and the whole sheet including the weld was cold rolled. What was the purpose of cold rolling?

14.2 It is known that bcc is less close-packed than fcc and thus has a higher diffusion coefficient. Are ferritic stainless steels (bcc at high temperatures) more or less susceptible to HAZ grain growth than austenitic stainless steels (fcc at high temperatures)? Explain why.

14.3 Do you expect grain growth during the welding of tantalum ($T_m = 2996°C$) to be more severe than during the welding of Al? Explain why.

15 Precipitation-Hardening Materials I: Aluminum Alloys

Aluminum alloys are more frequently welded than any other types of nonferrous alloys because of their widespread applications and fairly good weldability. In general, higher strength aluminum alloys are more susceptible to (i) hot cracking in the fusion zone and the PMZ and (ii) losses of strength/ductility in the HAZ. Aluminum–lithium alloys and PM (powder metallurgy) aluminum alloys can be rather susceptible to porosity in the fusion zone. Table 15.1 summarizes typical problems in aluminum welding and recommended solutions. The problems associated with the fusion zone and the PMZ have been discussed previously. In this chapter, we shall focus on the HAZ phenomena in heat-treatable aluminum alloys, which are strengthened through precipitation hardening. Table 15.2 shows the designation for aluminum alloys. As shown, the 2000, 6000, and 7000 series are heat treatable, while the rest are non–heat treatable.

15.1 BACKGROUND

Aluminum–copper alloys are a typical example of precipitation-hardening materials. As shown in the Al-rich side of the Al–Cu phase diagram in Figure 15.1, the solubility of Cu in the α phase increases with increasing temperature—a *necessary criterion for precipitation hardening*. Consider the precipitation hardening of Al–4% Cu as an example. Step 1, solution heat treating, is to heat treat the alloy in the α-phase temperature range until it becomes a solid solution. Step 2, quenching, is to rapidly cool the solid solution to room temperature to make it supersaturated in Cu. Step 3, aging, is to allow the strengthening phase to precipitate from the supersaturated solid solution. Aging by heating (e.g., at 190°C) is called *artificial aging* and aging without heating is called *natural aging*. In the heat-treating terminology, T6 and T4 refer to a heat-treatable aluminum alloy in the artificially aged condition and the naturally aged condition, respectively.

Five sequential structures can be identified during the artificial aging of Al–Cu alloys:

$$\text{Supersaturated solid solution} \rightarrow \text{GP} \rightarrow \theta'' \rightarrow \theta' \rightarrow \theta\,(\text{Al}_2\text{Cu}) \quad (15.1)$$

TABLE 15.1   Typical Welding Problems in Aluminum Alloys

Typical Problems	Alloy Type	Solutions	Sections in book
Porosity	Al-Li alloys (severe)	Surface scraping or milling	3.2
		Thermovacuum treatment	10.3
		Variable-polarity keyhole PAW	
	Powder-metallurgy alloys (severe)	Thermovacuum treatment	3.2
		Minimize powder oxidation and hydration during atomization and consolidation	
	Other types (less severe)	Clean workpiece and wire surface	3.2
		Variable-polarity keyhole PAW	
Solidification cracking in FZ	Higher-strength alloys (e.g., 2014, 6061, 7075)	Use proper filler wires and dilution	11.4
		In autogenous GTA welding, use arc oscillation or less susceptible alloys (2219)	11.4 7.6
Hot cracking and low ductility in PMZ	Higher-strength alloys	Use low heat input[a]	13.1
		Use proper filler wires	13.2
		Low-frequency arc oscillation	
Softening in HAZ	Work-hardened materials	Use low-heat input	14.2 14.3
	Heat-treatable alloys	Use low-heat input	15.2
		Postweld heat treating	15.3

[a] Low heat input processes (e.g., EBW, GTAW) or multiple-pass welding with low-heat input in each pass and low interpass temperature.

where θ (Al_2Cu) is the equilibrium phase with a body-centered-tetragonal (bct) structure. The GP zones (Guinier–Preston, sometimes called GP1), the θ'' phase (sometimes called GP2), and the θ' phase are metastable phases. Figure 15.2 shows the solvus curves of these metastable phases, which represent the highest temperatures these phases can exist (1, 2).

The GP zones are coherent with the crystal lattice of the α solid solution. They consist of disks a few atoms thick (4–6 Å) and about 80–100 Å in diameter, formed on the {100} planes of the solid solutions (3). Since a Cu atom is about 11% smaller than an Al atom in diameter and the GP zones are richer in Cu than the solid solution, the crystal lattice is strained around the GP zones. The strain fields associated with the GP zones allow them to be detected in the electron microscope. The θ'' phase is also coherent with the crystal lattice

TABLE 15.2 Designation of Wrought Aluminum Alloys

	Not Heat Treatable	Heat Treatable	Not Heat Treatable	Not Heat Treatable	Not Heat Treatable	Heat Treatable	Heat Treatable
Series	1000	2000	3000	4000	5000	6000	7000
Major alloying elements	None	Cu	Mn	Si	Mg	Mg/Si	Zn
Advantages	Electrical/thermal conductivity	Strength	Formability	Filler wires	Strength after welding	Strength, extrudability	Strength
Example	1100	2219	3003	4043	5052	6061	7075

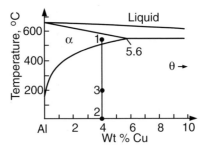

Figure 15.1 Aluminum-rich side of Al–Cu phase diagram showing the three steps of precipitation hardening.

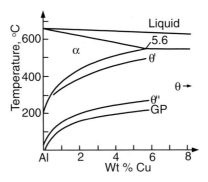

Figure 15.2 Metastable solvus curves for GP, θ'', and θ' in Al–Cu phase diagram. From Hornbogeni (1) and Beton and Rollason (2).

of the solid solution, its size ranging from 10 to 40 Å in thickness and 100 to 1000 Å in diameter. The θ' phase, on the other hand, is semicoherent with the lattice of the solid solution. It is not related to the GP zones or the θ'' phase; it nucleates heterogeneously, especially on dislocations. The size of the θ' phase ranges from 100 to 150 Å in thickness and 100 to 6000 Å or more in diameter depending on the time and temperature of aging (4). Figure 15.3 shows a transmission electron micrograph of the θ' phase in 2219 aluminum (Al–6Cu) (5). Finally, the θ phase, which can either form from θ' or directly from the solid solution, is incoherent with the lattice of the solid solution.

Figure 15.4 shows the correlation of these structures with the hardness of Al–4Cu (6). The maximum hardness (or strength) occurs when the amount of θ'' (or GP2) is at a maximum, although some contribution may also be provided by θ' (6). As θ' grows in size and increases in amount, the coherent strains decrease and the alloy becomes overaged. As aging continues even further, the incoherent θ phase forms and the alloy is softened far beyond its maximum-strength condition. As shown schematically in Figure 15.5, the lattice strains are much more severe around a coherent precipitate

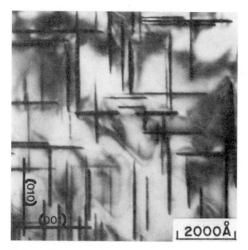

Figure 15.3 Transmission electron micrograph of a 2219 aluminum heat treated to contain θ' phase. From Dumolt et al. (5).

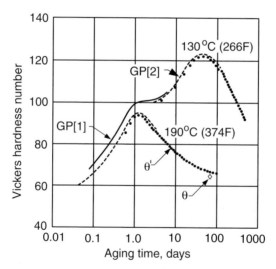

Figure 15.4 Correlation of structures and hardness of Al–4Cu at two aging temperatures. From Silcock et al. (6).

(Figure 15.5*b*) than around an incoherent one (Figure 15.5*c*). The severe lattice strains associated with the coherent precipitate make the movement of dislocations more difficult and, therefore, strengthen the material to a greater extent.

Similar to Al–Cu alloys, the precipitation structure sequence may be represented as follows for other alloy systems (8):

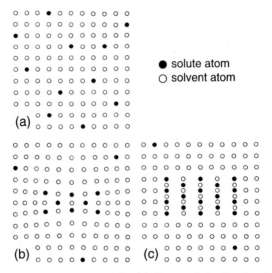

Figure 15.5 Three types of structure in Al–Cu precipitation hardening: (*a*) supersaturated solid solution; (*b*) coherent metastable phase; (*c*) incoherent equilibrium phase. From Guy (7).

TABLE 15.3   Compositions of Some Heat-Treatable Aluminum Alloys

Alloy	Si	Cu	Mn	Mg	Cr	Ni	Zn	Ti
2014	0.8	4.4	0.8	0.5	—	—	—	—
2024	—	4.4	0.6	1.5	—	—	—	—
2219	—	6.3	0.3	—	—	—	—	0.06
6061	0.6	0.3	—	1.0	0.2	—	—	—
7005	—	—	0.4	1.4	0.1	—	4.5	0.04
7039	0.3	0.1	0.2	2.8	0.2	—	4.0	0.1
7146	0.2	—	—	1.3	—	—	7.1	0.06

Source: *Aluminum Standards and Data* (10).

$$\text{Al–Cu–Mg (e.g., 2024):} \quad \text{SS} \rightarrow \text{GP} \rightarrow \text{S}'\,(\text{Al}_2\text{CuMg}) \rightarrow \text{S}\,(\text{Al}_2\text{CuMg}) \quad (15.2)$$

$$\text{Al–Mg–Si (e.g., 6061):} \quad \text{SS} \rightarrow \text{GP} \rightarrow \beta'\,(\text{Mg}_2\text{Si}) \rightarrow \beta\,(\text{Mg}_2\text{Si}) \quad (15.3)$$

$$\text{Al–Zn–Mg (e.g., 7005):} \quad \text{SS} \rightarrow \text{GP} \rightarrow \eta'\,(\text{Zn}_2\text{Mg}) \rightarrow \eta\,(\text{Zn}_2\text{Mg}) \quad (15.4)$$

where SS denotes the supersaturated solid solution. Table 15.3 shows the compositions of the commercial aluminum alloys mentioned above (10). It should be pointed out, however, that coherency strains are not observed in the GP zones or β' transition stages of precipitation in Al–Mg–Si alloys such as 6061.

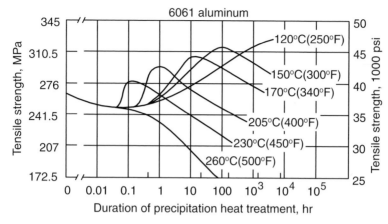

Figure 15.6 Aging characteristics for 6061-T4 aluminum (9). Modified from *Metals Handbook*, vol. 2, 8th edition, American Society for Metals, 1964, p. 276.

Therefore, it has been suggested that precipitation hardening in such aluminum alloys is due to the increased energy required for the dislocations to break the Mg–Si bonds as they pass through the precipitate, rather than due to coherency strains (4).

Figure 15.6 shows the precipitation-hardening curves of 6061 aluminum (8). The alloy has been naturally aged at room temperature (T4) before heat treating. The initial strength decrease is due to reversion (dissolution) of the GP zones formed in natural aging. As shown, the higher the temperature, the faster overaging occurs and the strength decreases.

15.2   Al–Cu–Mg AND Al–Mg–Si ALLOYS

15.2.1   Welding in Artificially Aged Condition

The 2000-series (Al–Cu–Mg) and 6000-series (Al–Mg–Si) heat-treatable alloys are known to have a tendency to overage during welding, especially when welded in the fully aged condition (T6).

Dumolt et al. (5) studied the HAZ microstructure of 2219 aluminum, a binary Al–6.3Cu alloy. Figure 15.7 shows the transmission electron micrographs of a 2219 aluminum plate artificially aged to contain only one metastable phase, θ', before welding and preserved in liquid nitrogen after welding to inhibit natural aging (5). Since the composition of alloy 2219 is beyond the maximum solid solubility, large θ particles are still present after heat treating, but the matrix is still α containing fine θ' precipitate. As shown in the TEM images, the volume fraction of θ' decreases from the base metal to the fusion boundary because of the reversion of θ' during welding. The reversion of θ' is accompanied by coarsening; that is, a few larger θ' particles

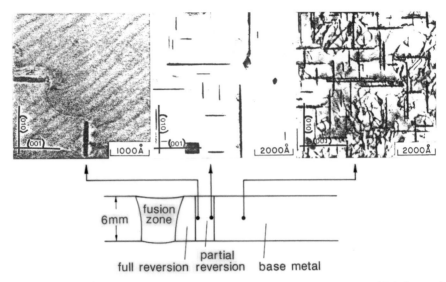

Figure 15.7 Transmission electron micrographs of a 2219 aluminum artificially aged to contain θ' before welding. From Dumolt et al. (5).

grow at the expense of many small ones (middle TEM image). The presence of such coarse precipitates suggests overaging and hence inability to recover strength by postweld artificial aging, as will be discussed later.

The microstructure in Figure 15.7 can be explained with the help of Figure 15.8. The base metal is heat treated to contain the θ' phase. Position 4 is heated to a peak temperature below the θ' solvus and thus unaffected by welding. Positions 2 and 3 are heated to above the θ' solvus and partial reversion occurs. Position 1 is heated to an even higher temperature and θ' is fully reverted. The cooling rate here is too high for reprecipitation of θ' to occur during cooling to room temperature. The θ' reversion causes the hardness to decrease in the HAZ, which is evident in the as-welded condition (AW). During postweld natural aging (PWNA), the GP zones form in the solutionized area near position 1, causing its hardness to increase and leaving behind a hardness minimum near position 2. During postweld artificial aging (PWAA), θ'' and some θ' precipitate near position 1 and cause its hardness to increases significantly. However, near position 2, where overaging has occurred during welding due to θ' coarsening, the hardness recovery is not as much.

A somewhat similar situation is welding a workpiece that has been heat treated to the T6 condition. Figure 15.9 shows the hardness profiles in a 3.2-mm-thick 6061 aluminum autogenous gas–tungsten arc welded in the T6 condition at 10 V, 110 A, and 4.2 mm/s (10 ipm) (11). A hardness minimum is evident after PWNA and especially after PWAA. Malin (12) welded 6061-T6 aluminum by pulsed GMAW with a filler metal of 4043 aluminum (essentially Al–5Si) and measured the HAZ hardness distribution after postweld natural

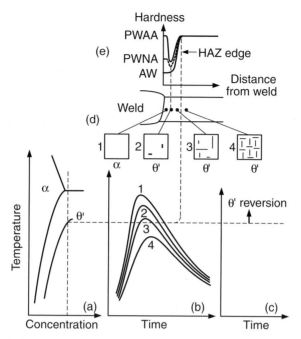

Figure 15.8   Al–Cu alloy heat treated to contain θ' before welding: (a) phase diagram; (b) thermal cycles; (c) reversion of θ'; (d) microstructure; (e) hardness distribution. θ in base metal not shown.

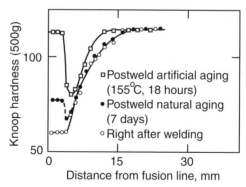

Figure 15.9   HAZ hardness profiles in a 6061 aluminum welded in T6 condition. From Kou and Le (11).

aging. He observed a hardness minimum similar to the PWNA hardness profile in Figure 15.9, where the peak temperature was 380°C (716°F) during welding and where failure occurred in tensile testing after welding. He pointed out that the precipitation range for the most effective strengthening phase β''

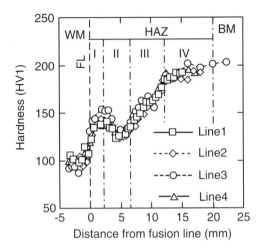

Figure 15.10   HAZ hardness profiles in a 2095 aluminum welded in T8 condition and measured after natural aging. Reprinted from Rading et al. (15). Courtesy of American Welding Society.

is 160–240°C (320–464°F) and that for the less effective strengthening phase β' is 240–380°C (464–716°F) (13, 14). He proposed that the losses of hardness and strength is a result of overaging due to β'' coarsening and β' formation. Malin also observed a sharp hardness decrease immediately outside the fusion boundary (PMZ) and speculated that this was caused by Mg migration into the Mg-poor weld.

Rading et al. (15) determined the fully naturally aged HAZ hardness profiles in a 9.5-mm-thick 2095 aluminum (essentially Al–4.3Cu–1.3Li) welded in the peak aged T8 condition with a 2319 filler metal (Al–6.3Cu), as shown in Figure 15.10. The T8 condition stands for solution heat treating, cold working, followed by artificially aging. The principal strengthening precipitate in the T8 condition is T_1 (Al$_2$CuLi) while in the naturally aged (T4) condition strengthening is provided mainly by δ' (Al$_3$Li) (16). The hardness profiles in Figure 15.10 are similar to the PWNA hardness profile in Figure 15.9 except for the sharp decrease near the fusion line (FL). According to TEM micrographs, the hardness minimum at 5 mm from the fusion line is due to overaging caused by T_1 coarsening during welding, while the hardness peak at 2 mm from the fusion line is caused by δ' precipitation during natural aging. It was proposed that the sharp hardness decrease near the fusion boundary is caused by diffusion of Li into the Li-poor weld, which reduces the propensity for δ' precipitation.

15.2.2 Welding in Naturally Aged Condition

Figure 15.11 shows TEM micrographs of a 2219 aluminum heat treated to contain GP zones alone before welding and preserved in liquid nitrogen after

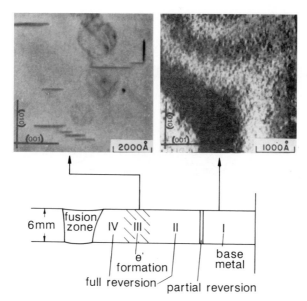

Figure 15.11 Transmission electron micrographs of a 2219 aluminum aged to contain GP zones before welding. From Dumolt et al. (5).

welding to inhibit natural aging (5). The GP zones in the HAZ are easily reverted during welding because of their small size. Precipitation of θ' occurs in the middle of the HAZ. Figure 15.12 shows the optical micrograph of a 2024 aluminum (Al–4.4Cu–1.5Mg) plate that was welded in the T4 (naturally aged) condition (17). The precipitation region in the HAZ is visible as a dark-etching band.

The microstructure in Figure 15.11 can be explained with the help of Figure 15.13. Positions 1–3 are heated to above the solvus of the GP zones, and GP zones are thus reverted. Since position 2 is heated to a maximum temperature within the precipitation temperature of θ', θ' precipitates and causes a small hardness peak right after welding (AW), as shown in Figure 15.13e. Precipitation of θ'', however, is not expected since the time typical of a welding cycle is not sufficient for its formation (18). During PWNA, the hardness increases slightly in the solutionized area at position 1 because of the formation of the GP zones. During PWAA, the hardness increases significantly in both this area and the base metal because of θ'' and θ' precipitation. The hardness recovery, however, is not as good near position 2, where some overaging has occurred during welding due to θ' precipitation.

Figure 15.14 shows the results of hardness measurements in a 3.2-mm-thick 6061 aluminum gas–tungsten arc welded in the T4 condition at 10 V, 110 A, and 4.2 mm/s (10 ipm) (11). A small peak appears in the as-welded condition, which is still visible after PWNA here but may be less clear in other cases. Similar results have been reported by Burch (19).

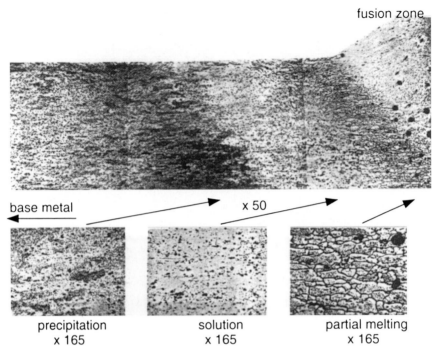

Figure 15.12   Precipitation zone in HAZ of a 2024 aluminum welded in naturally aged condition (weld metal at upper right corner). Reprinted from Arthur (17). Courtesy of American Welding Society.

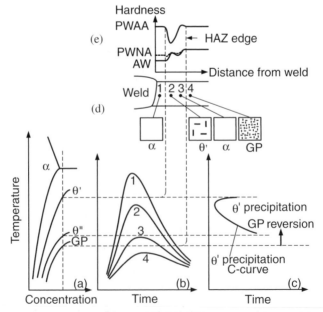

Figure 15.13   Al–Cu alloy heat treated to contain GP zones before welding: (*a*) phase diagram; (*b*) thermal cycles; (*c*) precipitation C curves; (*d*) microstructure; (*e*) hardness distribution. *θ* in base metal not shown.

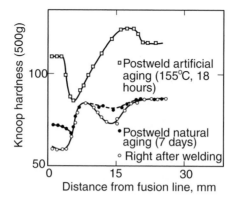

Figure 15.14   HAZ hardness profiles in a 6061 aluminum gas–tungsten arc welded in T4 condition. From Kou and Le (11).

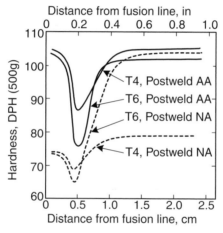

Figure 15.15   HAZ hardness profiles in 6061 aluminum welded in T4 or T6 and postweld naturally or artificially aged. Reprinted from Metzger (20). Courtesy of American Welding Society.

Figure 15.15 shows the results of hardness measurements in a 6061 aluminum (20). It suggests that welding a heat-treatable 6000- or 2000-series alloy in the T6 (artificially aged) condition can result in severe loss of strength (hardness) due to overaging. For this reason, welding in the T4 condition is often preferred to welding in the T6 condition (19, 20).

15.2.3   Effect of Welding Processes and Parameters

The loss of strength in the HAZ can be significantly affected by the welding process and by the heat input and welding speed. Figure 15.16 shows that as

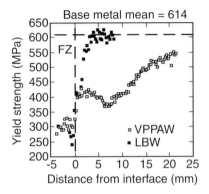

Figure 15.16   HAZ strength distributions in 2195-T8 aluminum made by LBW and PAW. Reprinted from Martukanitz and Howell (21).

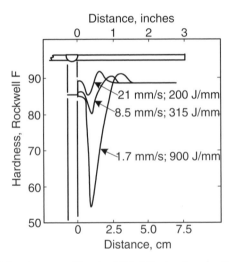

Figure 15.17   HAZ hardness profiles in 6061-T4 aluminum after postweld artificial aging. Reprinted from Burch (19). Courtesy of American Welding Society.

compared to variable-polarity PAW, LBW results in significantly less strength loss in the HAZ of a 2195-T8 aluminum (Al–Cu–Li) (21). Again, T8 stands for solution heat treating, cold working, followed by artificially aging. As shown in Figure 15.17, the higher the heat input per unit length of the weld (the higher the ratio of power input to welding speed), the wider the HAZ and the more severe the loss of strength (19). Therefore, the heat input should be limited when welding heat-treatable aluminum alloys.

Apparently, full strength can be recovered if the entire workpiece is solutionized, quenched, and artificially aged after welding. The preweld condition can be either solution heat treated or fully annealed, although the latter is somewhat inferior due to its poor machinability (18). However, for large welded structures heat-treating furnaces may not be available. In addition, distortion of the welded structure developed during postweld solution heat treating and quenching may be unacceptable.

15.3   Al–Zn–Mg ALLOYS

The age hardening of Al–Zn–Mg alloys is quite different from that of either 6000- or 2000-series aluminum alloys. As shown in Figure 15.18, alloy 7005 (Al–4.5Zn–1.2Mg) ages much more slowly than alloy 2014 (Al–4.5Cu–0.6Mg) (22). In fact, Al–Zn–Mg alloys, such as 7005, 7039, and 7146, age much more slowly than either 6000- or 2000-series aluminum alloys. As a result, Al–Zn–Mg alloys have a much smaller tendency to overage during welding than the other alloys. Furthermore, unlike 6000- or 2000-series aluminum alloys, Al–Zn–Mg alloys recover strength slowly but rather significantly by natural aging. For these reasons Al–Zn–Mg alloys are attractive when post-weld heat treatment is not practical. Figure 15.19 shows the hardness profiles in alloy 7005 after welding (23). Welding such an alloy in its naturally aged condition is most ideal since the strength in the HAZ can be recovered almost

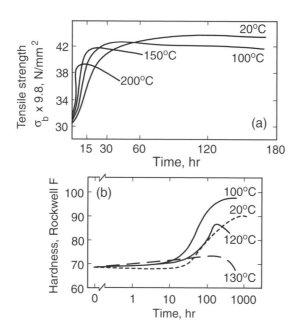

Figure 15.18 Aging characteristics of heat-treatable aluminum alloys: (*a*) Al–4.5Cu–0.6Mg quenched from 500°C; (*b*) Al–4.5Zn–1.2Mg quenched from 450°C (22).

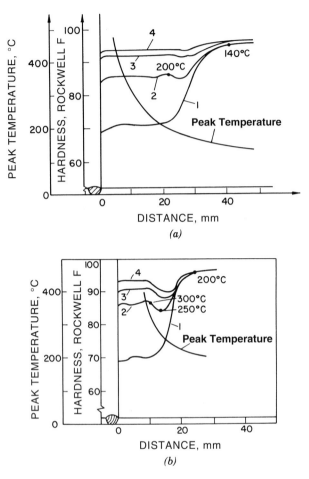

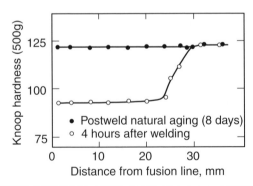

Figure 15.19 HAZ hardness profiles in Al–4.5Zn–1.2Mg alloy: (*a*) naturally aged before welding (1:3 hour, 2:4 days, 3:30 days, 4:90 days); (*b*) artificially aged at 130°C for 1 h before welding. From Mizuno (23).

Figure 15.20 HAZ hardness profiles in 7146 aluminum naturally aged before welding. From Kou and Le (11).

completely by postweld natural aging. Figure 15.20 shows similar results in alloy 7146 (Al–7.1Zn–1.3Mg) (11).

Similar to the welding of 6000- or 2000-series alloys, excessive heat inputs should be avoided in welding Al–Zn–Mg alloys in the artificially aged condition. Figure 15.21 shows the hardness profiles in the HAZ of alloy 7039 artificially aged before welding (24). As shown, the loss of strength can be reduced significantly by increasing the number of passes (thus decreasing the heat input in each pass) and maintaining a low interpass temperature.

Welding a heat-treatable aluminum alloy in the annealed condition is almost the opposite of welding it in the aged condition. Figure 15.22 shows the hardness profiles in an annealed heat-treatable aluminum alloy after welding (11). The base metal is relatively soft because of annealing. The HAZ is solutionized during welding and thus gets stronger by solution strengthening, as evident from the hardness increase in the HAZ after welding. The further

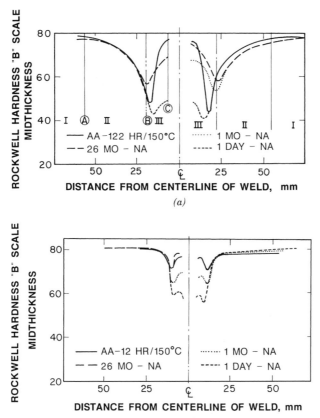

Figure 15.21 HAZ hardness profiles in 3-cm-thick artificially aged 7039 aluminum: (*a*) 4 passes, continuous welding; (*b*) 16 passes, 150°C interpass temperature. Reprinted from Kelsey (24). Courtesy of American Welding Society.

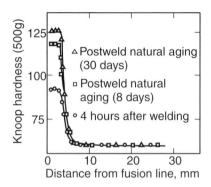

Figure 15.22   HAZ hardness profiles in alloy 7146 welded in annealed condition. From Kou and Le (11).

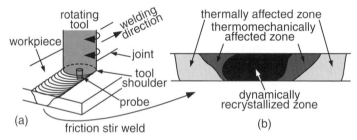

Figure 15.23   Friction stir welding: (*a*) process; (*b*) transverse cross section of resultant weld.

hardness increases in the HAZ after natural aging are consistent with the remarkable ability of Al–Zn–Mg alloys to gain strength by natural aging.

15.4   FRICTION STIR WELDING OF ALUMINUM ALLOYS

Friction stir welding is a solid-state joining process developed at the Welding Institute (25). As shown in Figure 15.23*a*, a rotating cylindrical tool with a probe is plunged into a rigidly clamped workpiece and traversed along the joint to be welded. Welding is achieved by plastic flow of frictionally heated material from ahead of the probe to behind it. For welding aluminum alloys, the tool is usually made of tool steel. As shown in Figure 15.23*b*, the resultant weld consists of three zones: thermally affected zone, thermomechanically affected zone, and dynamically recrystallized zone. In the thermally affected zone the grain structure is not affected by welding. In the thermomechanically affected zone, however, the grains are severely twisted. In the dynamically recrystallized zone, which is also called the weld nugget, all old grains

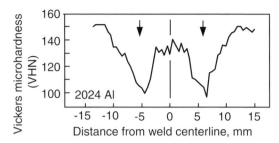

Figure 15.24   Hardness profile across friction stir weld of an artificially aged alloy 2024 (26). The arrows indicate the weld nugget.

disappear and numerous small new grains recrystallize. As in the HAZ in fusion welds of precipitation-hardened aluminum alloys, heating during welding can cause considerable loss of strength. Reversion of precipitate, overaging, and solutionizing occur during welding, from near the base metal to the weld centerline. Figure 15.24 shows a typical hardness distribution in precipitation-hardened aluminum alloys (26).

REFERENCES

1. Hornbogen, E., *Aluminum*, **43**(part 11): 9, 1967.

2. Beton, R. H., and Rollason, E. C., *J. Inst. Metals*, **86:** 77, 1957–58.

3. Nutting, J., and Baker, R. G., *The Microstructure of Metals*, Institute of Metals. London, 1965, pp. 65, 67.

4. Smith, W. F., *Structure and Properties of Engineering Alloys*, McGraw-Hill, New York, 1981.

5. Dumolt, S. D., Laughlin, D. E., and Williams, J. C., in *Proceedings of the First International Aluminum Welding Conference*, Welding Research Council, New York, p. 115.

6. Silcock, J. M., Heal, J. J., and Hardy, H. K., *J. Inst. Metals*, **82:** 239, 1953.

7. Guy, A. G., *Elements of Physical Metallurgy*, Addison-Wesley, Reading, MA, 1959.

8. Hundicker, H. Y., in *Aluminum*, Vol. 1, American Society for Metals, Metals Park, OH, 1967, Chapter 5, p. 109.

9. Metals Handbook, vol. 2, 8th edition, American Society for Metals, Metals Park, OH, 1964, p. 276.

10. *Aluminum Standards and Data*, Aluminum Association, New York, 1976, p. 15.

11. Kou, S., and Le. Y., unpublished research, Carnegie-Mellon University, Pittsburgh, PA, 1982.

12. Malin, V., *Weld. J.*, **74:** 305s, 1995.

13. Panseri, C., and Federighi, T., *J. Inst. Metals*, **94:** 94, 1966.

14. Miyauchi, T., Fujikawa, S., and Hirano, K., *J. Jpn. Inst. Light Metals*, **21:** 595, 1971.

15. Rading, G. O., Shamsuzzoha, M., and Berry, J. T., *Weld. J.*, **77:** 411s, 1998.

16. Langan, T. J., and Pickens, J. R., in Aluminum-Lithium Alloys, Vol. II, Eds. T. H. Sanders, Jr. and E. A. Starke, Jr., Materials and Component Engineering Publications, Birmingham, UK, p. 691.

17. Arthur, J. B., *Weld. J.*, **34:** 558s, 1955.

18. *Introductory Welding Metallurgy*, American Welding Society, Miami, FL, 1968, p. 65.

19. Burch, W. L., *Weld. J.*, **37:** 361s, 1958.

20. Metzger, G. E., *Weld. J.*, **46:** 457s, 1967.

21. Martukanitz, R. P., and Howell, P. R., in *Trends in Welding Research*, Eds. H. B. Smartt, J. A. Johnson, and S. A. David, ASM International, Materials Park, OH, 1996, p. 553.

22. *Principles and Technology of the Fusion Welding of Metals*, Vol. 2, Mechanical Engineering Publishing Co., Peking, China, 1981 (in Chinese).

23. Mizuno, M., Takada, T., and Katoh, S., J. Japanese Welding Society, vol. 36, 1967, pp. 74–81.

24. Kelsey, R. A., *Weld. J.*, **50:** 507s, 1971.

25. Dawes, C. J., *Friction Stir Welding of Aluminum*, IIW-DOC XII-1437-96, 1996, pp. 49–57.

26. Murr, L. E., Li, Y., Trillo, E. A., Nowak, B. M., and McClure, J. C., *Alumin. Trans.*, **1**(1), 141–154, 1999.

27. A. Umgeher and H. Cerjak, in *Recent Trends in Welding Science and Technology*, Eds. S. A. David and J. M. Vitek, ASM International, Materials Park, OH, 1990, p. 279.

FURTHER READING

1. *Aluminum*, edited by J. E. Hatch, American Society for Metals, Metals Park, OH, 1984.
2. Mondolfo, L. F., *Aluminum Alloys: Structure and Properties*, Butterworths, London, 1976.
3. Polmear, I. J., *Light Alloys*, Edward Arnold, London, 1981.
4. Smith, W. F., *Structure and Properties of Engineering Alloys.* McGraw-Hill, New York, 1981.

PROBLEMS

15.1 A reciprocal relationship between the tensile strength of 2219 aluminum (essentially Al–6.3Cu) weldments and the heat input per unit length of weld per unit thickness has been observed. Explain why.

15.2 **(a)** Based on the results of mechanical testing for 2219 aluminum given in Table P15.2, comment on the effect of postweld heat treatment on the tensile strength.

TABLE P15.2   2219 Aluminum

Test Specimen	Procedure	Tensile Strength (N/mm²)	Bend Angle (deg)
Base metal	SS	320	180
	SS + AA	412	121
Weldment	SS + welding	254	64
	SS + AA + welding	287	54
	SS + welding + AA	300	44
	SS + welding + SS + AA	373	84
	AN + welding + SS + AA	403	93

Abbreviations: SS, solid solution; AA, artificial aging; AN, annealing.

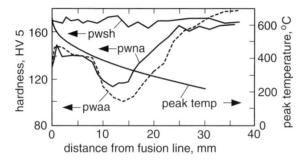

Figure P15.4

(b)   The bend angle (i.e., the maximum angle the specimen can be bent prior to failure) is an indication of the ductility of the material. Does the weld ductility seem to recover as effectively as the tensile strength? Explain why or why not.

15.3   A 12.7-mm-thick plate of 6061-T6 aluminum (T_L = 652°C) was gas–tungsten arc welded with DC electrode negative. The welding parameters were I = 222 A, E = 10.4 V, and V = 5.1 mm/s. Microhardness measurements after welding indicated that softening due to overaging starts about 5.3 mm from the fusion line and gradually increases as the fusion line is approached. Thermal measurements during welding revealed a peak temperature of about 300°C at the position where softening started. Calorimetric measurements showed that the arc efficiency was around 80%. How does the width of the HAZ compare with that predicted from Adams's equation (Chapter 2)?

15.4   Figure P15.4 show the hardness distributions measured by Umgeher and Cerjak (27) in 7075 aluminum after natural aging three months at room temperature (pwna), after artificial aging (pwaa), and after full postweld heat treatment of solutionizing, quenching, and then artificial aging (pwsh). Indicate the temperature ranges in which overaging and solutionizing occur during welding. Explain how these hardness distributions compare with each other.

15.5   Al–Li–Cu alloy 2095 was welded by LBW, GTAW, and GMAW and the HAZ hardness profiles of the resultant welds were measured. Rank the welds in the order of increasing hardness in the HAZ.

15.6   Al–Li–Cu alloy 2090 was welded with various filler metals such as 2319, 2090, 4047, and 4145. Joint efficiencies up to 65% of base-metal strength were obtained in the as-welded condition. After postweld solution heat treatment and artificial aging, joint efficiencies up to 98% were obtained. Explain why.

15.7   Sketch the hardness profiles in the following aluminum alloys after friction stir welding: (*a*) artificially aged 2219; (*b*) work-hardened 5083.

16 Precipitation-Hardening Materials II: Nickel-Base Alloys

Because of their high strength and good corrosion resistance at high temperatures, Ni-base alloys have become the most extensively used high-temperature alloys. Table 16.1 summarizes typical welding problems in Ni-base alloys and recommended solutions. The problems associated with the fusion zone and the PMZ have been discussed previously. In this chapter, we shall focus on weakening the HAZ and postweld heat treatment cracking in heat-treatable Ni-base alloys.

16.1 BACKGROUND

Table 16.2 shows the chemical compositions of several representative heat-treatable Ni-base alloys (1, 2). Aluminum and Ti are the two major precipitation-hardening constituents in heat-treatable Ni-base alloys, equivalent to Cu in heat-treatable Al–Cu alloys. As can be seen in Figure 16.1, the solubility of Ti or Al in the γ phase increases significantly with increasing temperature—a necessary criterion for precipitation hardening as in Al–Cu alloys (3).

Like heat-treatable Al alloys, the precipitation hardening of heat-treatable Ni-base alloys can be obtained by solutionizing at temperatures above the solvus, followed first by water quenching and then by artificial aging in the precipitation temperature range. In mill practice, however, the alloys are usually air cooled from the solutionizing temperature (usually in the range 1040–1180°C, or 1900–2150°F) to an intermediate aging temperature and held there for a number of hours before being further air cooled to a final aging temperature of about 760°C (1400°F). After aging at this final temperature for about 16h, the alloys can be air cooled to room temperature. For best results some alloys are aged at two, rather than one, intermediate temperatures. For example, both Udimet 700 and Astroloy are sometimes aged first at 980°C (1975°F) for 4h, then at 815°C (1500°F) for 24h before the 16-h final aging at 760°C (1400°F). For applications at low temperatures, the aging operation can be carried out solely at 760°C (1400°F) to avoid grain boundary carbide precipitation (2, 4).

The precipitation reaction to form the strengthening phase γ' can be written as follows (2):

TABLE 16.1   Typical Problems in Welding Nickel-Base Alloys

Typical Problems	Alloy Types	Solutions	Sections in Book
Low strength in HAZ	Heat-treatable alloys	Resolution and artificial aging after welding	16.2
Reheat cracking	Heat-treatable alloys	Use less susceptible grade (Inconel 718) Heat treat in vacuum or inert atmosphere Welding in overaged condition (good for Udimet 500) Rapid heating through critical temperature range, if possible	16.3
Hot cracking in PMZ	All types	Reduce restraint Avoid coarse-grain structure and Laves phase	13.1 13.2

TABLE 16.2   Composition of Heat-Treatable Nickel-Base Superalloys

Alloy	C	Cr	Co	W	Mo	Al	Ti	Others
Inconel X-750	0.04	16	—	—	—	0.6	2.5	7Fe, 1Cb
Waspaloy	0.07	19	14	—	3	1.3	3.0	0.1Zr
Udimet 700	0.10	15	19	—	5.2	4.3	3.5	0.02B
Inconel 718	0.05	18	—	—	3	0.6	0.9	18Fe, 5Cb
Nimonic 80A	0.05	20	<2	—	—	1.2	2.4	<5Fe
Mar-M200	0.15	9	10	12	—	5.0	2.0	1Cb, 2Hf
Rene 41	0.1	20	10	—	10	1.5	3.0	0.01B

Source: Owczarski (1) and Sims (2).

$$\underbrace{(\text{Ni, Cr, Co, Mo, Al, Ti})}_{\gamma} \rightarrow \underbrace{\text{Ni}_3(\text{Al, Ti})}_{\gamma'} + \underbrace{(\text{Cr, Co, Mo})}_{\substack{\text{matrix components}}} \quad (16.1)$$

where γ is a fcc matrix while γ', the precipitate, is an ordered fcc intermetallic compound. The $\text{Ni}_3(\text{Al,Ti})$ is only the abbreviation of γ'; the exact composition of γ' can be much more complicated. For example, the compositions of γ' in Inconel 713C and IN-731 have been determined to be (5)

$$(\text{Ni}_{0.980}\text{Cr}_{0.004}\text{Mo}_{0.004})_3(\text{Al}_{0.714}\text{Cb}_{0.099}\text{Ti}_{0.048}\text{Mo}_{0.038}\text{Cr}_{0.103})\gamma' \quad \text{in Inconel 713C}$$

and

$$(\text{Ni}_{0.884}\text{Co}_{0.070}\text{Cr}_{0.032}\text{Mo}_{0.008}\text{V}_{0.003})_3(\text{Al}_{0.632}\text{Ti}_{0.347}\text{V}_{0.013}\text{Cr}_{0.006}\text{Mo}_{0.002})\gamma' \quad \text{in IN-731}$$

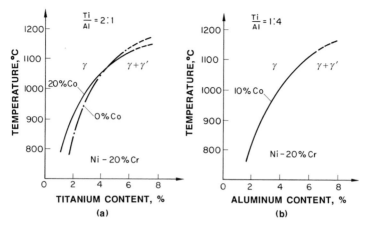

Figure 16.1   Effect of alloying elements on the solvus temperature of γ': (*a*) Ti; (*b*) Al. From Betteridge (3).

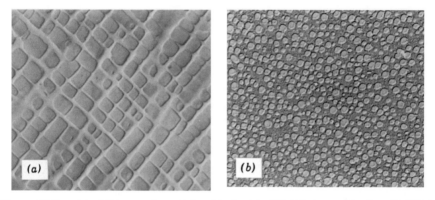

Figure 16.2 γ' in Ni-base alloys: (*a*) cubical γ' in IN-100 (magnification 13,625×); (*b*) spherical and cooling γ' in U500 (magnification 5450×). From Decker and Sims (6).

The precipitate γ' can assume several different shapes, such as spherical, cubical, and elongated. Figure 16.2 shows examples of cubical and spherical γ' precipitates in two different Ni-base alloys (6). It is interesting to note that, after the general precipitation of the dominant γ' particles, very fine γ' particles can further precipitate during cooling to room temperature. Such γ', called "cooling γ'," often generates roughening of the γ matrix. It should be pointed out here that the precipitation of γ' depletes the surrounding γ matrix of Al and Ti and results in a decrease in the lattice parameter of the matrix. This decrease creates the so-called aging contraction, which has been reported to be of the order of 0.1% (0.001 in./in.) in Rene 41 and 0.05% (0.0005 in./in.) in Inconel X-750 (7). As will be mentioned later in this chapter, the contraction strains so created hinder the relaxation of residual stresses in the HAZ and, therefore, promote the chance of postweld heat treatment cracking.

In most Ni-base alloys the high-temperature carbide MC can react with the γ matrix and form lower carbides, such as $M_{23}C_6$ and M_6C, according to the following reactions (6):

$$\underbrace{(Ti, Mo)C}_{MC} + \underbrace{(Ni, Cr, Al, Ti)}_{\gamma} \rightarrow \underbrace{Cr_{21} Mo_2 C_6}_{M_{23}C_6} + \underbrace{Ni_3(Al, Ti)}_{\gamma'} \qquad (16.2)$$

and

$$\underbrace{(Ti, Mo)C}_{MC} + \underbrace{(Ni, Co, Al, Ti)}_{\gamma} \rightarrow \underbrace{Mo_3(Ni, Co)_3 C}_{M_6C} + \underbrace{Ni_3(Al, Ti)}_{\gamma'} \qquad (16.3)$$

In alloys such as Udimet 700 the $M_{23}C_6$ carbide formed by reaction (16.2) appears as blocky carbide lining the grain boundaries. The γ' phase, on the other hand, envelops the $M_{23}C_6$ carbide along the grain boundaries, as shown schematically in Figure 16.3. If the $M_{23}C_6$ carbide develops in a brittle, cellular form rather than a hard, blocky form, the ductility and rupture life of the alloys are reduced. Since alloys that generate profuse γ' at grain boundaries appear to be resistant to cellular $M_{23}C_6$, grain boundary γ' formed by reaction (16.2) may play an important role in blocking its growth. The hard, blocky $M_{23}C_6$ carbide may initially strengthen the grain boundary beneficially. Ultimately, however, such $M_{23}C_6$ particles are the sites of the initiation of rupture fracture (2). In alloys such as Nimonic 80A and Inconel-X, the grain boundary γ' has not been noted as a product of reaction (16.2). In fact, as shown in Figure 16.3, the areas adjacent to the grain boundary are depleted of γ'. This can be caused by diffusion of Cr to form grain boundary carbides. Since these areas are depleted in Cr, their solubility for Ni and Al increases, thus causing the disappearance of γ' (2).

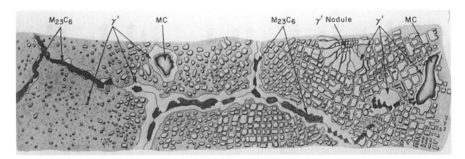

Figure 16.3 Schematic sketch of microstructure observed in some Ni-base superalloys. From Decker and Sims (6).

From the above discussion it is clear that the high strength of heat-treatable Ni-base alloys is due primarily to the precipitation hardening of γ' and the resistance to grain boundary sliding provided by carbides. However, Inconel 718 is an exception. It utilizes niobium (Nb) as its primary strengthening alloying element, and γ'' (an ordered bct intermetallic compound of composition Ni_3Nb) rather than γ' is responsible for precipitation hardening during aging.

In addition to γ', γ'', and carbides, a group of phases called *topologically close packed phases* can also be present in certain Ni-base alloys where composition control has not been carefully watched (6). Such phases, for instance, σ and μ, often appear as hard, thin plates and thus promote lowered rupture strength and ductility. However, in most Ni-base alloys such undesired phases do not usually appear, unless significant alteration of the matrix composition has occurred as a result of extensive exposure in the aging temperature range. Therefore, they are not of great concern during welding (4). Nevertheless, it should be pointed out that the presence of the *Laves phase*, which is also a topologically close packed phase, has been reported to promote hot cracking in Inconel 718 and A-286 due to its lower melting point (8–10).

Like heat-treatable Al alloys, heat-treatable Ni-base alloys can also *overage*. As seen in Figure 16.4a, the optimum aging temperature for Inconel X-750 is around 760°C (1400°F), above which it tends to overage. The aging characteristics of several heat-treatable Ni-base alloys at this temperature are shown in Figure 16.4b. Inconel 718, which is precipitation hardened by γ'', ages much more slowly than other alloys. As will be discussed later in this chapter, the slow aging characteristic of Inconel 718 makes it more resistant to cracking during postweld heat treatment.

16.2 REVERSION OF PRECIPITATE AND LOSS OF STRENGTH

Consider welding a heat-treatable Ni-base alloy in the aged condition, as shown in Figure 16.5. The area adjacent to the weld is heated above the precipitation temperature range of γ'. Reversion of γ' ranges from partial reversion near the edge of the HAZ (point 2) to full reversion near the fusion boundary (point 1). This γ' reversion causes loss of hardness or strength in the HAZ.

16.2.1 Microstructure

Owczarski and Sullivan (13) studied the reversion of the strengthening precipitates in the HAZ of Udimet 700 during welding. This material was in the full aged condition produced by the following heat treatment before welding:

1165°C/4 h + air cool (solution)
1080°C/4 h + air cool (primary age)

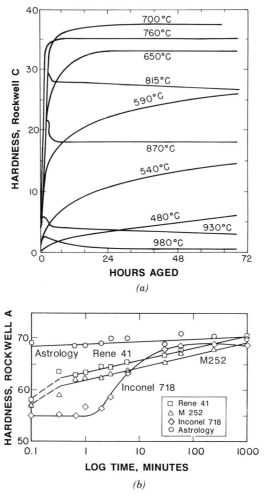

Figure 16.4 Aging characteristics of Ni-base alloys. (*a*) Inconel X-750. From Eiselstein (11). (*b*) Some other Ni-base alloys. Reprinted from Wilson and Burchfield (12). Courtesy of American Welding Society.

845°C/4 h + air cool (intermediate age)
760°C/16 h + air cool (final age)

The resultant HAZ microstructure is shown in Figure 16.6. The unaffected base metal consists of coarse angular γ' and fine spherical γ' between the coarse γ' (Figure 16.6*a*). The initial stage of reversion just inside the HAZ is characterized by the disappearing of the fine γ' and the rounding of the coarse angular γ' (Figure 16.6*b*). Further reversion of coarse γ' is evident in the middle of the HAZ (Figure 16.6*c*). Since this material is highly alloyed with

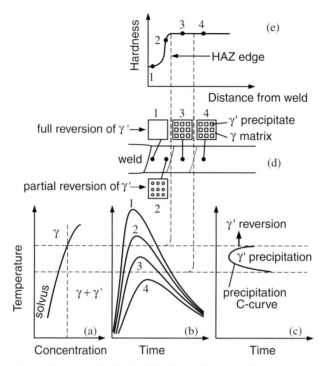

Figure 16.5   Reversion of γ' in HAZ: (*a*) phase diagram; (*b*) thermal cycles; (*c*) precipitation C curve; (*d*) microstructure; (*e*) hardness distribution.

Ti and Al, the areas near the reverted coarse γ' become so supersaturated with Ti and Al that finer γ' reprecipitates during cooling. This localized supersaturation is due to the fact that the retention time at high temperatures is too short to allow homogenization to occur in this region. Reversion continues toward completion as the fusion boundary is approached (Figure 16.6*d*). The prior sites of coarse γ' particles are marked by a periodic pattern of very fine γ' reprecipitated during cooling. The ultimate solution and distribution of γ' occur in the weld metal itself (Figure 16.6*e*). The weld metal contains a uniform distribution of fine γ' precipitate.

Figure 16.7 shows the HAZ microstructure after 16h postweld heat treatment at 760°C (1400°F). Reprecipitation of very fine γ' occurs in the region where coarse γ' has begun to dissolve during welding (Figure 16.6*b*). This is caused by the precipitation of the elements that were taken into solution during welding.

16.2.2   Hardness Profiles

Lucas and Jackson (14) and Hirose et al. (15) measured hardness profiles across welds of Inconel 718. Figure 16.8 shows the hardness distributions of

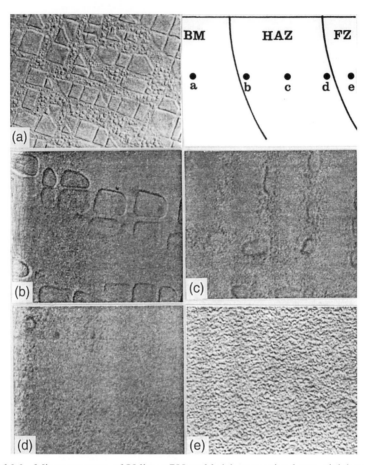

Figure 16.6   Microstructure of Udimet 700 weld: (*a*) as-received material (magnification 10,000×); (*b*) initial solution of fine γ'; (*c*) further solution of coarse γ'; (*d*) advanced stage of solution of coarse γ'; (*e*) weld metal containing fine γ'. (*b–d*). Magnification 15,000×. Reprinted from Owczarski and Sullivan (13). Courtesy of American Welding Society. Reduced to 84% in reproduction.

Hirose et al. (15) in Inconel 718 laser and gas–tungsten arc welded in the as-welded condition. The solutionized and laser-welded (SL) workpiece was not much affected by welding. The workpiece solutionized, aged, and laser welded (AL), however, became much softer in the HAZ. This is because of precipitate reversion in the HAZ and the fusion zone. The HAZ is much narrower in the laser weld (AL) than in the gas–tungsten arc weld (AT) because of the lower heat input used in the former. Aging after welding helped achieve the maximum hardness, either aged after welding or solutionized and then aged after welding, as shown in Figure 16.9.

Figure 16.7   HAZ of Udimet 700 showing reprecipitation of fine γ' in region where coarse γ' begins to dissolve. Magnification 16,000×. Reprinted from Owczarski and Sullivan (13). Courtesy of American Welding Society.

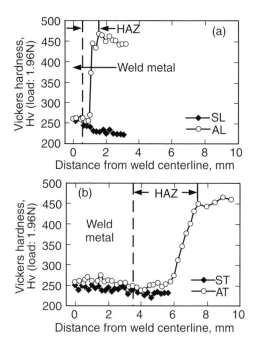

Figure 16.8   Hardness profiles in Inconel 718 welds in as-welded condition: (*a*) laser welds; (*b*) gas–tungsten arc welds. S: solutionized; A: aged after solutionization; L: laser welded; T: gas–tungsten arc welded. Broken line indicates fusion line. From Hirose et al. (15).

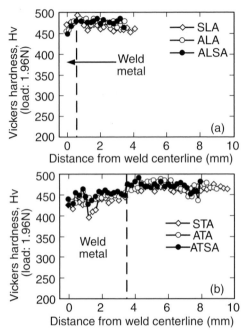

Figure 16.9 Hardness profiles in Inconel 718 welds after postweld heat treating: (*a*) laser welds; (*b*) gas–tungsten arc welds. S: solutionized; A: aged after solutionization; L: laser welded; T: gas–tungsten arc welded. Broken line indicates fusion line. From Hirose et al. (15).

16.3 POSTWELD HEAT TREATMENT CRACKING

Cracking can occur during the postweld heat treatment of heat-treatable Ni-base alloys. Hot cracking in both the fusion zone and the partially melted zone of heat-treatable Ni-base alloys (8, 14, 16–28) are similar to those in other materials (Chapters 11–13) and will not be discussed separately here.

16.3.1 Reasons for Postweld Heat Treatment

Heat-treatable Ni-base alloys are often postweld heat treated for two reasons: (i) to relieve stress, and (ii) to develop the maximum strength. To develop its maximum strength, the weldment is first solutionized and then aged. During solutionization the residual stresses in the weldment are also relieved. The problem is that aging may occur in the weldment while it is being heated up to the solutionization temperature because the aging temperature range is below the solutionization temperature. Since this aging action occurs before the residual stresses are relieved, it can cause cracking during postweld heat treatment. Such postweld heat treatment cracking is also called *strain-age cracking* or simply *reheat cracking*. The term strain-age cracking arises from the fact that cracking occurs in highly restrained weldments, as they are heated through the temperature range in which aging occurs.

16.3.2   Development of Cracking

Figure 16.10 shows the development of postweld heat treatment cracking. The precipitation temperature range is from T_1 to T_2 (Figure 16.10a). To relieve the residual stresses after welding, the workpiece is brought up to the solutionization temperature (Figure 16.10b). It passes through the precipitation temperature range. Unless the heating rate is high enough to avoid intersecting the precipitation C curve, precipitation and hence cracking will occur (Figure 16.6c). The microstructural changes in the HAZ are illustrated in Figures 16.10d and e.

Postweld heat treatment cracks usually, though not always, initiate in the HAZ. However, as shown in the weld circle-patch test in Figure 16.11a, it can propagate into regions unaffected by the welding heat (4). Such a test, which is often used for evaluating the strain-age cracking tendency of a material, is achieved by welding the circle-patch specimen to a stiffener (strong back) so that the combination can be heat treated without relaxation (stress relief) due to mechanical factors. As shown in Figure 16.11b, the cracks in the HAZ are intergranular (29).

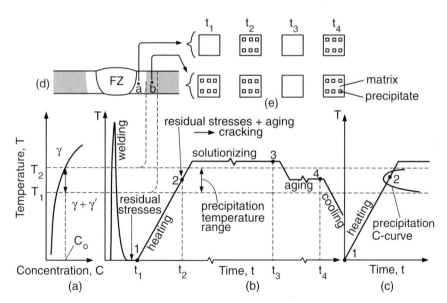

Figure 16.10 Postweld heat treatment cracking: (*a*) phase diagram; (*b*) thermal cycles during welding and heat treating; (*c*) precipitation C curve; (*d*) weld cross-section; (*e*) changes in microstructure.

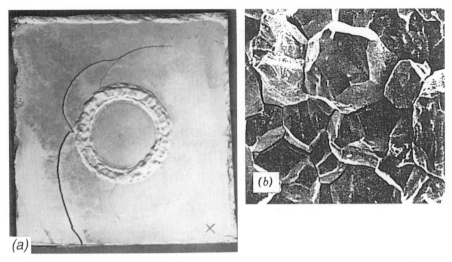

Figure 16.11   Postweld heat treatment cracking. (*a*) Circular-patch specimen of Rene 41. From Prager and Shira (4). (*b*) Scanning electron micrograph showing intergranular cracking. Reprinted from McKeown (29). Courtesy of American Welding Society.

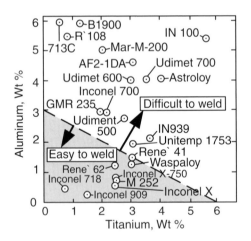

Figure 16.12   Effect of Al and Ti contents on postweld heat treatment cracking. Modified from Kelly (30).

16.3.3   Effect of Composition

Figure 16.12 shows the effect of Al and Ti contents on the postweld heat treatment cracking tendency in Ni-base alloys (30). Such a plot was first proposed by Prager and Shira (4). As can be seen, the γ'-strengthened Ni-base alloys with high Al and Ti contents are particularly difficult to weld because of high susceptibility to cracking. This is because such alloys tend to age harden rather rapidly and because their ductility is low.

16.3.4   Proposed Mechanisms

Postweld heat treatment cracking in Ni-base alloys is a result of low ductility and high strains in the HAZ (31, 32). So far several mechanisms have been proposed for the causes of low ductility in the HAZ, including, for example, embrittlement of the grain boundary due to liquidation or solid-state reactions during welding (33–36), embrittlement of the grain boundary by oxygen during heat treatment (37–39), and a change in deformation mode from transgranular slip to grain boundary sliding (14, 19, 22). The causes of high strains in the HAZ, on the other hand, can be the welding stresses and the thermal expansion and contraction of the material. In heat-treatable Ni-base alloys, the precipitation of strengthening phases results in contraction during aging. This aging contraction, in fact, has been proposed by several investigators (7, 22, 33, 40) to be a contributing factor to postweld heat treatment cracking in heat-treatable Ni-base alloys.

16.3.5   Remedies

Several different methods of avoiding postweld heat treatment cracking have been recommended. Most of these methods are based on experimentally observed crack susceptibility C curves. A crack susceptibility C curve is a curve indicating the onset of postweld heat-treatment cracking in a temperature–time plot. It is usually obtained by isothermal heat treating of welded circle patches at different temperatures for different periods of time and checking for cracking. Such a curve usually resembles the shape of a "C" and, therefore, is called a crack susceptibility C curve. Since the aging rate at the lower end of the aging temperature range is relatively slow, the occurrence of cracking approaches asymptotically some lower temperature limit. Likewise, since residual stresses are relaxed at the higher temperature end of the aging temperature range and precipitates formed at lower temperature are dissolved, the occurrence of cracking approaches asymptotically some higher temperature limit. At any temperature between the upper and lower asymptotic limits, there exists a minimum time, prior to which no cracking is possible and beyond which cracking is certain to occur (37).

Figure 16.13 shows the crack susceptibility C curves of Waspaloy and Inconel 718 (1). Inconel 718 ages much more sluggishly than γ'-strengthened Ni-base alloys (Figure 16.4*b*). As a result, the C curve of Inconel 718 is far to the right of the C curve of Waspaloy, suggesting that the former is much more resistant to postweld heat treatment cracking. In fact, Inconel 718 is a material designed specifically to minimize postweld heat treatment cracking and should be considered when postweld heat treatment cracking is of major concern. However, other alloys may still have to be used because of specific

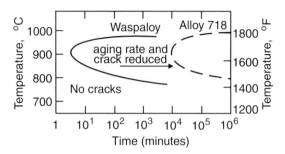

Figure 16.13 Crack susceptibility C curves for Waspaloy and Inconel 718 welds. Reprinted, with permission, from Owczarski (1).

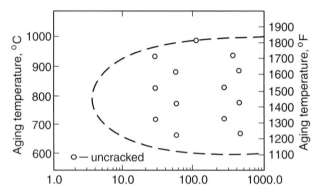

Figure 16.14 Crack susceptibility C curve for a Rene 41 solution annealed before welding and crack test data (O) from a Rene 41 overaged before welding. Reprinted from Berry and Hughes (36). Courtesy of American Welding Society.

requirements involved, and the following approaches to reduce the cracking susceptibility are worth considering.

Preweld overaging, either by multistep-type overaging or simply by cooling slowly from the solutionizing temperature, appears to significantly reduce postweld heat treatment cracking in Rene 41 (36, 37, 41), Figure 16.14 being an example (36). The base metal of an overaged workpiece is more ductile and does not age contract during postweld heat treatment, and this helps prevent severe residual stresses in the HAZ. But, as Franklin and Savage (32) pointed out, the overaged precipitate is dissolved in the HAZ during welding, and consequently the HAZ may still harden and age contract during postweld heat treatment.

The effect of preweld solution annealing appears to be controversial. Gottlieb (42) reported that solution-annealed Rene 41 (solutionized at 1080°C, or 1975°F, for half an hour followed by water quenching) is in fact more susceptible to postweld heat treatment cracking than fully age hardened Rene 41

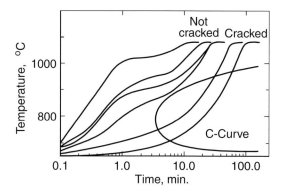

Figure 16.15 Effect of heating rate on postweld heat treatment cracking of a Rene 41 solution annealed before welding. Reprinted from Berry and Hughes (36). Courtesy of American Welding Society.

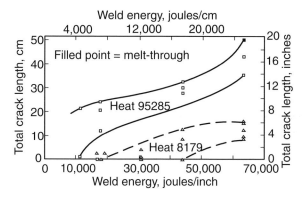

Figure 16.16 Effect of welding heat input on postweld heat treatment cracking of Rene 41. Reprinted from Thompson et al. (37). Courtesy of American Welding Society.

[solutionized at 1080°C for half an hour, water quenched, aged at 760°C (1400°F) for 4–16 h and then air cooled]. Berry and Hughes (36), on the other hand, reported that fully age hardened Rene 41 behaves essentially the same as solution-annealed Rene 41 during postweld heat treatment.

It is true that the base metal of a solution-annealed workpiece can age contract as well as harden during postweld heat treatment, and effective stress relief can be difficult. However, if the same weldment is heated rapidly during postweld heat treatment, effective stress relief can be achieved before it has a chance to harden and age contract, thus avoiding cracking (36, 37). As shown in Figure 16.15, no cracking occurs if the weldment is heated rapidly to avoid intersecting the crack susceptibility C curve (36). This approach is feasible when the welded structure can be heated rapidly in a furnace and when distortions due to nonuniform heating are not excessive.

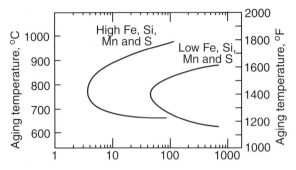

Figure 16.17 Effect of composition on postweld heat treatment cracking of Rene 41. Reprinted from Berry and Hughes (36). Courtesy of American Welding Society.

Another way of avoiding postweld heat treatment cracking is to use vacuum or inert atmospheres for heat treating (4, 36–38). It has been postulated that this technique is effective since there is no oxygen present to embrittle the grain boundary during postweld heat treatment (37–39, 43). Other recommended approaches include using low welding heat inputs (Figure 16.16), using small grain-size materials (36, 39), and controlling the composition (Figure 16.17). Of course, using low-restraint joint designs is helpful.

Finally, reheat cracking has also been reported in other alloys including $\frac{1}{2}$Cr–$\frac{1}{2}$Mo–$\frac{1}{4}$V steel (a creep-resistant ferritic steel), A517F (T1) steel (a structural steel), and 18Cr–12Ni–1Nb steel (a Nb-stabilized stainless steel) (44). The reheat cracking of creep-resistant ferritic steels will be discussed in Chapter 17.

REFERENCES

1. Owczarski, W. A., in *Physical Metallurgy of Metal Joining*, Eds. R. Kossowsky and M. E. Glicksman, Metallurgical Society of AIME, New York, 1980, p. 166.
2. Sims, C. T., *J. Metals*, **18:** 1119, October 1966.
3. Betteridge, W., *The Nimonic Alloys*, Arnold, London, 1959.
4. Prager, M., and Shira, C. S., *Weld. Res. Council Bull.*, **128:** 1968.
5. Mihalisin, J. R., and Pasquzne, D. L., Structural Stability in Superalloys, vol. 1, Joint ASTM-ASME, 1968, pp. 134–170.
6. Decker, R. F., and Sims, C. T., in *The Superalloys*, Eds. C. T. Sims and W. C. Hagel, Wiley, New York, 1972, p. 33.
7. Blum, B. S., Shaw, P., and Wickesser, A., Technical Report ASD-TRD-63-601, Republic Aviation Corp., Farmingdale, NY, June 1963.
8. Thompson, E. G., *Weld. J.*, **48:** 70s, 1969.
9. Thompson, E. G., unpublished research, University of Alabama, Birmingham, 1981.
10. Brooks, J. A., *Weld. J.*, **53:** 517s, 1974.

11. Eiselstein, H. L., *Adv. Technol. Stainless Steels*, Special Technical Publication No. 369.

12. Wilson, R. M. Jr., and Burchfield, L. W. G., *Weld. J.*, **35:** 32s, 1956.

13. Owczarski, W. A., and Sullivan, C. P., *Weld. J.*, **43:** 393s, 1964.

14. Lucas, M. J. Jr., and Jackson, C. E., *Weld. J.*, **49:** 46s, 1970.

15. Hirose, A., Sakata, K., and Kobayahi, K. F., in *Solidification Processing 1997*, Eds. J. Beech and H. Jones, Department of Engineering Materials, University of Sheffield, Sheffield, United Kingdom, 1997, p. 675.

16. Savage, W. F., Nippes, E. F., and Goodwin, G. M., *Weld. J.*, **56:** 245s, 1977.

17. Savage, W. F., and Krantz, B. M., *Weld. J.*, **50:** 29s, 1971.

18. Savage, W. F., and Krantz, B. M., *Weld. J.*, **45:** 13s, 1966.

19. Owczarski, W. A., Duvall, D. S., and Sullivan, C. P., *Weld. J.*, **45:** 145s, 1966.

20. Duvall, D. S., and Owczarski, W. A., *Weld. J.*, **46:** 423s, 1967.

21. Yeniscavich, W., *Weld. J.*, **45:** 344s, 1966.

22. Wu, K. C., and Herfert, R. E., *Weld. J.*, **46:** 32s, 1967.

23. Yeniscavich, W., in *Proceedings of the Conference on Methods of High-Alloy Weldability Evaluation*, Welding Research Council, New York, 1970, p. 2.

24. Gordine, J., in *Proceedings of the Conference on Methods of High-Alloy Weldability Evaluation*, Welding Research Council, New York, 1970, p. 28.

25. Owczarski, W. A., in *Proceedings of the Conference on Effects of Minor Elements on the Weldability of High-Nickel Alloys*, Welding Research Council, New York, 1967, p. 6.

26. Canonico, D. A., Savage, W. F., Werner, W. J., and Goodwin, G. M., in *Proceedings of the Conference on Effects of Minor Elements on The Weldability of High-Nickel Alloys*, Welding Research Council, New York, 1967, p. 68.

27. Valdez, P. J., and Steinman, J. B., in *Proceedings of the Conference on Effects of Minor Elements on The Weldability of High-Nickel Alloys*, Welding Research Council, New York, 1967, p. 93.

28. Grotke, G. E., in *Proceedings of the Conference on Effects of Minor Elements on The Weldability of High-Nickel Alloys*, Welding Research Council, New York, 1967, p. 138.

29. McKeown, D., *Weld. J.*, **50:** 201s, 1971.

30. Kelly, T. J., in *Weldability of Materials*, Eds. R. A. Patterson and K. W. Mahin, ASM International, Materials Park, OH, 1990, p. 151.

31. Baker, R. G., and Newman, R. P., *Metal Construction Br. Weld. J.*, **1:** 4, 1969.

32. Franklin, J. G., and Savage, W. F., *Weld. J.*, **53:** 380s, 1974.

33. Chang, W. H., Report DM58302 (58AD-16), General Electric, October 1958.

34. Hughes, W. P., and Berry, T. F., *Weld. J.*, **46:** 361s, 1967.

35. Morris, R. J., *Metal Prog.* **76:** 67, 1959.

36. Berry, T. F., and Hughes, W. P., *Weld. J.*, **46:** 505s, 1969.

37. Thompson, E. G., Nunez, S., and Prager, M., *Weld. J.*, **47:** 299s, 1968.

38. Carlton, J. B., and Prager, M., *Weld. Res. Council Bull.*, **150:** 13, 1970.

39. Prager, M., and Sines, G., *Weld. Res. Council Bull.*, **150:** 24, 1970.

40. Schwenk, W., and Trabold, A. F., *Weld. J.*, **42:** 460s, 1963.

41. Fawley, R. W., and Prager, M., *Weld. Res. Council Bull.*, **150:** 1, 1970.
42. Gottlieb, T.: unpublished Rocketdyne data.
43. Prager, M. A., and Thompson, E. G., Report R-71 11, Rocketdyne, September 1967.
44. Nichols, R. W., *Weld. World*, **7**(4)**:** 1969, p. 245.

FURTHER READING

1. Sims, C. T., and Hagel, W. C., Eds., *The Superalloys*, Wiley, New York, 1972.
2. Thamburaj, R., Wallace, W., and Goldak, J. A., *Int. Metals Rev.*, **28:** 1, 1983.

PROBLEMS

16.1 Explain why the susceptibility of Rene 41 welded in the solution anneal condition to postweld heat treatment cracking increases with increasing welding heat input.

16.2 It has been observed that the temperature of solution heat treatment before welding can significantly affect the susceptibility of Rene 41 to postweld heat treatment cracking. For instance, specimens subjected to 2150°F preweld solution heat treatment have been found to be more susceptible than those subjected to a 1975°F treatment. Explain why.

16.3 It has been reported that, in developing strain-age cracking C curves for Rene 41, water quenching following isothermal heat treatment of the welded circle patches often results in cracking. (a) Do you expect the C curves so developed to be reliable? (b) It has been suggested that at the end of isothermal heat treatment the furnace temperature be raised to 1975°F and kept there for 30 min and that the welded circle patches then be allowed to furnace cool at a rate of about 3–8°F/min (1.7–4.4°C/min). Cracking during cooling has been eliminated this way. Explain why. Do you expect the cracking C curves so obtained to be more reliable than those mentioned earlier?

16.4 Two rules are often quoted in postweld heat treatment of nickel-base alloys. First, never directly age weldments of heat-treatable nickel-base alloys. Second, the aging temperatures should exceed the service temperatures of the weldments. Explain why.

17 Transformation-Hardening Materials: Carbon and Alloy Steels

Carbon and alloy steels are more frequently welded than any other materials because of their widespread applications and good weldability. In general, carbon and alloy steels with higher strength levels are more difficult to weld because of the risk of hydrogen cracking. Table 17.1 summarizes some typical welding problems in carbon and alloy steels and their solutions. The problems associated with the fusion zone and the partially melted zone have been discussed in previous chapters. This chapter deals with basic HAZ phenomena in selected carbon and low-alloy steels.

17.1 PHASE DIAGRAM AND CCT DIAGRAMS

The HAZ in a carbon steel can be related to the *Fe–C phase diagram*, as shown in Figure 17.1, if the kinetic effect of rapid heating during welding on phase transformations is neglected. The HAZ can be considered to correspond to the area in the workpiece that is heated to between the lower critical temperature A_1 (the eutectoid temperature) and the peritectic temperature. Similarly, the PMZ can be considered to correspond to the areas between the peritectic temperature and the liquidus temperature, and the fusion zone to the areas above the liquidus temperature.

The Fe–C phase diagram and the *continuous-cooling transformation (CCT) diagrams* for heat treating carbon steels can be useful for welding as well, but some fundamental differences between welding and heat treating should be recognized. The thermal processes during the welding and heat treating of a carbon steel differ from each other significantly, as shown in Figure 17.2. First, in welding the peak temperature in the HAZ can approach 1500°C. In heat treating, however, the maximum temperature is around 900°C, which is not much above the upper critical temperature A_3 required for austenite (γ) to form. Second, the heating rate is high and the retention time above A_3 is short during most welding processes (electroslag welding being a notable exception). In heat treating, on the other hand, the heating rate is much slower and the retention time above A_3 is much longer. The A_1 and A_3 temperatures during heating (chauffage) are often referred to as the Ac_1 and Ac_3 temperatures, respectively.

TABLE 17.1 Typical Welding Problems and Practical Solution in Carbon and Alloy Steels, and Their Locations in the Text

Typical Problems	Alloy Types	Solutions	Locations
Porosity	Carbon and low-alloy steels	Add deoxidizers (Al, Ti, Mn) in filler metal	3.2 3.3
Hydrogen cracking	Steels with high carbon equivalent	Use low-hydrogen or austenitic stainless steel electrodes Preheat and postheat	3.2 17.4
Lamellar tearing	Carbon and low-alloy steels	Use joint designs that minimize transverse restrain Butter with a softer layer	17.6
Reheat cracking	Corrosion and heat-resisting steels	Use low heat input[a] to avoid grain growth Minimize restraint and stress concentrations Heat rapidly through critical temperature range, if possible	17.5
Solidification cracking	Carbon and low-alloy steels	Keep proper Mn/S ration	11.4
Low HAZ toughness due to grain growth	Carbon and low-alloy steels	Use carbide and nitride formers to suppress grain growth Use low heat input[a]	17.2 17.3
Low fusion-zone toughness due to coarse columnar grains	Carbon and low-alloy steels	Grain refining Use multipass welding to refine grains	7.6 17.2

[a] Low heat input processes (GMAW and SMAW vs. SAW and ESW) or multipass welding with low heat input in each pass.

For kinetic reasons the Ac_1 and Ac_3 temperatures tend to be higher than the equilibrium A_1 and A_3 temperatures, respectively, and they tend to increase with increasing heating rate during welding (1, 2). Kinetically, phase transformations require diffusion (the transformation to martensite is a well-known exception) and diffusion takes time. Consequently, upon rapid heating during welding, phase transformations may not occur at the equilibrium A_1 and A_3 temperatures but at higher temperatures Ac_1 and Ac_3. For steels containing greater amounts of carbide-forming elements (such as V, W, Cr, Ti, and Mo),

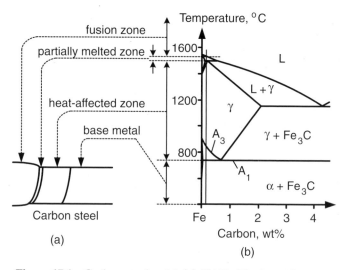

Figure 17.1   Carbon steel weld: (*a*) HAZ; (*b*) phase diagram.

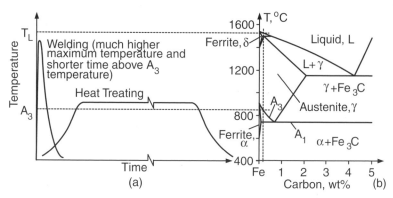

Figure 17.2   Comparison between welding and heat treating of steel: (*a*) thermal processes; (*b*) Fe–C phase diagram.

the effect of the heating rate becomes more pronounced. This is because the diffusion rate of such elements is orders of magnitude lower than that of carbon and also because they hinder the diffusion of carbon. As a result, phase transformations are delayed to a greater extent.

The combination of high heating rates and short retention time above Ac_3 in welding can result in the formation of inhomogeneous austenite during heating. This is because there is not enough time for carbon atoms in austenite to diffuse from the prior pearlite colonies of high carbon contents to prior ferrite colonies of low carbon contents. Upon rapid cooling, the former can transform into high-carbon martensite colonies while the latter into low-carbon ferrite colonies. Consequently, the microhardness in the HAZ can scatter over a wide range in welds made with high heating rates.

As a result of high peak temperatures during welding, grain growth can take place near the fusion boundary. The slower the heating rate, the longer the retention time above Ac_3 is and hence the more severe grain growth becomes. In the heat treating, however, the maximum temperature employed is only about 900°C in order to avoid grain growth.

The CCT diagrams (Chapter 9) for welding can be obtained by using a weld thermal simulator (Chapter 2) and a high-speed dilatometer that detects the volume changes caused by phase transformations (3–6). However, since CCT diagrams for welding are often unavailable, those for heat treating have been used. These two types of CCT diagrams can differ from each other because of kinetic reasons. For instance, grain growth in welding can shift the CCT diagram to longer times favoring transformation to martensite. This is because grain growth reduces the grain boundary area available for ferrite and pearlite to nucleate during cooling. However, rapid heating in welding can shift the CCT diagram to shorter times, discouraging transformation to martensite. *Carbide-forming elements* (such as Cr, Mo, Ti, V, and Nb), when they are dissolved in austenite, tend to increase the hardenability of the steel. Because of the sufficient time available in heat treating, such carbides dissolve more completely and thus enhance the hardenability of the steel. This is usually not possible in welding because of the high heating rate and the short high-temperature retention time encountered in the HAZ (7).

17.2 CARBON STEELS

According to the American Iron and Steel Institute (AISI), carbon steels may contain up to 1.65 wt % Mn, 0.60 wt % Si, and 0.60 wt % Cu in addition to much smaller amounts of other elements. This definition includes the *Fe–C steels* of the 10XX grades (up to about 0.9% Mn) and the *Fe–C–Mn steels* of the 15XX grades (up to about 1.7% Mn). The last two digits in the alloy designation number denote the nominal carbon content in weight percent, for instance, about 0.20% C in a 1020 and about 0.41% C in a 1541 steel. Manganese is an inexpensive alloying element that can be added to carbon steels to help increase hardenability.

17.2.1 Low-Carbon Steels

These steels, in fact, include both carbon steels with up to 0.15% carbon, called low-carbon steels, and those with 0.15–0.30% carbon, called mild steels (8). For the purpose of discussion 1018 steel, which has a nominal carbon content of 0.18%, is used as an example. Figure 17.3 shows the micrographs of a gas–tungsten arc weld of 1018 steel. The base metal consists of a light-etching

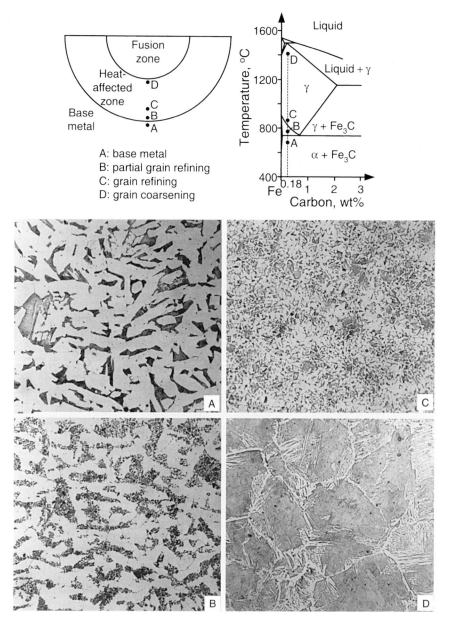

Figure 17.3 HAZ microstructure of a gas–tungsten arc weld of 1018 steel (magnification 200×).

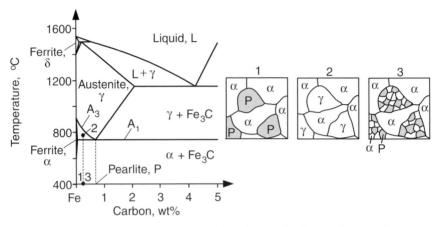

Figure 17.4 Mechanism of partial grain refining in a carbon steel.

ferrite and a dark-etching pearlite (position A). The HAZ microstructure can be divided into essentially three regions: partial grain-refining, grain-refining, and grain-coarsening regions (positions B–D). The peak temperatures at these positions are indicated in the phase diagram.

The *partial grain-refining region* (position B) is subjected to a peak temperature just above the effective lower critical temperature, Ac_1. As explained in Figure 17.4, the prior pearlite (P) colonies transform to austenite (γ) and expand slightly into the prior ferrite (F) colonies upon heating to above Ac_1 and then decompose into extremely fine grains of pearlite and ferrite during cooling. The prior ferrite colonies are essentially unaffected. The *grain-refining region* (position C) is subjected to a peak temperature just above the effective upper critical temperature Ac_3, thus allowing austenite grains to nucleate. Such austenite grains decompose into small pearlite and ferrite grains during subsequent cooling. The distribution of pearlite and ferrite is not exactly uniform because the diffusion time for carbon is limited under the high heating rate during welding and the resultant austenite is not homogeneous. The *grain-coarsening region* (position D) is subjected to a peak temperature well above Ac_3, thus allowing austenite grains to grow. The high cooling rate and large grain size encourage the ferrite to form side plates from the grain boundaries, called the Widmanstatten ferrite (9).

Grain coarsening near the fusion boundary results in coarse columnar grains in the fusion zone that are significantly larger than the HAZ grains on the average. As shown in Figure 17.5, in multiple-pass welding of steels the fusion zone of a weld pass can be replaced by the HAZs of its subsequent passes (10). This grain refining of the coarse-grained fusion zone by multiple-pass welding has been reported to improve the weld metal toughness.

Although martensite is normally not observed in the HAZ of a low-carbon steel, *high-carbon martensite* can form when both the heating and the cooling

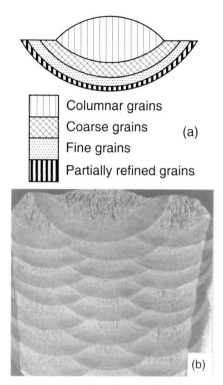

Figure 17.5 Grain refining in multiple-pass welding: (*a*) single-pass weld; (*b*) microstructure of multiple-pass weld. Reprinted from Evans (10). Courtesy of American Welding Society.

rates are very high, as in the case of some laser and electron beam welding. Figure 17.6 shows the HAZ microstructure in a 1018 steel produced by a high-power CO_2 laser beam (11). At the bottom of the HAZ (position B) high-carbon martensite (and, perhaps, a small amount of retained austenite) formed in the prior-pearlite colonies. The high-carbon austenite formed in these colonies during heating did not have time to allow carbon to diffuse out, and it transformed into hard and brittle high-carbon martensite during subsequent rapid cooling. Hard, brittle martensite embedded in a much softer matrix of ferrite can significantly degrade the HAZ mechanical properties. Further up into the HAZ (positions C and D), both the peak temperature and the diffusion time increased. As a result, the prior-pearlite colonies expanded while transforming into austenite and formed martensite colonies of lower carbon contents during subsequent cooling.

High-carbon martensite can also form in the HAZ of an as-cast low-carbon steel, where *microsegregation* during casting causes high carbon contents in the interdendritic areas. Aidun and Savage (12) have studied the repairing of

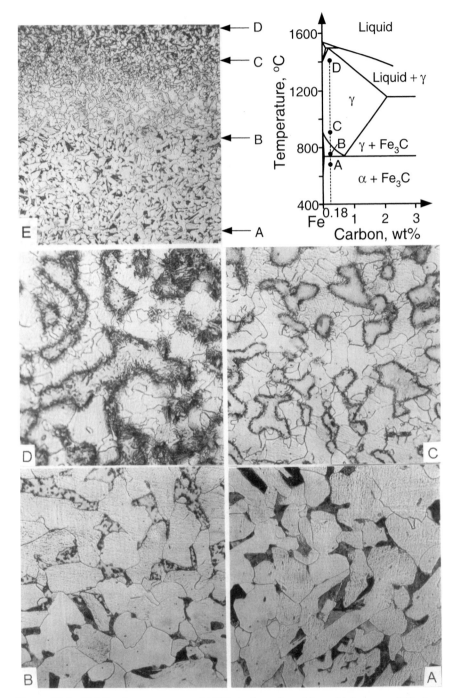

Figure 17.6   HAZ microstructure of 1018 steel produced by a high-power CO_2 laser. Magnification of (A)–(D) 415× and of (E) 65×. From Kou et al. (11).

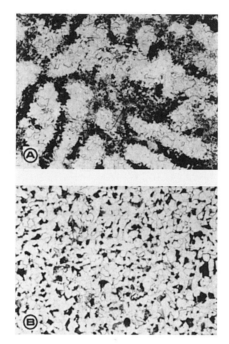

Figure 17.7 Microstructure of a carbon steel: (*a*) as-cast condition; (*b*) after homogenization. Reprinted from Aidun and Savage (12). Courtesy of American Welding Society.

cast steels used in the railroad industry. A series of materials with 0.21–0.31% C, 0.74–1.57% Mn, 0.50% Si, and up to about 0.20% Cr and Mo were spot welded using covered electrodes E7018. As a result of the microsegregation of carbon and alloying elements during casting, continuous networks of inter-dendritic pearlite nodules with carbon contents ranging from about 0.5 to 0.8% were present in the as-cast materials, as shown in Figure 17.7a. The resultant HAZ microstructure is shown in Figure 17.8. During welding of the as-cast materials, the continuous networks of pearlite nodules formed continuous networks of high-carbon austenite upon heating, which in turn transformed to continuous networks of high-carbon martensite upon cooling (region E). The islands scattered in the networks are untransformed ferrite. These networks of interdendritic pearlite nodules can be eliminated by homogenizing at 954°C for 2 h, as shown in Figure 17.7b.

17.2.2 Higher Carbon Steels

These steels include carbon steels with 0.30–0.50% carbon, called medium-carbon steels, and those with 0.50–1.00% carbon, called high-carbon steels (8). Welding of higher carbon steels is more difficult than welding lower

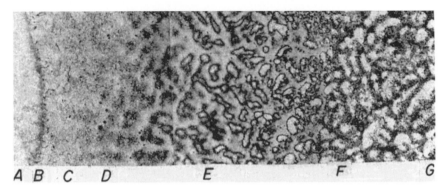

Figure 17.8   HAZ microstructure of a stationary repair weld of an as-cast carbon steel. Fusion zone: A, B; HAZ: C–F; base metal: G. Reprinted from Aidun and Savage (12). Courtesy of American Welding Society.

carbon steels because of the greater tendency of martensite formation in the HAZ and hence hydrogen cracking. For the purpose of discussion, 1040 steel, which has a nominal carbon content of 0.40%, is used as an example.

Figure 17.9 shows the micrographs of a gas-tungsten arc weld of 1040 steel. The base metal of the 1040 steel weld consists of a light-etching ferrite and a dark-etching pearlite (position A), as in the 1018 steel weld discussed previously. However, the volume fraction of pearlite (C rich with 0.77 wt % C) is significantly higher in the case of the 1040 steel because of its higher carbon content. As in the case of the 1018 steel, the HAZ microstructure of the 1040 steel weld can be divided essentially into three regions: the partial grain-refining, grain-refining, and grain-coarsening regions (positions B–D). The CCT diagram for the heat treating of 1040 steel shown in Figure 17.10 can be used to explain qualitatively the HAZ microstructure (13). In the grain coarsening region (position D), both the high cooling rate and the large grain size promote the formation of martensite. The microstructure is essentially martensite, with some dark-etching bainite (side plates) and pearlite (nodules). In the grain-refining region (position C), on the other hand, both the lower cooling rate and the smaller grain size encourage the formation of pearlite and ferrite. The microstructure is still mostly martensitic but has much smaller grains and more pearlite. Some ferrite and bainite may also be present at grain boundaries.

Because of the formation of martensite, *preheating* and control of *interpass temperature* are often required when welding higher carbon steels. For 1035 steel, for example, the recommended preheat and interpass temperatures are about 40°C for 25-mm (1-in.) plates, 90°C for 50-mm (2-in.) plates, and 150°C for 75-mm (3-in.) plates (assuming using low-hydrogen electrodes). For 1040 steel, they are about 90°C for 25-mm (1-in.) plates, 150°C for 50-mm (2-in.) plates, and 200°C for 75-mm (3-in.) plates (14). The reason for

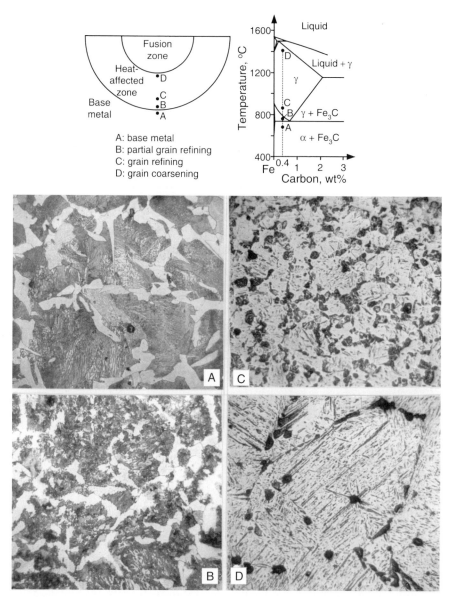

Figure 17.9 HAZ microstructure of a gas–tungsten arc weld of 1040 steel (magnification 400×).

more preheating for thicker plates is because for a given heat input the cooling rate is higher in a thicker plate (Chapter 2). In addition to the higher cooling rate, a thicker plate often has a slightly higher carbon content in order to ensure proper hardening during the heat-treating step of the steel-making process.

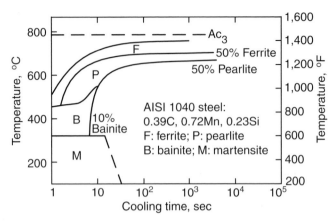

Figure 17.10 Continuous cooling transformation diagram for 1040 steel. Modified from *Atlas of Isothermal Transformation and Cooling Transformation Diagrams* (13).

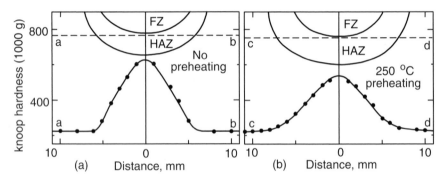

Figure 17.11 Hardness profiles across HAZ of a 1040 steel; (*a*) without preheating; (*b*) with 250°C preheating.

The hardness profile of the HAZ of the 1040 steel weld is shown in Figure 17.11*a*. When welded with preheating, the size of the HAZ increases but the maximum hardness decreases, as shown in Figure 17.11*b*. Examination of the HAZ microstructure near the fusion boundary of the preheated weld reveals more pearlite and ferrite but less martensite. This is because the cooling rate decreases significantly with preheating (Chapter 2).

17.3   LOW-ALLOY STEELS

Three major types of low-alloy steels will be considered here: high-strength, low-alloy steels; quenched-and-tempered low-alloy steels; and heat-treatable low-alloy steels.

17.3.1   High-Strength, Low-Alloy Steels

High-strength, low-alloy (HSLA) steels are designed to provided higher strengths than those of carbon steels, generally with minimum yield strengths of 275–550 MPa (40–80 ksi). Besides manganese (up to about 1.5%) and silicon (up to about 0.7%), as in carbon steels, HSLA steels often contain very small amounts of niobium (up to about 0.05%), vanadium (up to about 0.1%), and titanium (up to about 0.07%) to ensure both grain refinement and precipitation hardening. As such, they are also called microalloyed steels. Typically the maximum carbon content is less than 0.2% and the total alloy content is less than 2%. Alloys A242, A441, A572, A588, A633, and A710 are examples of HSLA steels, and their compositions are available elsewhere (15).

Niobium (Nb), vanadium (V), and titanium (Ti) are strong *carbide* and *nitride formers*. Fine carbide or nitride particles of these metals tend to hinder the movement of grain boundaries, thus reducing the grain size by making grain growth more difficult. The reduction in grain size in HSLA steels increases their strength and toughness at the same time. This is interesting because normally the toughness of steels decreases as their strength increases. Among the carbides and nitrides of Nb, V, and Ti, titanium nitride (TiN) is most stable; that is, it has the smallest tendency to decompose and dissolve at high temperatures. This makes it most effective in limiting the extent of grain growth in welding.

The higher the heat input during welding, the more likely the carbide and nitride particles will dissolve and lose their effectiveness as grain growth inhibitors. The low toughness of the coarse-grain regions of the HAZ is undesirable. It has been reported that steels containing titanium oxide (Ti_2O_3) tend to have better toughness (16, 17). The Ti_2O_3 is more stable than TiN and does not dissolve even at high heat inputs. The undissolved Ti_2O_3 particles do not actually stop grain growth but act as effective nucleation sites for acicular ferrite. Consequently, acicular ferrite forms within the coarse austenite grains and improves the HAZ toughness (16).

The HSLA steels are usually welded in the as-rolled or the normalized condition, and the weldability of most HSLA steels is similar to that of mild steel. Since strength is often the predominant factor in the applications of HSLA steels, the filler metal is often selected on the basis of matching the strength of the base metal (15). Any common welding processes can be used, but low-hydrogen consumables are preferred.

The preheat and interpass temperatures required are relatively low. For most alloys they are around 10°C for 25-mm (1-in.) plates, 50°C for 50-mm (2-in.) plates, and 100°C for 75-mm (3-in.) plates. For alloy A572 (grades 60 and 65) and alloy A633 (grade E), they are about 50°C higher (15). The amount of preheating required increases with increasing carbon and alloy content and with increasing steel thickness.

17.3.2   Quenched-and-Tempered Low-Alloy Steels

The quenched-and-tempered low-alloy (QTLA) steels, usually containing less than 0.25% carbon and less than 5% alloy, are strengthened primarily by quenching and tempering to produce microstructures containing martensite and bainite. The yield strength ranges from approximately 345 to 895 MPa (50 to 130 ksi), depending on the composition and heat treatment. Alloys A514, A517, A543, HY-80, HY-100, and HY-130 are some examples of QTLA steels, and their compositions are available elsewhere (15).

Low carbon content is desired in such alloys for the following two reasons: (i) to minimize the hardness of the martensite and (ii) to raise the M_s (martensite start) temperature so that any martensite formed can be tempered automatically during cooling. Due to the formation of low-carbon auto-tempered martensite, both high strength and good toughness can be obtained. Alloying with Mn, Cr, Ni, and Mo ensures the hardenability of such alloys. The use of Ni also significantly increases the toughness and lowers the ductile–brittle transition temperature in these alloys.

Any common welding processes can be used to join QTLA steels, but the weld metal hydrogen must be maintained at very low levels. Preheating is often required in order to prevent hydrogen cracking. The preheat and interpass temperatures required are higher than those required for HSLA steels but still not considered high. For HY130, for example, the preheat and interpass temperatures are about 50°C for 13-mm (0.5-in.) plates, 100°C for 25-mm (1-in.) plates, and 150°C for 38-mm (1.5-in.) plates (15). However, too high a preheat or interpass temperature is undesirable. It can decrease the cooling rate of the weld metal and HAZ and cause austenite to transform to either ferrite or coarse bainite, both of which lack high strength and good toughness. Postweld heat treatment is usually not required.

Excessive heat input can also decrease the cooling rate and produce unfavorable microstructures and properties. High heat input processes, such as ESW or multiple-wire SAW, should be avoided. Figure 17.12 shows the CCT curves of T1 steel, that is, A514 and A517 grade F QTLA steel (18). Curves p, f, and z represent the critical cooling rates for the formation of pearlite, ferrite, and bainite, respectively. The hatched area represents the region of optimum cooling rates. If the cooling rate during welding is too low, for instance, between curve p and the hatched area indicated, a substantial amount of ferrite forms. This can, in fact, be harmful since the ferrite phase tends to reject carbon atoms and turn its surrounding areas into high-carbon austenite. Such high-carbon austenite can in turn transform to high-carbon martensite and bainite during cooling, thus resulting in a brittle HAZ. Therefore, the heat input and the preheating of the workpiece should be limited when welding quenched-and-tempered alloy steels.

On the other hand, if the cooling rate during welding is too high, to the left of curve z in Figure 17.12, insufficient time is available for the auto-tempering of martensite. This can result in hydrogen cracking if hydrogen is

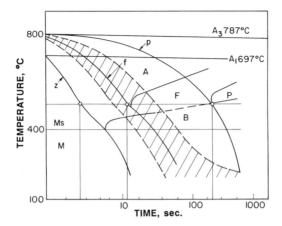

Figure 17.12   CCT curves for A514 steel. From Inagaki et al. (18).

present. Therefore, low-hydrogen electrodes or welding processes and a small amount of preheating are recommended. The hatched area in the figure represents the region of best cooling rates for welding this steel.

To meet the requirements of both limited heat inputs and proper preheating, multiple-pass welding is often used in welding thick sections of QTLA steels. In so doing, the interpass temperature is maintained at the same level as the preheat temperature. Multiple-pass welding with many small stringer beads improves the weld toughness as a result of the grain-refining and tempering effect of successive weld passes. The martensite in the HAZ of a weld pass is tempered by the heat resulting from deposition in subsequent passes. As a result, the overall toughness of the weld metal is enhanced. Figure 17.13 shows the effect of bead tempering (19). The HAZ of bead E is tempered by bead F and is, therefore, softer than the HAZ of bead D, which is not tempered by bead F of any other beads.

17.3.3   Heat-Treatable Low-Alloy Steels

The heat-treatable low-alloy (HTLA) steels refer to medium-carbon quenched-and-tempered low-alloy steels, which typically contain up to 5% of total alloy content and 0.25–0.50% carbon and are strengthened by quenching to form martensite and tempering it to the desired strength level (15). The higher carbon content promotes higher hardness levels and lower toughness and hence a greater susceptibility to hydrogen cracking than the quenched-and-tempered low-alloy steels discussed in the previous section. Alloys 4130, 4140, and 4340 are examples of HTLA steels.

The HTLA steels are normally welded in the annealed or overtempered condition except for weld repairs, where it is usually not feasible to anneal or overtemper the base metal before welding. Immediately after welding, the

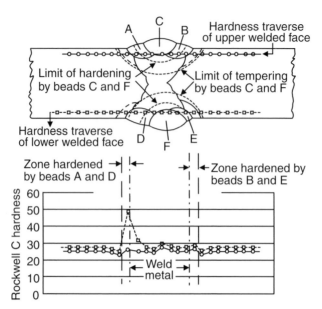

Figure 17.13 Tempering bead technique for multiple-pass welding of a butt joint in a quenched-and-tempered alloy steel. Reprinted form Linnert (19). Courtesy of American Welding Society.

entire weldment is heat treated, that is, reaustenized and then quenched and tempered to the desired strength level, or at least stress relieved or tempered to avoid hydrogen cracking. Any of the common welding processes can be used to join HTLA steels. To avoid hydrogen cracking, however, the weld metal hydrogen must be maintained at very low levels, proper preheat and interpass temperatures must be used, and preheat must be maintained after welding is completed until the commencement of postweld heat treatment.

In applications where the weld metal is required to respond to the same postweld heat treatment as the base metal in order to match the base metal in strength, a filler metal similar to the base metal in composition is used. In repair welding where it is possible to use the steel in the annealed or overtempered condition, the filler metal does not have to be similar to the base metal in composition. The weldment is stress relieved or tempered after welding. In some repair welding where neither annealing or overtempering the steel before welding nor stress relieving the weld upon completion is feasible, electrodes of austenitic stainless steels or nickel alloys can be used. The resultant weld metal has lower strength and greater ductility than the quenched-and-tempered base metal, and high shrinkage stresses during welding can result in plastic deformation of the weld metal rather than cracking of the HAZ.

High preheat and interpass temperatures are often required for welding HTLA steels. For alloy 4130, for instance, they are around 200°C for 13-mm

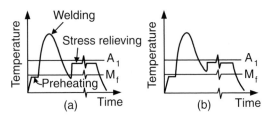

Figure 17.14 Thermal history during welding and postweld stress relieving of a heat-treatable alloy steel: (*a*) desired; (*b*) undesired.

(0.5-in.) plates, 250°C for 25-mm (1-in.) plates, and almost 300°C for 50-mm (2-in.) plates. For alloys 4140 and 4340, they are even higher (15).

In addition to the use of preheating, the weldment of heat-treatable alloy steels is often immediately heated for stress-relief heat treatment before cooling to room temperature. During the stress-relief heat treatment, martensite is tempered and, therefore, the weldment can be cooled to room temperature without danger of cracking. After this, the weldment can be postweld heat treated to develop the strength and toughness the steel is capable of attaining.

A sketch of the thermal history during welding and postweld stress relieving is shown in Figure 17.14*a*. The preheat temperature and the temperature immediately after welding are both maintained slightly below the martensite-finish temperature M_f. Stress relieving begins immediately after welding, at a temperature below the A_1 temperature.

This can be further explained using a 25-mm (1-in.) 4130 steel as an example. Based on the isothermal transformation diagram for 4130 steel (13) shown in Figure 17.15, the ideal welding procedure for welding 4130 steel is to use a filler metal of the same composition, preheat to 250°C (about 500°F), weld while maintaining a 250°C interpass temperature using low-hydrogen electrodes, and stress-relief heat treat at about 650°C (1200°F) immediately upon completion of welding (19). The postweld heat treatment of the entire weldment can be done in the following sequence: austenitizing at about 850°C (about 1600°F), quenching, and then tempering in the temperature range 400–600°C (about 800–1100°F).

As shown in Figure 17.14*a*, the HAZ should be cooled to a temperature slightly below the *martensite-finish temperature* M_f before being heated for stress relieving. If this temperature is above M_f, as shown in Figure 17.14*b*, there can be untransformed austenite left in the HAZ and it can decompose into ferrite and pearlite during stress relieving or transform to untempered martensite upon cooling to room temperature after stress relieving.

In the event that stress-relief heat treatment cannot be carried out immediately upon completion of welding, the temperature of the completed weldment can be raised to approximately 400°C (750°F), which is the vicinity of the bainite "knee" for 4130 and most other heat-treatable alloy steels. By

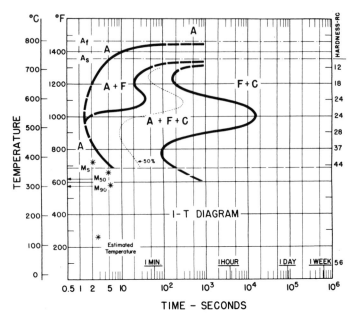

Figure 17.15   Isothermal transformation curves for 4130 steel. Reprinted from *Atlas of Isothermal Transformation and Cooling Transformation Diagrams* (13).

holding at this temperature for about 1 h or less, the remaining austenite can transform to bainite, which is more ductile than martensite. Therefore, when the weldment is subsequently cooled to room temperature, no cracking should be encountered. Further heat treatment can be carried out later in order to optimize the microstructure and properties of the weldment.

In cases where heat-treatable low alloy steels cannot be postweld heat treated and must be welded in the quenched-and-tempered condition, HAZ softening as well as hydrogen cracking can be a problem. To minimize softening, a lower heat input per unit length of weld (i.e., a lower ratio of heat input to welding speed) should be employed. In addition, the preheat, interpass, and stress-relief temperatures should be at least 50°C below the tempering temperature of the base metal before welding. Since postweld heat treating of the weldment is not involved, the composition of the filler metal can be substantially different from that of the base metal, depending on the strength level of the weld metal required.

17.4   HYDROGEN CRACKING

Several aspects of hydrogen cracking have been described previously (Chapter 3), including the various sources of hydrogen during welding, the hydrogen

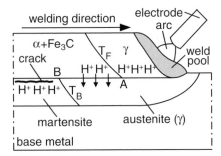

Figure 17.16 Diffusion of hydrogen from weld metal to HAZ during welding. Modified from Granjon (20).

levels in welds made with various welding processes, the solubility of hydrogen in steels, the methods for measuring the weld hydrogen content, and the techniques for reducing the weld hydrogen content have been described previously in Chapter 3.

17.4.1 Cause

Hydrogen cracking occurs when the following four factors are present simultaneously: hydrogen in the weld metal, high stresses, susceptible microstructure (martensite), and relatively low temperature (between −100 and 200°C). High stresses can be induced during cooling by solidification shrinkage and thermal contraction under constraints (Chapter 5). Martensite, especially hard and brittle high-carbon martensite, is susceptible to hydrogen cracking. Since the martensite formation temperature M_s is relatively low, hydrogen cracking tends to occur at relatively low temperatures. For this reason, it is often called "cold cracking." It is also called "delayed cracking," due to the incubation time required for crack development in some cases.

Figure 17.16 depicts the diffusion of hydrogen from the weld metal to the HAZ during welding (20). The terms T_F and T_B are the austenite/(ferrite + pearlite) and austenite/martensite transformation temperatures, respectively. As the weld metal transforms from austenite (γ) into ferrite and pearlite ($\alpha +$ Fe_3C), hydrogen is rejected by the former to the latter because of the lower solubility of hydrogen in ferrite than in austenite. The weld metal is usually lower in carbon content than the base metal because the filler metal usually has a lower carbon content than the base metal. As such, it is likely that the weld metal transforms from austenite into ferrite and pearlite before the HAZ transforms from austenite into martensite (M). The build-up of hydrogen in the weld metal ferrite causes it to diffuse into the adjacent HAZ austenite near the fusion boundary, as indicated by the short arrows in the figure. As shown in Figure 17.17, the diffusion coefficient of hydrogen is much higher in ferritic materials than austenitic materials (21). The high diffusion coefficient

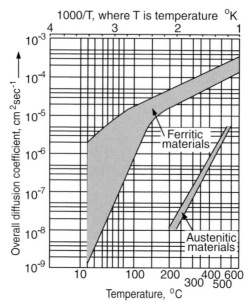

Figure 17.17   Diffusion coefficient of hydrogen in ferritic and austenitic materials as a function of temperature. Modified from Coe (21).

of hydrogen in ferrite favors this diffusion process. On the contrary, the much lower diffusion coefficient of hydrogen in austenite discourages hydrogen diffusion from the HAZ to the base metal before the HAZ austenite transforms to martensite. This combination of hydrogen and martensite in the HAZ promotes hydrogen cracking.

The mechanism for hydrogen cracking is still not clearly understood, though numerous theories have been proposed. It is not intended here to discuss these theories since this is more a subject of physical metallurgy than welding metallurgy. For practical purposes, however, it suffices to recognize that Troiano (22) proposed that hydrogen promotes crack growth by reducing the cohesive lattice strength of the material. Petch (23) proposed that hydrogen promotes crack growth by reducing the surface energy of the crack. Beachem (24) proposed that hydrogen assists microscopic deformation ahead of the crack tip. Savage et al. (25) explained weld hydrogen cracking based on Troiano's theory. Gedeon and Eagar (26) reported that their results substantiated and extended Beachem's theory.

17.4.2   Appearance

Figure 17.18 is a typical form of hydrogen crack called "underbead crack" (27). The crack is essentially parallel to the fusion boundary. Hydrogen cracking can

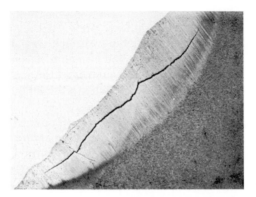

Figure 17.18 Underbead crack in a low-alloy steel HAZ (magnification 8×). Reprinted from Bailey (27).

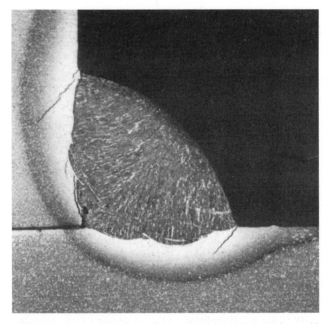

Figure 17.19 Hydrogen cracking in a fillet weld of 1040 steel (magnification 4.5×). Courtesy of Buehler U.K., Ltd., Coventry, United Kingdom.

be accentuated by stress concentrations. Figure 17.19 shows cracking at the junctions between the weld metal surface and the workpiece surface of a fillet weld of 1040 steel (28). This type of crack is called "toe crack." The same figure also shows cracking at the root of the weld, where lack of fusion is evident. This type of crack is called "root cracks."

17.4.3   Susceptibility Tests

There are various methods for testing the hydrogen cracking susceptibility of steels, such as the implant test (20, 29), the Lehigh restraint test (30), the RPI augmented strain cracking test (31), the controlled thermal severity test (32), and the Lehigh slot weldability test (33). Due to the limitation in space, only the first two tests will be described here.

Figure 17.20 is a schematic of the implant test. In this test, a cylindrical specimen is notched and inserted in a hole in a plate made from a similar material. A weld run is made over the specimen, which is located in such a way that its top becomes part of the fusion zone and its notch lies in the HAZ. After welding and before the weld is cold, a load is applied to the specimen and the time to failure is determined. As an assessment of hydrogen cracking susceptibility, the stress applied is plotted against the time to failure, as shown in Figure 17.21 for a high-strength, low-alloy pipeline steel (33). In this case loading was applied to the specimen when the weld cooled down to 125°C. As

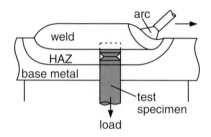

Figure 17.20   Implant test for hydrogen cracking.

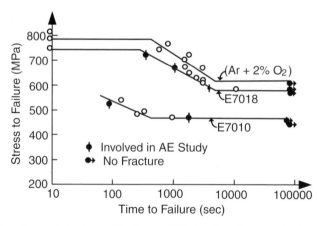

Figure 17.21   Implant test results for a HSLA pipe line steel. Reprinted from Vasudevan et al. (33). Courtesy of American Welding Society.

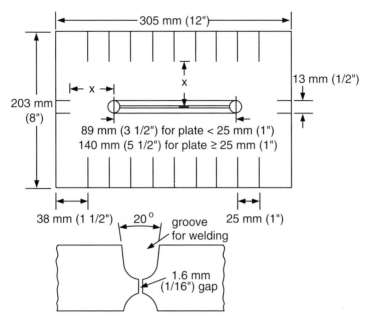

Figure 17.22 Lehigh restraint specimen. Reprinted from Stout et al. (30). Courtesy of American Welding Society.

shown, the welds made with low-hydrogen electrodes (E7018, basic limestone-type covering) have a higher threshold stress below which no cracking occurs and a longer time to cracking than the welds made with high-hydrogen electrodes (E7010, cellulose-type covering). In other words, the former is less susceptible to hydrogen cracking than the latter. The gas–metal arc welds made with $Ar + 2\% \ O_2$ as the shielding gas are least susceptible. Obviously, no electrode covering is present to introduce hydrogen into these welds.

Figure 17.22 shows the Lehigh restraint specimen (30). The specimen is designed with slots cut in the sides (and ends). The longer the slots, the lower the degree of plate restraint on the weld is. A weld run is made in the root of the joint, and the length of the slots required to prevent hydrogen cracking is determined. Cracking is detected visually or by examining transverse cross sections taken from the midpoint of the weld.

17.4.4   Remedies

A. *Control of Welding Parameters*

A.1. Preheating   As already described in the previous section, the use of the proper preheat and interpass temperatures can help reduce hydrogen cracking. Figure 17.23 shows such an example (26). Two general approaches have

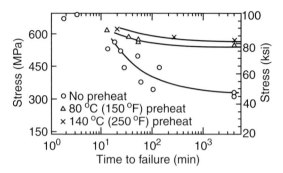

Figure 17.23   Effect of preheating on hydrogen cracking of a high-strength steel. Reprinted from Gedeon and Eagar (26). Courtesy of American Welding Society.

been used to select the most appropriate temperatures. One approach is to use empirically derived tables that list the steels and the recommended welding procedures, including the preheat and interpass temperatures, for instance, for various grades of carbon steels (14) and low-alloy steels (15). The other approach is to relate the cracking tendency to the hardenability of steels based on the carbon equivalence. One such formula for the carbon equivalence of alloy steel is as follows (34):

$$\text{Carbon equivalence} = \%C + \frac{\%Mn}{6} + \frac{\%Si}{24} + \frac{\%Ni}{40} + \frac{\%Cr}{5} + \frac{\%Mo}{4} \quad (17.1)$$

Figure 17.24 shows the recommended preheat and interpass temperatures based on this formula (34).

A.2. Postweld Heating   Postweld heat treatment, as described in the previous section, can be used to stress relieve the weld before it cools down to room temperature. In the event that stress relief heat treatment cannot be carried out immediately upon completion of welding, the completed weldment can be held at a proper temperature to allow austenite to transform into a less susceptible microstructure than martensite. Postweld heating can also help hydrogen diffuse out of the workpiece (Chapter 3). For most carbon steels, the postweld heat treatment temperature range is 590–675°C (1100–1250°F) (14).

A.3. Bead Tempering   Bead tempering in multiple-pass welding can also be effective in reducing hydrogen cracking. This has been described previously.

B. Use of Proper Welding Processes and Materials

B.1. Use of Low-Hydrogen Processes and Consumables   The use of lower hydrogen welding processes (such as GTAW or GMAW insead of SMAW or

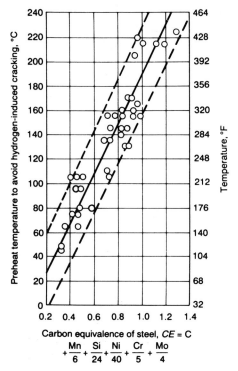

Figure 17.24   Effect of carbon equivalent on preheat requirement to prevent hydrogen cracking. Reprinted from Lesnewich (34).

FCAW) or consumables (such as basic rather than cellulosic SMAW electrodes) reduces the amount of hydrogen present in the welding zone (Chapter 3). The electrodes should be kept in sealed containers. If they are exposed, they may need to be baked around 300–400°C (600–800°F) to dry them out.

B.2. Use of Lower Strength Filler Metals   The use of filler metals of lower strength than the base metal can help reduce the stress levels in the HAZ and, hence, the chance of hydrogen cracking.

B.3. Use of Austenitic Stainless Steel Filler Metals   Austenitic stainless steels and Ni-based alloys have been used as the filler metal for welding HTLA steels. The diffusion coefficient of hydrogen is much lower in austenite (which includes Ni-based alloys) than ferrite (Figure 17.17). As such, hydrogen is trapped in the austenitic weld metal and unable to reach the HAZ to cause cracking. Furthermore, the good ductility of austenitic weld metal can help prevent the buildup of excessively high residual stresses in the HAZ, thus reducing the susceptibility of hydrogen cracking.

17.5   REHEAT CRACKING

Reheat cracking is a well-recognized problem in low-alloy ferritic steels that contain chromium, molybdenum, and sometimes vanadium and tungsten to enhance corrosion resistance and elevated-temperature strength. These alloy steels, sometimes called creep-resistant ferritic steels, are frequently used for elevated-temperature service for nuclear and fossil energy applications.

Creep-resistant ferritic steels are reheated, often to about 550–650°C, after welding to relieve stresses and hence reduce susceptibility to hydrogen cracking or stress corrosion cracking. However, cracking can occur in the HAZ during reheating. Examples of ferritic steels susceptible to reheat cracking include 0.5Cr–0.5Mo–0.25V, 0.5Cr–1Mo–1V, and 2.25Cr–1Mo. Nakamura et al. (35) have proposed the following formula for the effect of alloying elements on crack susceptibility:

$$CS = \%\,Cr + 3.3 \times (\%\,Mo) + 8.1 \times (\%\,V) - 2 \qquad (17.2)$$

When the value of CS is equal to or greater than zero, the steel may be susceptible to reheat cracking.

17.5.1   Appearance

Figure 17.25a shows reheat cracking in a CrMoV steel (27). Cracks are along the prior austenite grain boundaries in the HAZ, as shown in Figure 17.25b. Figure 17.26 shows a scanning electron micrograph of the fracture surface in

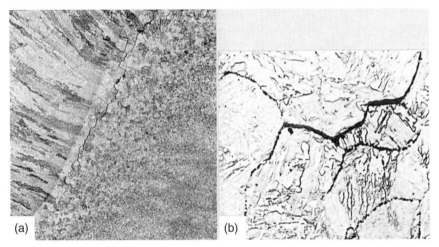

(a)   (b)

Figure 17.25   Reheat cracking in a CrMoV steel: (*a*) macrostructure (magnification ×35); (*b*) microstructure (magnification ×1000). The cracks are along the prior austenite grain boundaries of the bainitic structure. From Bailey (27).

Figure 17.26 SEM micrograph of the fracture surface of a 2.25Cr–1Mo ferritic steel caused by reheat cracking. Reprinted from Nawrocki et al. (36). Courtesy of American Welding Society.

a 2.25Cr–1Mo steel subjected to a Gleeble thermal weld simulation and a tensile stress that was maintained during cooling and throughout reheating at an elevated temperature (36). Intergranular cracking is evident.

17.5.2   Cause

The mechanism of reheat cracking (37–41) is explained as follows. During welding the HAZ near the fusion line is heated to temperatures very high in the austenite phase field, where preexisting carbides of Cr, Mo, and V dissolve and the austenite grains grow. Subsequent rapid cooling does not allow enough time for carbides to reprecipitate, resulting in supersaturation of these alloying elements as the austenite transforms to martensite (if hardenability is sufficient). When the coarse-grained HAZ is reheated to elevated temperature for stress relieving, fine carbides precipitate at dislocations in the prior austenite grain interiors and strengthen them before stresses are relieved. Since the grain interiors are strengthened more than the grain boundaries and since this occurs before stresses are relieved, cracking can occur along the grain boundaries.

17.5.3   Susceptibility Tests

The reheat cracking susceptibility can be evaluated by measuring the extent of crack growth in a compact tension specimen (42) or simply a Charpy V-notch specimen (43), which is loaded and kept at the reheat temperature (say around 600°C). The Vinckier test shown in Figure 17.27 has also been used (44). The test specimens are made by welding two pieces of 50-mm-thick plates together. The ends of the test specimens are welded to a stainless steel block. Upon reheating, the test specimens are subjected to tensile loading caused by the higher thermal expansion coefficient of the stainless steel block.

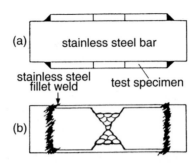

Figure 17.27   Vinckier test for reheat cracking. From Glover et al. (44).

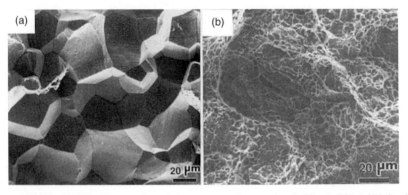

Figure 17.28   SEM micrographs of fracture surfaces of a 2.4Cr–1.5W–0.2V ferritic steel: (*a*) single-pass sample; (*b*) multiple-pass sample. Reprinted from Nawrocki et al. (36, 45). Courtesy of American Welding Society.

The following formula (44) can be used to determine the overall strain in the test specimen:

$$\varepsilon_1 = \frac{\sigma_1}{E_1} = \frac{(\alpha_2 - \alpha_1)T}{(E_1 A_1 / E_2 A_2) + 1} \tag{17.3}$$

where ε is the overall strain in the test specimen, α the thermal expansion co-efficient, E Young's modulus, T the reheat temperature, and A the cross-sectional area. Subscripts 1 and 2 refer to the test specimen and the stainless steel block, respectively. Figure 17.25a, in fact, is the macrograph of the HAZ of a Vinckier test specimen.

17.5.4   Remedies

Multiple-pass welding has been reported to reduce reheat cracking in creep-resistant ferritic steels (42, 43, 45). Figure 17.28 shows the results of a new

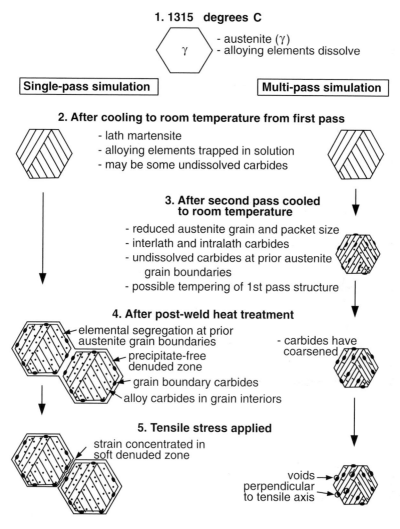

1. 1315 degrees C

- austenite (γ)
- alloying elements dissolve

| Single-pass simulation | Multi-pass simulation |

2. After cooling to room temperature from first pass

- lath martensite
- alloying elements trapped in solution
- may be some undissolved carbides

3. After second pass cooled to room temperature

- reduced austenite grain and packet size
- interlath and intralath carbides
- undissolved carbides at prior austenite grain boundaries
- possible tempering of 1st pass structure

4. After post-weld heat treatment

- elemental segregation at prior austenite grain boundaries
- precipitate-free denuded zone
- grain boundary carbides
- alloy carbides in grain interiors

- carbides have coarsened

5. Tensile stress applied

strain concentrated in soft denuded zone

voids perpendicular to tensile axis

Figure 17.29 Microstructural changes and failure mode of single- and multiple-pass samples of a 2.4Cr–1.5W–0.2V ferritic steel. Modified from Nawrocki et al. (45). Courtesy of American Welding Society.

ferritic steel, 2.4Cr–1.5W–0.2V, that was subjected to a Gleeble weld thermal simulation and a tensile stress that was maintained during cooling and throughout reheating to elevated temperatures (45). With a single simulated weld pass (36), reheat cracking is evident from the brittle intergranular failure mode shown in Figure 17.28a. With two simulated weld passes (45), however, reheat cracking is avoided, as evident from the microvoid coalescence failure mode shown in Figure 17.28b and the increase in the time to failure. Figure 17.29

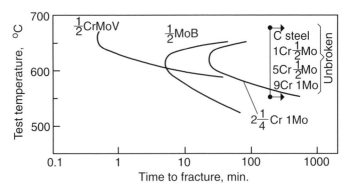

Figure 17.30 Temperature vs. time to fracture in ferritic steels. From Murray (46).

explains the effect of the multiple-pass weld procedure (45). With a single-pass weld procedure, the grains are coarse. As mentioned previously, during reheating fine carbides precipitate at dislocations in the grain interior. Meanwhile, coarse carbides can also form at the grain boundaries and deplete the adjacent region of carbide-forming alloying elements. This can cause a precipitation-free denuded zone to form along the grain boundaries. As such, fine carbides strengthen the grain interiors and the precipitation-free denuded zone, if it is present, can weaken the grain boundaries. In any case, since the grain interiors are strengthened more than the grain boundaries, intergranular cracking occurs along the grain boundaries. With a multiple-pass weld procedure, however, the coarse grains are refined, fine grain interior carbides coarsen, and there is no longer a precipitation-free denuded zone along the grain boundaries.

Reducing the level of residual stresses, which can be achieved by minimizing the restraint during welding, can also reduce reheat cracking in creep-resistant ferritic steels. Stress concentrations, such as fillet toes and prior liquation cracks, tend to accentuate the problem. As in heat-treatable nickel-base alloys, rapid heating through the carbide precipitation range, if possible, can also be considered. Based on the crack susceptibility C curves shown in Figure 17.30, this seems more feasible with 2.25Cr–1Mo than with 0.5Cr–Mo–V (46).

17.6 LAMELLAR TEARING

17.6.1 Cause

The cause of lamellar tearing in welded structures has been described by several investigators (47–52). In brief, it is the combination of high localized stresses due to weld contraction and low ductility of the base metal in its through-thickness direction due to the presence of elongated stringers of nonmetallic inclusions parallel to its rolling direction. Tearing is triggered by decohesion of such nonmetallic inclusions (usually silicates and sulfides) near the weld in or just outside the HAZ.

Figure 17.31 shows the typical microstructure of steels susceptible to lamellar tearing (51). The alignment of inclusions in the rolling direction is evident. Figure 17.32 shows lamellar tearing under a T butt weld in a C–Mn steel that shows a stepwise path in the rolling direction—typical of lamellar tearing.

Besides the content and morphology of inclusions, other factors such as hydrogen, preheating, and electrode strength have also been reported to affect the susceptibility of steels to lamellar tearing (53). For example, welds made with E7010 cellulosic electrodes, which supply a high hydrogen potential, have been found significantly more susceptible than those made with the GMAW process. The effect of hydrogen appears to be associated with embrittlement rather than with cold cracking. Preheating has been found to reduce the

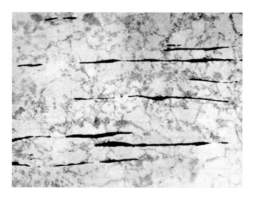

Figure 17.31 Microstructure of a lamellar-tearing susceptible steel. From Dickinson and Nichols (51).

Figure 17.32 Lamellar tearing near a C–Mn steel weld. Reproduced by permission of TWI Ltd.

susceptibility to lamellar tearing, especially when high hydrogen potentials were present (53). It should be noted that preheating, if added to the final contraction strains in a restrained joint in service, may not be effective in reducing the susceptibility to lamellar tearing. As can be expected, electrodes of lower strength levels provide the weld metal the ability to accommodate for contraction strains and, therefore, improve the resistance to lamellar tearing. The strength level of the electrode used, of course, cannot be lower than that required by design purposes.

17.6.2 Susceptibility Tests

The lamellar tearing susceptibility of steels can be evaluated in several ways, for instance, the Lehigh cantilever lamellar tearing test (53), the Cranefield lamellar tearing test (54), and the tensile lamellar tearing test (55). Figure 17.33 shows the Lehigh cantilever lamellar tearing test (53). As shown, a multiple-pass weld is made in a V-groove between the test specimen and a beveled 50-mm- (2-in.-) thick beam set at a right angle to it.

17.6.3 Remedies

A practical way of avoiding lamellar tearing is to employ joint designs that allow the contraction stresses to act more in the rolling direction of the susceptible material and less in the transverse (through-thickness) direction. Figure 17.34 shows the use of one such joint design.

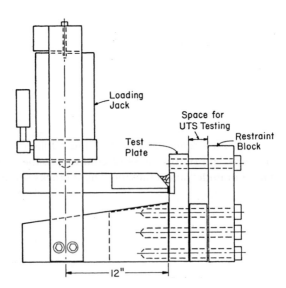

Figure 17.33   The Lehigh cantilever lamellar tearing test. Reprinted from Kaufman et al. (53). Courtesy of American Welding Society.

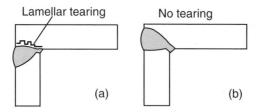

Figure 17.34   Lamellar tearing of a corner joint: (*a*) improper design; (*b*) improved design.

17.7   CASE STUDIES

17.7.1   Failure of Socket/Pipe Welds (56)

Figure 17.35*a* shows the fracture region of a saddle socket/pipe weldment. The steel pipe (150 mm outside diameter) has 0.51% C and 0.75% Mn, and the steel socket has less than 0.1% C. As shown in Figure 17.35*b*, underbead cracking is evident in the pipe. The microstructure in the area, as shown in Figure 17.35*c*, is martensitic and its hardness level is around 700–800 HV. No preheating or postheating was used, and the failure occurred during cooling after welding.

17.7.2   Failure of Oil Engine Connecting Rod (57)

Figure 17.36*a* shows the top end of one of the connecting rods of a vertical four-cylinder engine. The rod was made from a medium-carbon or low-alloy steel in the hardened-and-tempered condition, according to the microstructure analysis of the base metal. The top end bearings of the connecting rods in this engine were originally fed with oil from the bottom ends by means of external pipes carried along the side flutes in the rods. Due to the subsequent modification in the method of lubrication, the holes at the top ends of the rods were sealed off with deposits of weld metal. The connecting rods failed about five years after the modification. Figures 17.36*b* shows underbead cracks in the HAZ. The HAZ exhibited a martensitic structure with a hardness of 450 VPN, which was measured with a diamond pyramid and a load of 3 kg. Because a small weld deposit was laid onto a mass of cold metal, the cooling rate was rather high, resulting in underbead cracking. These cracks served as the starting points of fatigue cracking, which led to the final rupture of the rods. The fracture surface, as shown in Figure 17.36*a* was smooth, conchoidal, and characteristic of that resulting from fatigue.

17.7.3   Failure of Water Turbine Pressure Pipeline (58)

Figure 17.37*a* shows a big crack along the submerged arc weld of a turbine pressure pipeline. The inside diameter of the pipeline was 1.2 m and the wall

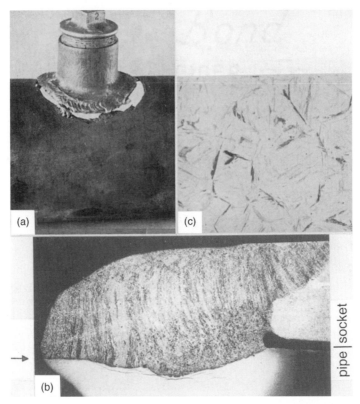

Figure 17.35 Failure in a socket/pipe weldment: (*a*) fracture region (magnification 0.4×); (*b*) underbead cracking (magnification 7.5×); (*c*) martensitic microstructure in the cracked region (magnification 370×). Reprinted, with permission, from Naumann (56).

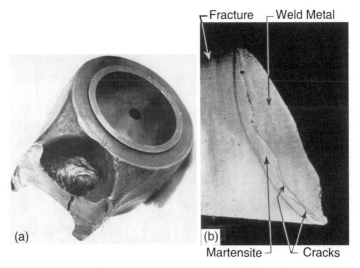

Figure 17.36 Failure of an oil engine connecting rod: (*a*) top end of connecting rod showing fracture surface and weld deposit; (*b*) section through weld deposit showing underbead cracking (magnification 3.5×). From Hutchings and Unterweiser (57).

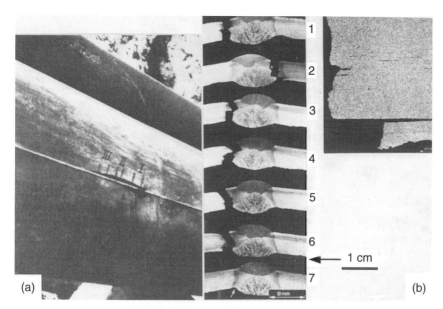

Figure 17.37   Failure of a water turbine pressure pipeline (58): (*a*) cracking in the pipeline; (*b*) weld microstructure along the crack showing lamellar tearing.

thickness varied from 6 to 11 mm. The pipeline was made from two flat sheets of steel containing 0.1% C, 0.4% Mn, 0.02% S, and 0.02% P. The pipe had two longitudinal seams. After the welding of the first seam between the two sheets, a cold-rolling operation followed. Then, the second seam was welded and the pipe was cold rolled again after welding for roundness. The pipe was mounted on saddle supports spaced 14 m apart.

Figure 17.37*b* shows the microstructure around the fracture area. As shown, lamellar tearing, possibly due to stringers of nonmetallic inclusions of sulfur and phosphorus, was evident. The cold-rolling operation, the gravity force acting on the part of the pipeline between the saddle supports, and the internal pressure cycles could have all contributed to the stresses required for lamellar tearing.

REFERENCES

1. Verein, Deutscher Eisenhuttenlcute, *Steel—A Handbook for Materials Research and Engineering*, Vol. 1: *Fundamentals*, Springer Verlag, Berlin, 1992, p. 175.

2. Albutt, K. J., and Garber, S., *J. Iron Steel Inst.,* **204:** 1217, 1966.

3. Nippes, E. F., and Savage, W. F., *Weld. J.,* **28:** 534s, 1949.

4. Nippes, E. F., and Nelson, E. C., *Weld. J.,* **37:** 30s, 1958.

5. Nippes, E. F., Savage, W. F., and Paez, J. M., *Weld. J.,* **38:** 475s, 1959.

6. D'Andrea, M. M. Jr., and Adams, C. M. Jr., *Weld. J.*, **42:** 503s, 1963.

7. Kou, S., *Welding Metallurgy*, 1st ed., Wiley, New York, 1987.

8. *Welding Handbook*, 7th ed, Vol. 4, American Welding Society, Miami, FL, 1982, p. 9.

9. Chadwick, G. A., *Metallography of Phase Transformations*, Crane, Russak and Co., New York, 1972.

10. Evans, G. M., *Weld. J.*, **59:** 67s, 1980.

11. Kou, S., Sun, D. K., and Le, Y., *Metall. Trans.*, **14A:** 643, 1983.

12. Aidun, D. K., and Savage, W. F., *Weld. J.*, **63:** 1984; **64:** 97s, 1985.

13. *Atlas of Isothermal Transformation and Cooling Transformation Diagrams*, American Society for Metals, Metals Park, OH, 1977.

14. Smith, R. B., in *ASM Handbook*, Vol. 6: *Welding, Brazing and Soldering*, ASM International, Materials Park, OH, 1993, p. 641.

15. Winsor, F. J., in *ASM Handbook*, Vol. 6: *Welding, Brazing and Soldering*, ASM International, Materials Park, OH, 1993, p. 662.

16. Homma, H., Ohkida, S., Matsuda, and Yamamoto, K., *Weld. J.*, **66:** 301s, 1987.

17. Yurika, N., paper preseuted at Fifth International Symposium, Japan Welding Society, Tokyo, April 1990.

18. Inagaki, M., et al., *Fusion Welding Processing*, Seibundo Shinko Sha, Tokyo, 1971 (in Japanese).

19. Linnert, G. E., *Welding Metallurgy*, 3rd ed., Vol. 2, American Welding Society. Miami, FL, 1967.

20. Granjon, H., in *Cracking and Fracture in Welds*, Japan Welding Society, Tokyo, 1972, p. IB1.1.

21. Coe, F. R., *Welding Steels without Hydrogen Cracking*, Welding Institute, Cambridge, 1973.

22. Troiano, A. R., *Trans. ASM*, **52:** 54, 1960.

23. Petch, H. J., *Nature*, **169:** 842, 1952.

24. Beachem, C. D., *Metal. Trans.*, **3:** 437, 1972.

25. Savage, W. F., Nippes, E. F., and Szekeres, E. S., *Weld. J.*, **55:** 276s, 1976.

26. Gedeon, S. A., and Eagar, T. W., *Weld. J.*, **69:** 213s, 1990.

27. Bailey, N., in *Residual Stresses*, Welding Institute, Cambridge, 1981, p. 28.

28. *Welding Metallography*, Metallurgical Services Laboratories, Betchworth, Surrey, United Kingdom 1970, p. 22.

29. Sawhill, J. M. Jr., Dix, A. W., and Savage, W. F., *Weld. J.*, **53:** 554s, 1974.

30. Stout, R. D., Tor, S. S., McGeady, L. J., and Doan, G. E., *Weld. J.*, **26:** 673s, 1947.

31. Savage, W. F., Nippes, E. F., and Husa, E. I., *Weld. J.*, **61:** 233s, 1982.

32. Cottrell, C. L. M., *Weld. J.*, **32:** 257s, 1953.

33. Vasudevan, R., Stout, R. D., and Pense, A. W., *Weld. J.*, **60:** 155s, 1981.

34. Lesnewich, A., in *ASM Handbook*, Vol. 6: *Welding, Brazing and Soldering*, ASM International, Materials Park, OH, 1993, p. 408.

35. Nakamura, H., Naiki, T., and Okabayashi, H., in *Proceedings of the First International Conference on Fracture*, Sendai, Japan, Vol. 2, 1966, p. 863.

36. Nawrocki, J. G., Dupont, J. N., Robino, C. V., and Marder, A. R., *Weld. J.,* **79:** 355s, 2000.

37. Swift, R. A., *Weld. J.,* **50:** 195s, 1971.

38. Swift, R. A., and Rogers, H. C., *Weld. J.,* **50:** 357s, 1971.

39. Meitzner, C. F., *WRC Bull.,* **211:** 1, 1975.

40. McPherson, R., *Metals Forum,* **3:** 175, 1980.

41. Dhooge, A., and Vinckier, A., *Weld. World,* **30:** 44, 1992.

42. Gooch, D. J., and King, B. L., *Weld. J.,* **59:** 10s, 1980.

43. Batte, A. D., and Murphy, M. C., *Metals Technol.,* **6:** 62, 1979.

44. Glover, A. G., Jones, W. K. C., and Price, A. T., *Metals Technol.,* **4:** 326, 1977.

45. Nawrocki, J. G., Dupont, J. N., Robino, C. v., and Marder, A. R., *Weld. J.,* **80:** 18s, 2001.

46. Murray, J. D., *Br. Weld. J.,* **14:** 447, 1967.

47. Kihara, H., Suzuki, H., and Ogura, N., *J. Jpn. Weld. Soc.,* **24:** 94, 1956.

48. Jubb, J. E. M., *Weld. Res. Council Bull.,* **168:** 1971.

49. Ganesh, S., and Stout, R. D., *Weld. J.,* **55:** 341s, 1976.

50. Skinner, D. H., and Toyama, M., *Weld. Res. Council Bull.,* **232:** 1977.

51. Dickinson, F. S., and Nichols, R. W., in *Cracking and Fractures in Welds*, Japan Welding Society, Tokyo, 1972, p. IA4.1.

52. Bailey, N., *Weldability of Ferritic Steels,* Arbington Publishing, Arbington Hall, Cambridge, England, and ASM International, Materials Park, OH, 1994, p. 103.

53. Kaufman, E. J., Pense, A. W., and Stout, R. D., *Weld. J.,* **60:** 43s, 1981.

54. Kanazawa, S., *Weld. Res. Abroad,* **21(5):** 2, 1975.

55. Wold, G., and Kristoffersen, T., *Sverseteknikk,* **27:** 33, 1972.

56. Naumann, F. K., *Failure Analysis: Case Histories and Methodology,* Riederer Verlag GmbH, Stuttgart, Germany, and American society for Metals, Metals Park, OH, 1983.

57. Hutchings, F. R., and Unterweiser, P. M., *Failure Analysis: The British Engine Technical Reports,* American Society of Metals, Metals Park, OH, 1981.

58. *Fatigue Fractures in Welded Constructions,* Vol. II, International Institute of Welding, London, 1979.

FURTHER READING

1. *ASM Handbook*, Vol. 6: *Welding, Brazing and Soldering*, ASM International, Materials Park, OH, 1993, pp. 70–87, 405–415, 641–676.

2. Easterling, K., *Introduction to the Physical Metallurgy of Welding*, 2nd ed., Butterworths, London, 1992.

3. Bailey, N., *Weldability of Ferritic Steels*, Arbington Publishing, Cambridge, 1994.

4. Evans, G. M., and Bailey, N., *Metallurgy of Basic Weld Metal*, Arbington Publishing, Cambridge, 1997.

PROBLEMS

17.1   In welding quenched-and-tempered alloy steels, the maximum allow-able heat input per unit length of weld can be higher, say by about 25%, in T-joints than in butt joints. Explain why.

17.2   Type E14018 electrodes have been recommended for SMAW of HY-130 steel. (a) A moisture level lower than 0.1% in the electrode covering is required. Explain why such a low moisture level is required. (b) It is suggested that in multipass welding an interpass temperature not less than 60°C (135°F) be maintained for 1 h or longer between two con-secutive passes. Explain why such a long waiting time is suggested.

17.3   Would you consider using the electroslag welding process for welding quenched-and-tempered alloy steels and heat-treatable alloy steels requiring postweld heat treatment? Why or why not?

17.4   It has been observed that 2.25Cr–1Mo steel specimens subjected to a simulated HAZ thermal cycle with a peak temperature of 1000°C exhibit intragranular cracking when reheated to 500–700°C under strain. However, under the same testing conditions the same specimens subjected to a simulated thermal cycle with a 1350°C peak temperature exhibit intergranular cracking typical of reheat cracking. Can you comment on these observations from the viewpoint of grain growth and carbide dissolution at high temperatures and carbide reprecipitation during reheating?

17.5   A fatigue failure was reported in the main leaf of a large laminated spring used in heavy motive machinery. The spring consisted of alloy steel sections heat treated to spring temper. It was found that fatigue cracking originated from the martensite under a weld spatter acciden-tally formed during welding repairs to the undercarriage. (a) Explain how martensite had formed. (b) Comment on the danger of spattering, careless arc striking, and accidentally touching some part of the struc-ture with the electrode.

17.6   The development of the high-speed ESW process has led to a signifi-cant improvement in weld toughness over the conventional ESW process. One technique of achieving high welding speed is through the addition of metal powder into the molten slag. In some structural steels the maximum grain size in the HAZ has been reduced from 500 to 250 μm, resulting in significantly better joint toughness. Explain why the reduction in HAZ grain size can be obtained by adding metal powder.

18 Corrosion-Resistant Materials: Stainless Steels

Stainless steels are widely used in various industries because of their resistance to corrosion. The welding of stainless steels, especially the austenitic grades, is important in energy-related systems, for instance, power generation and petrochemical refining systems. Table 18.1 summarizes typical welding problems in stainless steels and some recommended solutions. The problems associated with the fusion zone and the partially-melted zone have been addressed in previous chapters. This chapter focuses on HAZ phenomena in stainless steels.

18.1 CLASSIFICATION OF STAINLESS STEELS

Stainless steels are a class of Fe-base alloys that are noted for their high corrosion and oxidation resistance. They usually contain from 12 to 27% Cr and 1 to 2% Mn by weight, with the addition of Ni in some grades. A small amount of carbon is also present, either deliberately added or as an unavoidable impurity. As shown in Table 18.2, stainless steels can be classified into three major categories based on the structure: ferritic, martensitic, and austenitic (1).

18.1.1 Ferritic Stainless Steels

These are composed mainly of a bcc phase. Chromium (bcc) tends to stabilize α-Fe (bcc) and its high-temperature counterpart δ-Fe (bcc), which merge to form the so-called closed γ loop, as shown in Figure 18.1 (2). It can be said that ferritic stainless steels have metallurgical characteristics similar to those of Fe–Cr alloys containing sufficient (more than about 12%) Cr to remain outside the γ loop. They are essentially ferritic over the whole solid-state temperature range. To avoid forming an excessive amount of the brittle σ phase, however, the maximum Cr content of ferritic stainless steels is in general kept below 27%.

18.1.2 Martensitic Stainless Steels

These stainless steels can be said to behave like Fe–Cr alloys containing less than about 12% Cr (inside the γ loop). These alloys solidify as δ-ferrite and

TABLE 18.1   Typical Welding Problems in Stainless Steels

	Typical Problems	Solutions	Sections in book
Austenitic	Solidification cracking	Use proper filler wires and dilution to keep 4–10% ferrite	11.4
	Weld decay	Use stabilized grades (321 347) or low-carbon grades (304L and 316L)	
		Postweld heat-treat to dissolve carbides, followed by quenching	18.2
	Knife-line attack (in stabilized grades)	Use low-carbon grades	
		Add La and Ce to stabilized grades	18.2
		Postweld heat-treat to dissolve carbides	
	Hot cracking in partially melted zone	Switch from 347 to 304 or 316	13.1
		Use low heat inputs	13.2
Ferritic	Low toughness due to HAZ grain growth and grain-boundary martensite	Use low heat inputs or add carbide and nitride formers to suppress grain growth	18.3
		Add Ti or Nb to reduce martensite	
Martensitic	Hydrogen cracking	Preheat and postheat	
		Use low-hydrogen or austenitic stainless steel electrodes	18.4

transform to austenite during cooling. When the cooling rate is sufficiently rapid, as in the case of welding, the austenite that forms transforms into martensite. It should be emphasized that the Fe–Cr phase diagram can be used only as a convenient basis for distinguishing the above two different structural categories of stainless steels. The presence of minor elements, especially carbon, can significantly displace the boundaries of the austenite and ferrite ranges. These effects will be considered later in this chapter.

18.1.3   Austenitic Stainless Steels

The addition of Ni (fcc) into Fe–Cr alloys, as shown in Figure 18.2, tends to widen the range over which austenite (fcc) exists and increase its stability at low temperatures (3). Generally speaking, austenitic stainless steels contain at least 15% Cr and enough Ni to maintain a stable austenitic structure over the temperature range from 1100°C to room temperature without the formation

TABLE 18.2 Nominal Composition of Wrought Stainless Steels

AISI Type	Nominal Composition (%)				
	C	Mn	Cr	Ni	Other
Austenitic grades					
301	0.15 max	2.0	16–18	6–8	
302	0.15 max	2.0	17–19	8–10	
304	0.08 max	2.0	18–20	8–12	
304L	0.03 max	2.0	18–20	8–12	
309	0.20 max	2.0	22–24	12–15	
310	0.25 max	2.0	24–26	19–22	
316	0.08 max	2.0	16–18	10–14	2–3% Mo
316L	0.03 max	2.0	16–18	10–14	2–3% Mo
321	0.08 max	2.0	17–19	9–12	$(5 \times \%C)$ Ti min
347	0.08 max	2.0	17–19	9–13	$(10 \times \%C)$ Nb–Ta min
Martensitic grades					
403	0.15 max	1.0	11.5–13		
410	0.15 max	1.0	11.5–13		
416	0.15 max	1.2	12–14		0.15% S min
420	0.15 min	1.0	12–14		
431	0.20 max	1.0	15–17	1.2–2.5	
440A	0.60–0.75	1.0	16–18		0.75% Mo max
440B	0.75–0.95	1.0	16–18		0.75% Mo max
440C	0.95–1.20	1.0	16–18		0.75% Mo max
Ferritic grades					
405	0.08 max	1.0	11.5–14.5		0.1–0.3% Al
430	0.15 max	1.0	14–18		
446	0.20 max	1.5	23–27		

Source: Fisher and Maciag (1).

of martensite (4). For instance, for a stainless steel bearing 15–16% Cr, about 6–8% Ni is required. The most widely used stainless steel is 304 stainless steel. It is also known as 18-8 stainless steel because of its composition of 18% Cr and 8% Ni.

18.2 AUSTENITIC STAINLESS STEELS

18.2.1 Weld Decay

The HAZ of austenitic stainless steels containing more than about 0.05% C can be susceptible to a form of intergranular corrosion called weld decay. Figure 18.3 shows the microstructure in the HAZ of a 0.05% C 304 austenitic stainless steel subjected to an accelerated corrosion test (5). The preferential attack at the grain boundaries is evident.

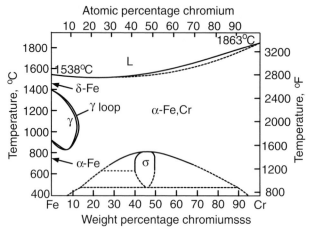

Figure 18.1 Fe–Cr phase diagram (2).

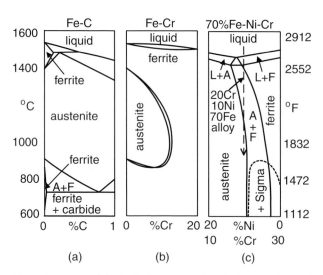

Figure 18.2 Phase diagrams: (*a*) Fe–C; (*b*) Fe–Cr; (*c*) Fe–Ni–Cr at 70% Fe. Reprinted from DeLong (3). Courtesy of American Welding Society.

A. Cause   Weld decay in austenitic stainless steels is caused by precipitation of Cr carbide at grain boundaries, which is called *sensitization*. Typically, the Cr carbide is Cr-enriched $M_{23}C_6$, in which M represents Cr and some small amount of Fe. Within the sensitization temperature range carbon atoms rapidly diffuse to grain boundaries, where they combine with Cr to form Cr carbide. Figure 18.4*a* shows a TEM micrograph of Cr carbide particles along a grain boundary of a 316 stainless steel that has been sensitized at 750°C for

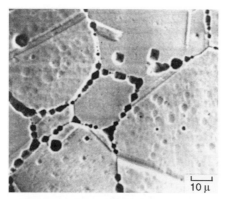

Figure 18.3 Intergranular corrosion in HAZ of a 304 stainless steel containing 0.05% C. From Fukakura et al. (5).

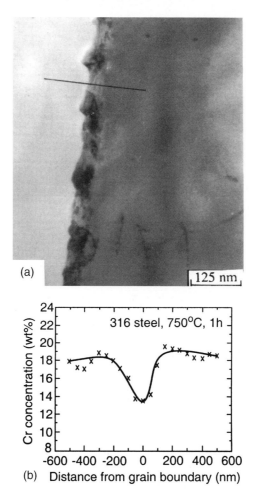

Figure 18.4 Chromium carbide precipitation at a grain boundary in 316 stainless steel: (*a*) TEM micrograph; (*b*) Cr concentration profile across grain boundary. From Magula et al. (6).

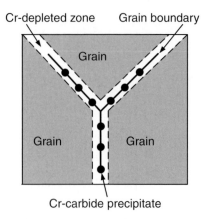

Figure 18.5   Grain boundary microstructure in sensitized austenitic stainless steel.

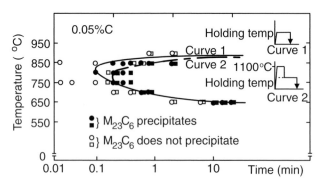

Figure 18.6   Isothermal precipitation curve for $Cr_{23}C_6$ in 304 stainless steel. From Ikawa et al. (7).

1 h (6). Composition analyses show that these particles contain 63–71 wt % Cr. The Cr concentration distribution measured along the straight line in Figure 18.4a is shown in Figure 18.4b, and it clearly indicates Cr depletion at the grain boundary. Because of Cr carbide precipitation at the grain boundary, the areas adjacent to the grain boundary are depleted of Cr, as shown schematically in Figure 18.5. These areas become anodic to the rest of the grain and hence are preferentially attacked in corrosive media, resulting in intergranular corrosion (Figure 18.3).

Figure 18.6 shows the isothermal precipitation or *time–temperature–sensitization* (TTS) curves of a 304 stainless steel containing 0.05% C (7). As shown, the sensitization temperature range of Cr carbide is about 600–850°C. Figure 18.7 shows the solvus curves of $Cr_{23}C_6$ and TiC (and NbC) in 18Cr–8Ni stainless steel (8). The marked areas indicate the precipitation temperature ranges

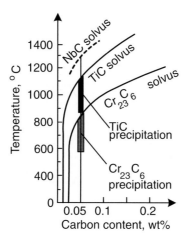

Figure 18.7 Solvus curves for $Cr_{23}C_6$ and TiC in 304 stainless steel. Modified from *Principle and Technology of the Fusion Welding of Metals* (8).

of the carbides, about 600–850°C for $Cr_{23}C_6$ and up to about 1100°C for TiC. Chromium carbide dissolves at temperatures above its solvus curve. Upon cooling slowly from above the solvus temperature, Cr carbide can precipitate again. However, if the cooling rate is high, Cr carbide may not have enough time to precipitate and the material can be supersaturated with free carbon.

Solomon and Lord (9–11) have reported TTS curves for 304 stainless steel with 0.077% C. It was also observed that deformation prior to welding or strain during cooling can enhance sensitization. According to Solomon (12), this is perhaps due to the fact that dislocations can increase the carbide nucleation rate and the diffusion rate.

Weld decay does not occur immediately next to the fusion boundary, where the peak temperature is highest during welding. Instead, it occurs at a short distance away from it, where the peak temperature is much lower. This phenomenon can be explained with the help of thermal cycles during welding, as shown in Figure 18.8. At position 1 near the fusion boundary, the material experiences the highest peak temperature and cooling rate (Chapter 2). Consequently, the cooling rate through the precipitation range is too high to allow Cr carbide precipitation to occur. At position 2, which is farther away from the fusion line, the retention time of the material in the sensitization temperature range is long enough for precipitation to take place. At position 3, outside the HAZ, the peak temperature is too low to allow any precipitation. Figure 18.9 shows thermal cycles measured during the welding of 304 stainless steel and the location of the resultant weld decay (13). Consider the microstructure of a 304 stainless steel weld shown previously in Figure 14.6. At position *c*, which is away from the fusion boundary, the thick, dark grain boundaries are the evidence of weld decay due to grain boundary precipita-

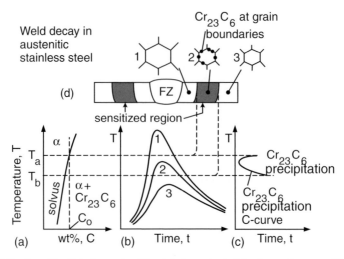

Figure 18.8 Sensitization in austenitic stainless steel: (*a*) phase diagram; (*b*) thermal cycle; (*c*) precipitation curve; (*d*) microstructure.

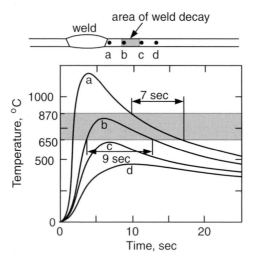

Figure 18.9 Thermal cycles and weld decay in 304 stainless steel weld. Modified from Fontana and Greene (13).

tion. At position *e*, which is immediately adjacent to the fusion boundary, there are no signs of grain boundary precipitation.

B. Effect of Carbon Content   The carbon content of the material can affect the degree of sensitization. As shown in Figure 18.10, sensitization takes place

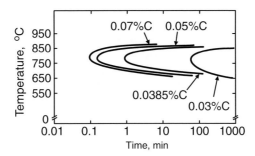

Figure 18.10 Effect of carbon content on isothermal precipitation of $Cr_{23}C_6$ in 304 stainless steel. From Ikawa et al. (14).

more rapidly in 304 stainless steel as its carbon content is increased (14). Ikawa et al. (14) showed that under a given heat input and welding speed weld decay in 304 stainless steel increases with increasing carbon content.

C. Effect of Heat Input Ikawa et al. (14) also showed that for a 304 stainless steel of a given carbon content weld decay increases with increasing heat input per unit length of weld. The higher the heat input, the wider the region of sensitization and the longer the retention time in the sensitization temperature range. It is interesting to point out that in the case of resistance spot welding the metal is rapidly heated by a momentary electric current followed by a naturally rapid cooling, As such, sensitization may not occur (13).

D. Remedies Weld decay in austenitic stainless steels can be avoided as follows (15)

D.1. Postweld Heat Treatment The weldment can be heat treated at 1000–1100°C followed by quenching. The high-temperature heat treatment dissolves the Cr carbide that has precipitated during welding, and quenching prevents its re-formation. This treatment is not always possible, however, because of the size and/or the quenching-induced distortion of the weldment.

D.2. Reduction of Carbon Content Low-carbon grades such as 304L and 316L stainless steels can be used. These stainless steels are designed to have less than 0.035% C to reduce susceptibility to weld decay (Table 18.2). Figure 18.11 shows that alloy 316L is much less susceptible to sensitization than alloy 316 (6). The TTS curve of the former is much farther to the right (longer time) than that of the latter.

D.3. Addition of Strong Carbide Formers Elements such as titanium (Ti) and niobium (Nb) have a higher affinity for C than Cr and thus form carbides more easily than Cr. Types 321 and 347 stainless steels are essentially identical to

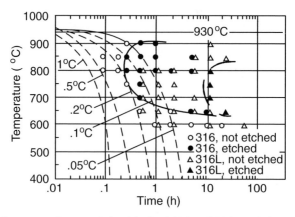

Figure 18.11   Intergranular attack (etching) of 316 and 316L stainless steels by oxalic acid. From Magula et al. (6).

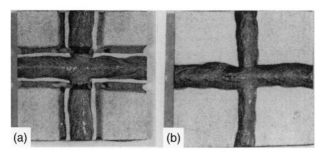

Figure 18.12   Welds in austenitic stainless steels: (*a*) weld decay in 304 stainless steel; (*b*) no weld decay in 321 stainless steel. From Linnert (16).

304 stainless steel except for the addition of Ti and Nb, respectively. As shown in Figure 18.12, the stabilized grade is more resistant to weld decay since Cr carbide precipitation is suppressed during welding (16).

18.2.2   Knife-Line Attack

Although stabilized austenitic stainless steels such as 321 and 347 are not susceptible to weld decay, they can be susceptible to a different kind of intergranular corrosion attack, called knife-line attack. Like weld decay, knife-line attack is also caused by precipitation of Cr carbide at grain boundaries. Knife-line attack differs from weld decay in two ways: (i) knife-line attack occurs in a narrow region immediately adjacent to the weld metal and (ii) knife-line attack occurs in stabilized-grade stainless steels (13, 17). Figure 18.13 shows knife-line attack next to the weld metal (left) in a 321 stainless steel.

Figure 18.13 Knife-line attack in a stabilized austenitic stainless steel. Reproduced by permission of TWI, Ltd.

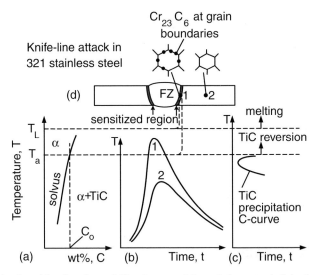

Figure 18.14 Sensitization in stabilized austenitic stainless steel: (*a*) phase diagram; (*b*) thermal cycle; (*c*) precipitation curve; (*d*) microstructure.

A. Cause Knife-line attack in stabilized austenitic stainless steels can be explained with the help of thermal cycles during welding, as shown in Figure 18.14. Position 1 is very close to the fusion boundary and is thus subjected to a high peak temperature and a high cooling rate during welding. Since this

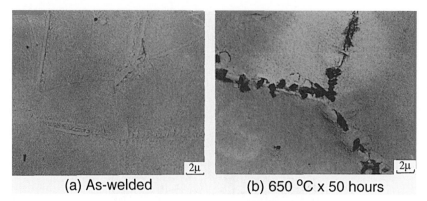

(a) As-welded   (b) 650 °C x 50 hours

Figure 18.15   Transmission electron micrographs near fusion boundary of 347 stainless steel. From Ikawa et al. (18).

peak temperature is above the solvus temperature of Ti carbide, Ti carbide dissolves in this area. Due to the rapid cooling rate through its precipitation temperature range, Ti carbide does not reprecipitate during cooling, thus leaving abundant free carbon atoms in this area. When the weld is reheated in the Cr carbide precipitation range (for stress-relief or in multiple-pass welding), Ti carbide does not form appreciably since the temperature level is not high enough. Consequently, Cr carbide precipitates at grain boundaries, and this area becomes susceptible to intergranular corrosion attack. Position 2, however, is not susceptible because of its low peak temperature. Because of the high temperate gradient near the fusion boundary (this is especially true for austenitic stainless steels due to the low thermal conductivity) and the high Ti carbide dissolution temperature, the region in which Ti carbide dissolves during welding is very narrow. As a result, subsequent intergranular corrosion attack occurs in a very narrow strip immediately adjacent to the fusion boundary, and thus the name knife-line attack.

Figure 18.15 shows the transmission electron micrographs in the area immediately outside the fusion boundary of a 347 stainless steel (18). Precipitation of Cr carbides is clearly visible at grain boundaries after postweld sensitizing heat treating at 650°C for 50h. Figure 18.16 shows the intergranular corrosion attack in the same area by a corrosive liquid for 15h.

B. Remedies   The knife-line attack of stabilized austenitic stainless steels can be avoided as follows.

B.1. Postweld Heat Treatment   Annealing in the temperature range of 1000–1100°C after welding helps Cr carbide dissolve. Since Ti carbide and Nb carbide form in this temperature range (Figure 18.7), subsequent quenching to avoid Cr carbide precipitation is not necessary.

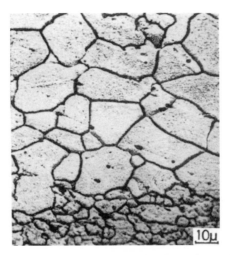

Figure 18.16 Intergranular corrosion near fusion boundary of 347 stainless steel. From Ikawa et al. (18).

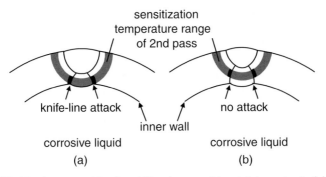

Figure 18.17 Dual-pass weld of stabilized austenitic stainless steel: (*a*) knife-line attack; (*b*) no attack. Modified from *Principle and Technology of the Fusion Welding of Metals* (8).

B.2. Using Low-Carbon Grades   As in the case of weld decay, low-carbon grades reduce the chance of Cr carbide precipitation.

B.3. Adjusting Welding Procedure   Sometimes, it is possible to avoid knife-line attack by modifying the welding procedure. Figure 18.17 shows one such example involving the dual-pass seam weld of a stainless steel pipe of the stabilized grade, the first pass on the inside of the pipe and the second pass on the outside (8). During the deposition of the first pass, Ti or Nb carbide near its fusion boundary dissolves. As shown in Figure 18.17*a*, the sensitization temperature range (600–850°C) of the second pass overlaps the fusion boundary

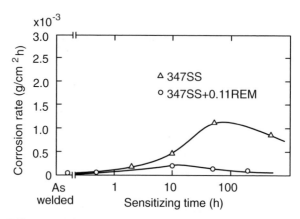

Figure 18.18   Effect of addition of REMs on knife-line attack sensitivity of 347 stainless steel. From Ikawa et al. (19).

of the first pass at the pipe inner wall and causes Cr carbide to precipitate, thus making the inner wall susceptible to attack by the corrosive liquid inside the pipe. To correct the problem, the size of the first pass should be increased and that of the second pass decreased so that the sensitization temperature range of the second pass overlaps the fusion boundary of the first pass away from the pipe inner wall, as shown in Figure 18.17b. The problem can also be corrected by making the first pass from the outside of the pipe and the second pass from the side if this is possible to do.

B.4. Adding Rare Earth Metals   It has been shown that adding rare earth metals (REMs) such as La and Ce can reduce the knife-line attack of stabilized stainless steels (18, 19). High-resolution electron micrographs reveal that carbide precipitation in the grain interior is accelerated in these materials, thus leaving less free carbon atoms for carbide precipitation at grain boundaries. Figure 18.18 shows the reduction in the rate of intergranular corrosion in a 347 stainless steel by addition of a small amount of REMs. As can be seen by comparing Figure 18.19 with Figure 18.15b, treating with REMs significantly reduces Cr carbide precipitation at grain boundaries. Similar results have been observed in REM-treated 321 stainless steel. Addition of small amounts of REMs in stabilized stainless steels does not appear to have an adverse effect on their mechanical properties (18).

C. Susceptibility Tests   The susceptibility of austenitic stainless steels to weld decay or knife-line attack can be evaluated by the Huey test (13). This test (ASTM A-262) consists of exposure of sensitized materials to boiling 65% nitric acid for five 48-h periods. A less time-consuming method is the Streicher test (13). This test (ASTM A-262–55T) consists of polishing a small specimen,

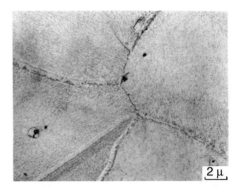

Figure 18.19 Transmission electron micrograph near fusion boundary of REM-treated 347 stainless steel. From Ikawa et al. (18).

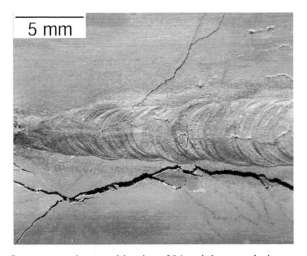

Figure 18.20 Stress corrosion cracking in a 304 stainless steel pipe gas–tungsten arc welded and exposed to a corrosive liquid.

through No. 000 emery paper, etching in 10% oxalic acid for 1.5 min under an applied current density of $1 \, A/cm^2$, and then examining the surface at 250–500 magnification. The specimen is the anode and a stainless steel beaker is used as the cathode.

18.2.3 Stress Corrosion Cracking

Due to the low thermal conductivity and high thermal expansion coefficient of austenitic stainless steels, severe residual stresses can be present in these materials after welding. When exposed to corrosive media, stress corrosion

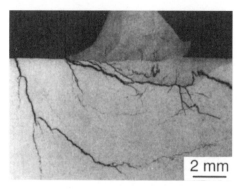

Figure 18.21   Stress corrosion cracking near the toe of a 316L weld. Reprinted from Brooks and Lippold (21).

cracking may occur in the HAZ. Cracking is usually transgranular and branching and can extend far into the base metal (20). Figure 18.20 shows stress corrosion cracking in a circumferential gas–tungsten arc weld in a 304 pipe of 100 mm (4 in.) diameter and 2.4 mm ($^3/_{32}$ in.) wall exposed to a corrosive liquid. In the lower half of the photograph there are many very fine cracks branching from the main crack that are covered with corrosion products. They are visible with the naked eye but not in the photograph. Figure 18.21 shows stress corrosion cracking that has initiated near the toe of a weld in 316L stainless steel (21).

When stress-relief heat treating is carried out, the possibility of Cr carbide precipitation should be carefully considered. Stress relief can be achieved while solutionizing austenitic stainless steels to dissolve Cr carbide at a temperature level of 1000°C. New residual stresses may develop, however, during subsequent quenching. To avoid this problem, low-carbon or stabilized grades can be used to allow the material to cool slowly after solutionizing.

18.3   FERRITIC STAINLESS STEELS

18.3.1   Phase Diagram

Ferritic stainless steels, although generally considered less weldable than austenitic stainless steels, have advantages such as lower costs and better resistance to stress corrosion cracking than austenitic stainless steels. The most commonly used ferritic stainless steel perhaps is the 430 type, which usually contains about 17% Cr and 0.05–0.12% C. Figure 18.22 is a vertical section of the Fe–Cr–C phase diagram at 17% Cr (4).

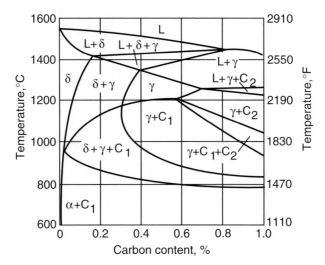

Figure 18.22 Vertical section of Fe–Cr–C phase diagram at 17% Cr. Reprinted, with permission, from Castro and De Cadenet (4). Copyright 1974 Cambridge University Press.

18.3.2 Sensitization

Sensitization of ferritic stainless steels (such as 430 and 446) is quite different from that of austenitic stainless steels (unstabilized, such as 304). The sensitizing range for ferritic stainless steels lies above 925°C, and immunity to intergranular corrosion is restored by annealing in the range of 650–815°C for about 10–60 min (15). These temperatures are essentially the opposite of those applying to austenitic stainless steels. Because of the high sensitization temperature range, the weld decay in ferritic stainless steels occurs close to the weld metal, rather than at a distance away, as in the case of austenitic stainless steels. Furthermore, unlike in austenitic stainless steels, lowering the carbon content is not very effective in preventing weld decay in ferritic stainless steels. In fact, a 430 stainless steel with a carbon content as low as 0.009% C was found still susceptible to weld decay. As in the case of austenitic stainless steels, addition of Ti or Nb was found helpful (15).

Because the diffusion rates of C and Cr are much higher in ferrite (bcc) than in austenite (fcc), rapid cooling from above 925°C during welding does not really suppress precipitation of Cr carbide at grain boundaries in ferritic stainless steels. For the same reason, lowering the carbon content does not effectively prevent Cr carbide from precipitating, unless the carbon content is extremely low (e.g., 0.002% in 446 stainless steel). According to Uhlig (15), postweld annealing in the range of 650–815°C encourages diffusion of Cr atoms to the Cr-depleted region adjacent to Cr carbide precipitates and thus helps reestablish a uniform Cr composition to resist intergranular corrosion.

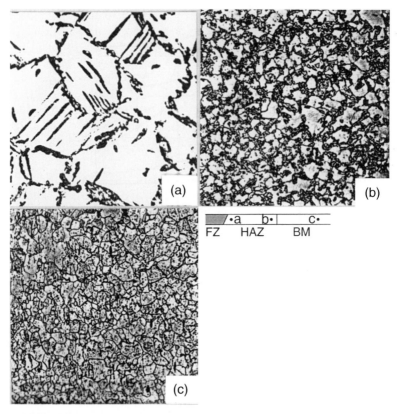

Figure 18.23   Microstructure of 430 stainless steel (magnification 212×): (*a*) near fusion boundary; (*b*) near base metal; (*c*) base metal. From Kou and Le (22).

18.3.3   Martensite Formation and Grain Growth

Figure 18.23 shows the microstructure of 430 stainless steel in the base metal and the HAZ (22). Position *b* (away from the fusion boundary) reaches the austenite formation ranges ($\delta + \gamma + C_1$ and $\delta + \gamma$) of the phase diagram (Figure 18.22 at about 0.08%C). Here, austenite formed along grain boundaries and, upon fast cooling during welding, transformed into martensite (Figure 18.23*b*). Little grain growth took place. Position *a* (near the fusion boundary), on the other hand, reaches the δ-ferrite range and excessive grain growth takes place at high temperatures (Figure 18.23*a*). During cooling through the austenite formation ranges, a considerable amount of austenite formed at the grain boundaries. The austenite appears needlelike because of the large grain size and the high cooling rate. During further rapid cooling to room temperature, the austenite transforms into needlelike martensite.

Because of excessive grain coarsening and formation of needlelike martensite at grain boundaries, the notch toughness of as-welded 430 stainless steel is poor. Postweld heat treatment at 800°C (martensite tempering) significantly improves the notch toughness. Suppressing martensite formation by adding 0.5% Ti or 1% Nb also helps (4). This is probably because both Ti and Nb increase the stability of δ-ferrite, thus suppressing the formation of austenite. Furthermore, as mentioned previously, they tend to form carbides at high temperatures. Therefore, the carbon content is reduced and the tendency to form martensite decreases. Excessive grain growth can be avoided, of course, by using lower welding heat inputs. It has also been suggested that nitride and carbide formers such as B, Al, V, and Zr can be added to ferritic stainless steels to suppress grain growth during welding (23).

18.4 MARTENSITIC STAINLESS STEELS

18.4.1 Phase Diagram

Martensitic stainless steels are employed because of the combination of high strength and good corrosion resistance. The commercial 13% Cr steel containing more than 0.08% C is the most widely used martensitic stainless steel. Figure 18.24 shows a vertical section of the Fe–Cr–C at 13% Cr (4). This diagram is somewhat similar to that shown in Figure 18.22 at 17% Cr. The temperature and composition ranges of the austenite phase (γ), however, are much wider in the case of 13% Cr steel.

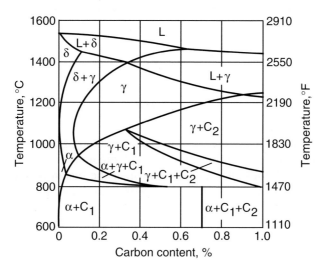

Figure 18.24 Vertical section of Fe–Cr–C phase diagram at 13% Cr. Reprinted, with permission, from Castro and De Cadenet (4). Copyright 1974 Cambridge University Press.

18.4.2   Underbead Cracking

The hardness of martensitic stainless steels increases rather significantly with increasing carbon content. For 12% Cr steel quenched from 1050°C, for instance, the hardness increases from 364 HV at 0.07% C to 620 HV at 0.60% C (4). Consequently, martensitic stainless steels with higher carbon contents are rather susceptible to underbead cracking, especially when hydrogen is present during welding. Because of the internal stresses induced by the volume increase associated with the austenite-to-martensite transformation, under-bead cracking can still occur even if relatively low restraint is employed during welding. As a result, martensitic stainless steels with a carbon content above 0.25–0.30% are not normally welded.

18.4.3   Remedies

To avoid underbead cracking in martensitic stainless steels, both preheating and postweld tempering (between 600 and 850°C) are usually employed. In addition, filler metals of austenitic stainless steels are often used. This is because the weld metal so produced remains ductile, thus reducing the chance of underbead cracking. Also, due to the greater solubility of hydrogen in austenite, hydrogen can dissolve in the weld metal and the possibility of under-bead cracking is thus reduced.

As in the case of welding heat-treatable alloy steels, martensitic stainless steels are usually not allowed to cool directly to room temperature upon the completion of welding in order to avoid underbead cracking. Figure 18.25 shows the postweld heat treatment for a creep-resistant 12% Cr steel (12% Cr/0.2% C/1% Mo/0.4% W/0.3% V) and the resultant microstructure (4).

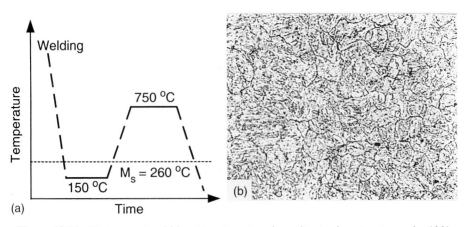

Figure 18.25   Proper postweld heat treatment and resultant microstructure of a 12% Cr martensitic steel (magnification 500×). Reprinted, with permission, from Castro and De Cadenet (4). Copyright 1974 Cambridge University Press.

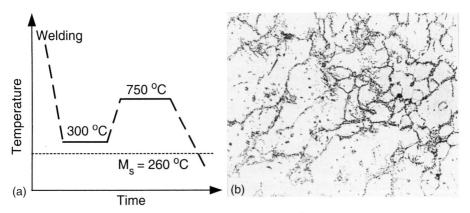

Figure 18.26 Improper postweld heat treatment and resultant microstructure of a 12% Cr martensitic steel (magnification 500×). Reprinted, with permission, from Castro and De Cadenet (4). Copyright 1974 Cambridge University Press.

Upon the completion of welding, the weldment is held at 150°C in order to avoid cracking. After about 1 h the weldment can be heat treated at 750°C for 4 h to temper the martensite that has formed. The resultant microstructure of the HAZ is essentially tempered martensite. Figure 18.26 shows a different postweld heat treatment and the resultant microstructure (4). From the viewpoint of completely avoiding the possibility of underbead cracking, it might seem better to hold the weldment at a temperature higher than M_S (say 300°C) so that no martensite can form after welding. During subsequent heat treating at 750°C, however, the untransformed austenite phase decomposes and the microstructure of the HAZ becomes ferritic with agglomerated carbides precipitated at the grain boundaries. Unfortunately, such microstructure is rather brittle and is, in fact, inferior to that of tempered martensite.

18.5 CASE STUDY: FAILURE OF A PIPE

Figure 18.27*a* shows a section from a failed 304 stainless steel pipe (24). The pipe, about 100 mm in diameter, was seam welded first from the inside and then from the outside. It was carrying hot, dilute nitric acid and it was in service only two months before it failed.

The transverse cross section of the seam weld and its surrounding area is shown in Figure 18.27*b*. The attack is about 1.5–3.0 mm from the inner weld pass, thus suggesting that weld decay is the cause of failure. Corrosion is intergranular judging from the missing grains in the microstructure shown in Figure 18.27*c*, and this is consistent with the intergranular characteristics of weld decay. During welding the corroded areas were heated up into the sensitization temperature range, causing Cr carbide to precipitate along grain boundaries.

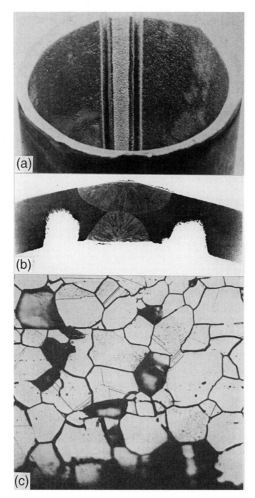

Figure 18.27 Failure of a 304 stainless steel pipe (10 cm diameter): (*a*) section from failed pipe; (*b*) transverse cross section of seam weld; (*c*) microstructure of weld decay area. Reprinted from *Welding Handbook* (24). Courtesy of American Welding Society.

REFERENCES

1. Fisher, G. J., and Maciag, R. J., in *Handbook of Stainless Steels*, Eds. D. Peckner and I. M. Bernstein, McGraw-Hill, New York, 1977, p. 1.
2. *Metals Handbook*, 8th ed., Vol. 8, American Society for Metals, Metals Park, OH, 1973, p. 291.
3. Delong, W. T., *Weld. J.*, **54:** 273s, 1974.
4. Castro, R., and De Cadenet, J. J., *Welding Metallurgy of Stainless and Heat Resisting Steels*, Cambridge University Press, London, 1974.

5. Fukakura, J., Kashiwaya, H., Mori, T., Iwasaki, S., and Arii, M., in *Weldments: Physical Metallurgy and Failure Phenomena*, Eds. R. J. Christoffel, E. F. Nippes, and H. D. Solomon, General Electric Co., Schenectady, NY, 1979, p. 173.

6. Magula, V., Liao, J., Ikeuchi, K., Kuroda, T., Kikuchi, Y., and Matsuda, F., *Trans. Japn. Weld. Res. Inst.*, **25:** 49, 1996.

7. Ikawa, H., Nakao, Y., and Nishimoto, K., *Technol. Repts. Osaka Univ.*, **28**(1434): 1978.

8. *Principle and Technology of the Fusion Welding of Metals*, Vol. 2, Mechanical Engineering Publishing Co., Peking, 1979 (in Chinese).

9. Solomon, H. D., *Corrosion*, **34:** 183, 1978.

10. Solomon, H. D., *Corrosion*, **36:** 356, 1980.

11. Solomon, H. D., and Lord, D. C., *Corrosion*, **36:** 395, 1980.

12. Solomon, H. D., in *Weldments: Physical Metallurgy and Failure Phenomena*, Eds. R. J. Christoffel, E. F. Nippes, and H. D. Solomon, General Electric Company, Schenectady, NY, 1979, p. 149.

13. Fontana, M. G., and Greene, N. D., *Corrosion Engineering*, 2nd ed., McGraw-Hill, New York, 1978.

14. Ikawa, H., Nakao, Y., and Nishimoto, K., *Technol. Repts. Osaka Univ.*, **29**(1648): 1979.

15. Uhlig, H. H., *Corrosion and Corrosion Control*, 2nd ed., Wiley, New York, 1971.

16. Linnert, G. E., in *Source Book on Stainless Steels*, American Society for Metals, Metals Park, OH, 1976, p. 237.

17. Ikawa, H., Shin, S., Nakao, Y., and Nishimoto, K., *Technol. Repts. Osaka Univ.*, **25**(1262): 1975.

18. Ikawa, H., Nakao, Y., and Nishimoto, K., *Technol. Repts. Osaka Univ.*, **28**(1402): 1978.

19. Ikawa, H., Nakao, Y., and Nishimoto, K., *Trans. Jpn. Weld. Soc.*, **8:** 9, 1977.

20. *Metals Handbook*, 8th ed., Vol. 7, American Society for Metals, Metals Park, OH, 1972, p. 135.

21. Brooks, J. A., and Lippold, J. C., in *ASM Handbook*, Vol. 6, ASM International, Materials Park, OH, 1993, p. 456.

22. Kou, S., and Le, Y., *Metall. Trans.*, **13A:** 1141, 1982.

23. Kah, D. H., and Dickinson, D. W., *Weld. J.*, **60:** 135s, 1981.

24. *Welding Handbook*, 7th ed., Vol. 4, American Welding Society, Miami, FL, 1982.

FURTHER READING

1. Fontana, M. G., and Greene, N. D., *Corrosion Engineering*, 2nd ed., McGraw-Hill, New York, 1978.

2. Uhlig, H. H., *Corrosion and Corrosion Control*, 2nd ed., Wiley, New York, 1971.

3. Peckner, D., and Bernstein, I. M., *Handbook of Stainless Steels*, McGraw-Hill, New York, 1977.

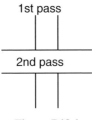

Figure P18.1

PROBLEMS

18.1 Two welds are made at 90° to each other in a stabilized-grade stainless steel (e.g., alloy 321), as shown in Figure P18.1. Indicate the locations where knife-line attack is expected.

18.2 Is sensitization expected to be more serious in SAW or GTAW of 304 stainless steel and why?

18.3 Sketch the concentration profiles of TiC and $M_{23}C_6$ across the HAZ from the fusion line in of a 321 stainless steel (Ti-stabilized) weld subjected to a postweld sensitization heat treatment at 650°C.

18.4 Rate 410, 420, and 440 stainless steels in the decreasing order of H-cracking susceptibility and explain why.

INDEX